COMBATING CLIMATE CHANGE

A ROLE FOR UK FORESTS

An assessment of the potential of the UK's trees and woodlands to mitigate and adapt to climate change

TSO
information & publishing solutions

First published in 2009 by The Stationery Office Limited, 26 Rutland Square, Edinburgh EH1 2BW.

ISBN: 978-0-11-497351-3

Recommended citation: The Read Report, or in full as below.

Read, D.J., Freer-Smith, P.H., Morison, J.I.L., Hanley, N., West, C.C. and Snowdon, P. (eds). 2009. *Combating climate change – a role for UK forests. An assessment of the potential of the UK's trees and woodlands to mitigate and adapt to climate change.* The Stationery Office, Edinburgh.

Foreword

In a world confronted with the possibility that human-induced warming of the climate will bring catastrophic consequences, it is incumbent upon each nation to maximise its own contribution to ameliorative activities. These actions can be of two kinds. The first involves the production of national targets for reducing emissions of carbon dioxide and the other greenhouse gases that drive the warming process, while the second seeks to increase the rates at which we contribute to the removal of these potentially damaging components from the atmosphere.

In this assessment we provide a scientific analysis of the potential for actions of this second type. We examine the abilities of the UK's trees, forests and forest products to absorb and retain the greenhouse gas carbon dioxide. What emerges is an urgent need to increase the extent of forest cover in the UK so that we can make an appropriate contribution to the global requirement for mitigation of greenhouse gas emissions.

By intercepting and trapping solar energy and carbon dioxide, tree canopies form effective sinks in which the principal agent of the warming process is removed from the atmosphere and sequestered in solid form as wood. The effectiveness of trees in these light- and carbon-scavenging processes is revealed both by the extent of shading and temperature amelioration observed at the forest floor, and by the rapid growth rates that they achieve, often on land too marginal for agricultural practice. These capabilities have long been recognised and exploited in Britain, particularly in the production of timber for strategic and economic purposes. Herein lies a problem of a historical nature that is not peculiar to our islands but which is particularly disadvantageous to those of us who now recognise the potential of trees to act as carbon 'sinks' in the British context. This is that, while our once bountiful indigenous resource was removed for the development of agriculture and effectively exploited for fuel and building materials, the UK has been far less effective in replanting forests. Essentially, harvesting along with long periods of little or no reforestation or afforestation have had

the consequence that, whereas our islands before human occupation had supported up to 80% forest cover, this figure had fallen to around 5% by the mid-1920s. Even today, with UK forest cover at around 12% of land area, we are one of the least well forested countries in Europe. While importation of timber and pulp has provided some solution to the demands of constructional and paper-producing industries of the UK, the problem confronting those who contemplate the use of home-grown trees and forests as frontline defences against climate change remains a challenging one! Here, we address this problem. We do so, not with a view to finding some 'quick fix' for an emerging short-term climate challenge but with the intention to identify an ongoing and progressively increasing contribution of UK forests to the mitigation of what will be a lasting legacy of excessive fossil fuel exploitation.

We ask and answer questions concerning the present and predicted abilities of UK forests to act as carbon sinks. We also examine how the substitution of wood for fossil fuels and its use in place of materials which require large greenhouse gas emission in their manufacture, can optimise the benefits of the UK forest resource. Ongoing climate change is impacting our woodlands now, influencing their ability to provide environmental, biodiversity, economic and social benefits. Some adaptation will occur naturally, but UK woodlands are managed ecosystems and forest management provides an opportunity to achieve adaptation in a time frame that will limit catastrophic impacts. Inevitably, planting for the future involves some risks and uncertainties but these are no justification for inaction. We are in a position to make maximum likelihood assessments of climate scenarios in mid century and are already armed with sufficient ecological knowledge to enable informed choices concerning the abilities of current as well as new species and provenances to grow well and meet our forestry objectives in the environments that will confront the next generation.

Analysis of the basic science confirms both that climate warming is occurring and that trees are already responding in their growth patterns to these changes. In view of the fact that emissions of greenhouse gases and the associated warming are predicted to increase further, we need to optimise the effectiveness of trees as natural agents for the capture of carbon from the atmosphere. Clearly such considerations cannot be made in isolation. Other pressures upon human populations and the environment, not least those surrounding the adequacy of food supply, will be increasingly to the fore. In such contexts, foresters will need, where possible, to avoid competition with agriculture by exploiting the abilities of trees to grow well, and without the need for fertiliser application, on land that is not suitable for food crops. One legacy of our history of deforestation is that land considered marginal or unsuitable for agriculture is abundantly available in the UK. If, in our quest to enhance carbon capture by trees, we are to exploit these resources, we must do so in a way that minimises carbon release from the soil and other adverse consequences. Changes of forest practice and of wood utilisation to accommodate these requirements are also considered in this text.

While fully recognising that under any foreseeable circumstance economic, social and local environmental factors will act to constrain our abilities to use trees and forests as agents for mitigation of climate change, our intention here is to expose the considerable potential of forestry to contribute to the emerging environmental challenges. Our analysis has been principally of UK forests, their current and potential benefits, however much of the science reviewed here has wider implications and the UK has considerable expertise in woodland creation and sustainable forest management. Climate change is a global problem and there are important international dimensions. We owe it to succeeding generations both here in the UK and elsewhere in the world to investigate, evaluate and employ every possible means to ameliorate the environmental impacts that are arising from the comfortable but profligate styles of life that many in the developed world are fortunate to enjoy.

Sir David Read FRS

Contents

Acknowledgements

It is a pleasure to acknowledge the support and encouragement of Tim Rollinson (Director General, Forestry Commission) who commissioned this report. Special thanks are due to Elaine Dick, our Production Editor, for her constructive inputs, as well as for the forbearance and fortitude that she showed throughout the production process, and to Peter Freer-Smith and Pat Snowdon for meticulously attending to editorial and administrative matters. We would also like to thank our two international reviewers William Hyde and Denis Loustau.

Contributors

Steering Group members and editors

Sir David Read FRS [1, 2]
Department of Animal & Plant Sciences, University of
Sheffield, Western Bank, Sheffield, S10 2TN
d.j.read@sheffield.ac.uk

Jan Bebbington [2]
School of Management, University of St Andrews,
The Gateway, St Andrews, Fife, KY16 9SS
jan.bebbington@st-andrews.ac.uk

Mark Broadmeadow [2]
Forestry Commission, Alice Holt Lodge, Wrecclesham,
Farnham, Surrey, GU10 4LH
mark.broadmeadow@forestry.gsi.gov.uk

Jeremy Eppel
Defra, Nobel House, 9 Milbank, 17 Smith Square, London,
SW1P 3JR
jeremy.eppel@defra.gsi.gov.uk

Peter Freer-Smith [1, 2]
Forest Research, Alice Holt Lodge, Wrecclesham, Farnham,
Surrey, GU10 4LH
peter.freer-smith@forestry.gsi.gov.uk

John Handley [2]
Centre for Urban and Regional Ecology, School of
Environment and Development, Humanities Building,
Bridgeford Street, University of Manchester, Oxford Road,
Manchester, M13 9PL
john.handley@manchester.ac.uk

Nick Hanley [1]
Economics Division, School of Management, University of
Stirling, Stirling, FK9 4LA
n.d.hanley@stir.ac.uk

Wilma Harper
Forestry Commission, Silvan House, 231 Corstorphine Road,
Edinburgh, EH10 7AT
wilma.harper@forestry.gsi.gov.uk

Simon Hodge
Forestry Commission Scotland, Silvan House, 231
Corstorphine Road, Edinburgh, EH10 7AT
simon.hodge@forestry.gsi.gov.uk

Paul Jarvis FRS, FRSE [2]
Institute of Atmospheric & Environmental Science, School
of GeoSciences, The University of Edinburgh, The King's
Buildings, Edinburgh EH9 3JN
margaretsjarvis@aol.com

Keith Kirby [2]
Natural England, Northminster House, Peterborough,
PE1 1UA
keith.kirby@naturalengland.org.uk

Bill Mason [2]
Forest Research, Northern Research Station, Roslin,
Midlothian, EH25 9SY
bill.mason@forestry.gsi.gov.uk

James Morison [1, 2]
Forest Research, Alice Holt Lodge, Wrecclesham, Farnham,
Surrey, GU10 4LH
james.morison@forestry.gsi.gov.uk

Gert-Jan Nabuurs
European Forest Institute, Torikatu 34, 80100 Joensuu,
Finland
gert-jan.nabuurs@efi.int

Martin Parry
Centre for Environmental Policy, Imperial College London,
London, SW7 2AZ
martin@mlparry.com

Pat Snowdon [1, 2]
Forestry Commission, Silvan House, 231 Corstorphine Road,
Edinburgh, EH10 7AT
pat.snowdon@forestry.gsi.gov.uk

Ed Suttie [2]
Building Research Establishment, Garston, Watford,
Hertfordshire, WD25 9XX
suttiee@bre.co.uk

Chris West [1, 2]
UK Climate Impacts Programme, School of Geography and
the Environment, University of Oxford, South Parks Road,
Oxford, OX1 3QY
chris.west@ukcip.org.uk

[1] Editorial team [2] Also authors

International Reviewers

William Hyde
Retired, formerly Professor at Duke and Virginia Tech
Universities, Visiting Professor at Gothenburg University. 1930
South Broadway, Grand Junction, CO 81507 USA
wfhyde@aol.com

Denis Loustau
Directeur de Recherches, Unité de Recherches Ecologie
Fonctionnelle et Physique de l'Environnement (EPHYSE) INRA,
Centre de Bordeaux – Aquitaine France
Loustau@pierroton.inra.fr

Authors

Pam Berry
Environmental Change Institute, School of Geography and the Environment, Oxford University, South Parks Road, Oxford, OX1 3QY
pam.berry@eci.ox.ac.uk

Nick Brown
Department of Plant Sciences, University of Oxford, South Parks Road, Oxford, OX1 3RB
nick.brown@plants.ox.ac.uk

Claudia Carter
Forest Research, Alice Holt Lodge, Wrecclesham, Farnham, Surrey, GU10 4LH
claudia.carter@forestry.gsi.gov.uk

Robert Clement
Institute of Atmospheric & Environmental Science, School of GeoSciences, The University of Edinburgh, The King's Buildings, Edinburgh EH9 3JN
robert.clement@ed.ac.uk

Susannah Gill
The Mersey Forest, Risley Moss Visitors Centre, Ordnance Avenue, Birchwood, Warrington, WA3 6QX
susannah.gill@merseyforest.org.uk

John Grace
Institute of Atmospheric & Environmental Science, School of GeoSciences, The University of Edinburgh, The King's Buildings, Edinburgh EH9 3JN
jgrace@ed.ac.uk

Anna Lawrence
Forest Research, Alice Holt Lodge, Wrecclesham, Farnham, Surrey, GU10 4LH
anna.lawrence@forestry.gsi.gov.uk

Katie Livesey
Building Research Establishment, Garston, Watford, Hertfordshire, WD25 9XX
liveseyk@bre.co.uk

Robert Matthews
Forest Research, Alice Holt Lodge, Wrecclesham, Farnham, Surrey, GU10 4LH
robert.matthews@forestry.gsi.gov.uk

Mike Morecroft
Natural England, John Dower House, Crescent Place, Cheltenham, GL50 3RA
michael.morecroft@naturalengland.org.uk

Bruce Nicoll
Forest Research, Northern Research Station, Roslin, Midlothian, EH25 9SY
bruce.nicoll@forestry.gsi.gov.uk

Maria Nijnik
Socio-Economic Research Group, The Macaulay Land Use Research Institute, Craigiebuckler, Aberdeen, AB15 8QH
m.nijnik@macaulay.ac.uk

Guillaume Pajot
Socio-Economics Research Group, Macaulay Land Use Research Institute, Craigiebuckler, Aberdeen, AB15 8QH
g.pajot@macaulay.ac.uk

Mike Perks
Forest Research, Northern Research Station, Roslin, Midlothian, EH25 9SY
mike.perks@forestry.gsi.gov.uk

Chris Quine
Forest Research, Northern Research Station, Roslin, Midlothian, EH25 9SY
chris.quine@forestry.gsi.gov.uk

Duncan Ray
Forest Research, Northern Research Station, Roslin, Midlothian, EH25 9SY
duncan.ray@forestry.gsi.gov.uk

Bill Slee
Socio-Economics Research Group, Macaulay Land Use Research Institute, Craigiebuckler, Aberdeen, AB15 8QH
b.slee@macaulay.ac.uk

Keith Smith
Institute of Atmospheric & Environmental Science, School of GeoSciences, The University of Edinburgh, The King's Buildings, Edinburgh EH9 3JN
keith.smith@ed.ac.uk

Gail Taylor
School of Biological Sciences, University of Southampton, Bassett Crescent East, Southampton, SO16 7PX
g.taylor@soton.ac.uk

Fergus Tickell
Northern Energy Developments, Parkhouse, Canonbie, Dumfriesshire DG14 0RA
fergus@energydevelopments.co.uk

Joan Webber
Forest Research, Alice Holt Lodge, Wrecclesham, Farnham, Surrey, GU10 4LH
joan.webber@forestry.gsi.gov.uk

EXECUTIVE SUMMARY

Key Findings

UK forests and trees have the potential to play an important role in the nation's response to the challenges of the changing climate. Substantial responses from the UK forestry sector will contribute both to mitigation by abatement of greenhouse gas (GHG) emissions and to adaptation, so ensuring that the multiple benefits of sustainable forestry continue to be provided in the UK.

A clear need for more woodlands

Forests remove CO_2 from the atmosphere through photosynthesis and, globally, could provide abatement equivalent to about 25% of current CO_2 emissions from fossil fuels by 2030, through a combination of reduced deforestation, forest management and afforestation. Analysis of woodland planting scenarios for the UK indicate that forestry could make a significant contribution to meeting the UK's challenging emissions reduction targets. Woodlands planted since 1990, coupled to an enhanced woodland creation programme of 23 200 ha per year (14 840 ha additional to the 8360 ha per year assumed in business as usual projections) over the next 40 years, could, by the 2050s, be delivering, on an annual basis, emissions abatement equivalent to 10% of total GHG emissions at that time. Such a programme would represent a 4% change in land cover and would bring UK forest area to 16% which would still be well below the European average.

Woodland creation provides highly cost-effective and achievable abatement of GHG emissions when compared with potential abatement options across other sectors. The Committee on Climate Change considered that abatement costing less than £100 per tonne of CO_2 was cost-effective. All the woodland creation options evaluated here met this criterion including a range of broadleaved woodlands. The two most cost-effective options were conifer plantations and rapidly growing energy crops, but mixed woodlands managed for multiple objectives can also deliver abatement at less than £25 per tonne CO_2.

An enhanced woodland creation programme would help to reverse the decline in the rate of atmospheric CO_2 uptake by forests that is reported in the UK's Greenhouse Gas Inventory. From a maximum of 16 $MtCO_2$ per year in 2004, the strength of the 'forest carbon sink' is projected to fall to 4.6 $MtCO_2$ per year by 2020, largely because of the age structure of UK forests and the maturation and harvesting of the woodlands created as a result of the afforestation programmes of the 1950s to 1980s. The decline in planting rates since the 1980s also contributes to this serious projected decline in the sink strength of UK forests.

The new woodlands would also deliver a range of co-benefits but would need to respect a range of other land-use objectives including biodiversity, food security, landscape and water supply.

An asset to be managed wisely

Existing UK forests, including soils, are both a large store of carbon (estimated at around 790 MtC) and a system removing CO_2 from the atmosphere (about 15 $MtCO_2$ per year in 2007). Sustainable forest management can maintain the carbon store of a forest at a constant level while the trees continue to remove CO_2 from the atmosphere and transfer a proportion of the carbon into long-term storage in forest products. The total carbon stored in the forest and its associated 'wood chain' therefore increases over time under appropriate management systems.

Impacts of climate change are beginning to become apparent in the UK's woodlands, including effects on productivity, tree condition, woodland soil function, woodland fauna and flora and forest hydrology.

There is increasing concern over the number of outbreaks of novel pests and diseases in forestry and arboriculture. Forest pests and diseases could compromise the ability of woodlands to adapt and contribute to meeting the challenge of climate change.

The regulatory framework and sustainability standards for UK forestry will need to be maintained and, in some cases, adapted to address climate change. A similar approach should be put in place for the management of urban trees. This will ensure that trees continue to deliver a wide range of ecosystem services.

The status quo is not an option

Since tree crops take many years to mature, the planning horizons for forestry are inherently long. Actions taken now may only prove their worth in 50–100 years time and must be appropriate for both the current and future climates. A move towards planned rather than reactive adaptation in woodland creation and management is therefore preferable.

The creation of new woodlands and the restocking programmes of existing forests present major opportunities for adapting forests to future climate change. Changes to the selection of species and provenances for particular sites using the current range of species are required now. These choices can be accommodated using the range of species currently in use. Over longer timeframes, and if greenhouse gas emissions do not decline, we will need to consider the introduction of new species, including those from continental Europe. However, further research is urgently needed to establish which species will be best suited to the changed environmental conditions. The preference for use of native tree species and local provenances under all circumstances will need to be reconsidered.

The changing climate raises difficult questions for conservation of woodland biodiversity. Current descriptions of native woodland communities based on species composition are unlikely to remain valid because some native members of the flora and fauna may struggle to survive.

Harvesting and use of wood increases forestry's mitigation potential

Harvesting of trees leads to transfers of their carbon into wood products where it is stored, often over long periods. These can be used to substitute for those materials the production of which involves high emissions of GHG. Wood products can also be used directly as sources of energy to replace fossil fuels.

Forests achieve their considerable productivities largely in the absence of nitrogen (N) fertilisation, thus avoiding the high fossil fuel costs of N fertiliser production, direct losses of the greenhouse gas nitrous oxide (N_2O) and the risk of pollutant-N loss to the environment as nitrate in water catchments.

Within the next five years sustainably-produced wood fuel has the potential to save the equivalent of approximately 7 $MtCO_2$ emissions per year by replacing fossil fuels in the UK. This contribution could be increased further as bioenergy, including energy derived from woody biomass, makes an increasing contribution to UK targets for renewable heat, power and liquid fuels. The use of biomass for heating provides one of the most cost-effective and environmentally acceptable ways of decreasing UK GHG emissions.

The estimated total quantity of carbon stored in wood-based construction products in the UK housing stock in 2009 is 19 Mt (equivalent to 70 $MtCO_2e$). If the market for wood construction products continues to grow at its current rate over the next 10 years there is the potential to store an estimated additional 10 Mt of carbon (equivalent to 36.7 $MtCO_2e$) in the UK's new and refurbished homes by 2019.

Part of the current failure to accept wood products for use in construction arises from conservatism in the construction industry. Outmoded attitudes need to be robustly challenged by drawing on the evidence and promoting the technical properties of wood.

Trees help people adapt

Trees have an important role in helping society to adapt to climate change, particularly in the urban environment, through providing shelter, cooling, shade and runoff control. Tree and woodland planting should be targeted to: (a) places where people live, especially the most vulnerable members of society, and (b) places where people gather (such as town and local centres) which currently have low tree cover.

Forestry practitioners should engage with the public to contribute to societal understanding and responses to climate change. The changes required will challenge both policy makers and managers to adopt a more flexible approach in response to the emerging body of evidence.

Policy incentives need to be re-designed so that adequate reward is given to the provision of the non-market benefits of forests, especially those relating to the climate change mitigation and adaptation functions of forests.

We conclude that further scientific and socio-economic analysis is required to enable the UK to achieve the full adaptation and mitigation potential of forestry that is identified in this first national assessment. Clear, robust, research programmes will be needed to underpin the changes of forestry policy and practice which are required to meet the new and challenging circumstances.

Forests are a multiple-purpose resource which make up almost a third of the Earth's land surface. Through their photosynthetic and respiratory activities they play a critical role in the global carbon cycle. While there have been numerous global- and continental-scale determinations of the contributions of forests to the planetary carbon cycle, few have considered these issues in depth at the national scale.

Here we present a synthesis of a scientifically-based analysis of the potential of UK forests and trees to play a role in the nation's response to the challenges of the changing climate.

To date the primary scientific input to global climate change negotiations is the Fourth Assessment Report of the Intergovernmental Panel on Climate Change (IPCC) published in 2007, and a summary of more recent developments was published in 2009. Chapter 9 of the Working Group 3 report on mitigation of the Fourth Assessment concluded that forestry could make a very significant contribution to a low-cost global mitigation portfolio that provided synergies with adaptation and sustainable development. The Stern Review commissioned by the UK Government in 2006 similarly concluded that curbing deforestation was a highly cost-effective way of reducing greenhouse gas (GHG) emissions and that action to preserve forest areas was urgently needed. In response to these two reviews the Forestry Commission (FC) hosted a meeting of all interested parties in London (November 2007) to consider the UK response and produced a climate change action plan. One of the four key actions identified was a national assessment of UK forestry's contribution to mitigating and adapting to climate change, to be published prior to the UNFCCC meeting in Copenhagen in December 2009. To achieve this the present independent expert assessment was commissioned with the remit to provide a better understanding of how UK forestry can adapt to and improve its contribution to mitigation of climate change, with the following specific objectives:

- review and synthesise existing knowledge on the impacts of climate change on UK trees, woodlands and forests;
- provide a baseline of the current potential of different mitigation and adaptation actions;
- identify gaps and weaknesses to help determine research priorities for the next five years.

The UK Government has taken a lead in climate change policy development and, while the current assessment was in progress, set challenging and legally binding targets leading to an emissions reduction of 80% of 1990 GHG emissions by 2050. A contribution to the targeted reductions in atmospheric GHG concentrations can be achieved by increasing the rates at which the gases are removed from the atmosphere through biological uptake.

In its White Paper, the UK Low Carbon Transition Plan (2009), the Government identified woodland creation as a cost-effective way of fighting climate change and recognised the urgency of action to support tree-planting initiatives. This pressure to act is further accentuated by the climate change projections (UKCP 2009) which use model simulations to provide probabilistic estimates of future climate. These indicate that the UK climate will continue to warm substantially through this century; that there will be changes in rainfall patterns and its seasonal distribution; and that considerable regional variations can be expected. These projections are timely since they provide the essential physical background needed to inform those adjustments of forestry policy and practice that will be required both to mitigate the impacts of the projected changes and to tailor adaptation measures to local conditions.

The science reviewed here and the general implications for policy advice which arise from it are thus presented at a critical time in the development of UK policies on woodland creation and of other actions designed to achieve adaptation and mitigation through UK forestry. Our assessment has yielded the overarching and strongly-held conviction that, confronted by climate change, substantial responses are required of the forestry sector. This evaluation of the science shows that the UK forestry sector can contribute significantly both to the abatement of emissions and to ensuring, through effective adaptation, that the multiple benefits of sustainable forest management continue to be provided.

Forests and atmospheric carbon

The largest impact of forests on atmosphere/land surface exchange arises through the net ecosystem exchange of carbon. Woodlands and forests are a net sink of CO_2, i.e. they remove CO_2 from the atmosphere, except during tree harvesting and for a relatively short period thereafter (the duration depending on soil type and other site factors). The strength of this sink (i.e. the rate of removal) has been quantified for general UK forest types (coniferous and broadleaved woodlands) and the impact of some management activities on CO_2 emissions and removal are known. While there are gaps in understanding, sufficient is known for the overall GHG balance to be calculated and for projections to be provided for both the existing UK forest cover and for a range of alternative forest management scenarios.

Measurements made over several years in a coniferous forest in Scotland show average annual removal from the atmosphere of around 24 tonnes of CO_2 per hectare per year (tCO_2 ha^{-1} yr^{-1}). Comparable measurements made in an oak forest in southern England indicate that

it removes c. 15 tCO_2 ha^{-1} yr^{-1} (Chapter 3, Figures 3.3 and 3.6, pages 30 and 32). Analysis shows that the rate at which carbon accumulates in forest stands from one crop of trees ('rotation') to the next rotation is influenced by site (particularly soil type), forest operations and by the extent of soil disturbance at planting and harvest. In considering emissions abatement it is important to note that, while forest carbon stocks will reach upper limits, the total abatement can continue to rise over successive rotations because of carbon storage in wood products and the substitution of wood for fossil fuel. The dynamics of the current and projected UK forest carbon sink are largely determined by historic planting patterns which involved extensive afforestation, mostly in the uplands, through the 1950s to 1980s. The c. 1 million hectares of coniferous forest planted in the UK, mainly on marginal land, over this period represents a major resource as both a carbon store and a carbon 'sink'. In contrast to agricultural crops, such forests achieve their considerable productivities largely without the application of nitrogen (N) fertilisers, thus avoiding the high fossil fuel costs of N fertiliser production, direct losses of the greenhouse gas nitrous oxide (N_2O) and the risk of nitrate (NO_3^-) pollution of water resources.

Although the UK's existing forest area has more than doubled over the past 80 years, at around 12% it is amongst the lowest of any country in Europe. Further, annual areas of new planting have declined sharply since 1989 (Chapter 1, Figure 1.1, page 6) and this has important consequences for the potential contribution that UK forests can make to mitigation of climate change. Current estimates show that, largely as a result of the earlier planting activities, the strength of the UK carbon sink increased from 12 $MtCO_2$ yr^{-1} in 1990 to a peak of 16 $MtCO_2$ yr^{-1} in 2004. In Scotland, where woodland cover is higher than in the rest of the UK and the population density is smaller, the removal of CO_2 by forests currently accounts for around 12% of emissions. However, the situation is very different in England, where the forest carbon sink equates to less than 1% of total GHG emissions.

Because of the age structure created by the planting history in UK forestry, as the harvesting of the forests created in the last century continues, falls in net CO_2 uptake by UK forests are expected even though harvesting is usually followed by restocking. Significantly, the afforestation rates of the post-war years have subsequently been greatly reduced, so the projected CO_2 uptake by UK forests shows a marked decline to as little as 4.6 $MtCO_2$ yr^{-1} by 2020 (Chapter 8, Figure 8.1, page 141). Such declines, which are reported as part of the UK's GHG inventory to the UNFCCC, have serious implications for our ability to meet the challenging targets for emission reductions outlined in the Climate Change Act and the UK Low Carbon Transition Plan. The rapid diminution of the carbon sink provides both a challenge and an opportunity for the forestry sector, and by making provision for reversal of this trend it has the potential to increase its total contribution to climate change mitigation.

Baseline and potential for mitigation in UK forests

The total carbon stock in UK forests (including their soils) is approximately 790 MtC and the stock in timber and wood products outside forests is estimated to be a further 80 MtC. Changes to the large forest carbon stocks can be achieved through management practices. More intensive Forest Management Alternatives (FMAs – see Chapter 6, Table 6.3, page 104) have the lowest standing carbon stocks, but by far the highest annual rates of carbon sequestration (uptake). Since these estimates consider only carbon in the forest, they do not include the further potential of wood to provide abatement of GHG emissions by substituting for fossil fuels as well as for materials which cause large GHG emissions in their production.

Models have been developed which simulate the full life cycle of carbon, including that retained in forest products and fossil fuel emissions avoided by the use of forest products. These enable us to provide, for the first time, a comprehensive evaluation of the abatement potential of UK forests. These model calculations demonstrate that a combination of increased new planting and the substitution benefits of wood fuel and wood products for fossil fuel intensive materials, has the potential to deliver significant abatement (i.e. reduction in net GHG emissions). New woodlands can be planted to deliver a range of forestry and other objectives. Different options for planting provide the potential for significant abatement either as sequestration or substitution, particularly over the longer term (Chapter 8, Figure 8.5, page 154). Energy forestry, in contrast to multi-purpose and conifer/mixed forestry, shows a larger contribution to abatement through substitution than through sequestration in the forest stand and soil.

Clearly, the extent to which abatement can be achieved will be directly proportional both to the increases in area over which the new planting takes place and the extent of fossil fuel substitution that can be gained. For example, woodlands planted since 1990, coupled to an enhanced woodland creation programme involving planting 23 200 ha (14 840 ha over and above the business as usual assumption of 8360 ha per year) of forest per year over the next 40 years, could deliver abatement of c. 15 MtCO$_2$ by the 2050s, providing the substitution benefits of wood and timber products are taken into account (see Chapter 8, Figure 8.4, page 148). This level of abatement would

equate to about 10% of total GHG emissions from the UK if recent emissions reduction commitments were achieved. Such a programme of woodland creation might incorporate energy forestry, conifer forests, farm and native broadleaved woodland and would establish nearly one million hectares of woodland, bringing total forest cover in the UK to approximately 3.8 million hectares. This rate of afforestation would represent both a major change in, and challenge to, the forestry sector. However it would represent only a 4% change in land use and result in UK woodland cover of 16% which would remain well below the European average. If any such changes of land-use practice were to be implemented it would be important to ensure that the regulatory framework and sustainability standards for woodland creation in the UK were maintained.

Much discussion has centred on the potential of changes in approaches to forest management to deliver emissions abatement. Delaying thinning and harvesting operations to increase in-forest carbon stocks do deliver abatement in the short term if sequestration in forest biomass alone is considered. However, when the abatement associated with wood and timber products substituting for fossil fuels is considered in the analysis, the additional carbon retention delivered through storage in forest biomass is rapidly negated by the lost ability to deliver abatement through substitution. Within a period of 40–50 years, total abatement potential is lower for management scenarios that aim to increase in-forest carbon stocks (see Chapter 8, Figures 8.4 and 8.6, pages 148 and 155). There is therefore a risk that measures that focus solely on increasing forest carbon stocks are likely to limit the abatement potential because of lost opportunities for fossil fuel and product substitution. It is also evident (see Figures 8.4 and 8.6) when a range of options are considered, that there is limited scope for changes in forest management, alone, to deliver significant levels of emission abatement, implying that woodland creation should be the initial focus of activity.

While it is evident from such analyses that the forestry sector has the potential to make a significant contribution to emissions reduction commitments, the economic viability of specific options for woodland creation has to be considered. In establishing the capacity for abatement to be delivered by all sectors, the first report of the Committee on Climate Change considered that abatement costing less than £100 per tonne CO$_2$ was potentially cost-effective. Here, a detailed analysis of new woodland planting scenarios shows that multi-purpose forestry provides highly cost-effective abatement (0 to £26 per tonne of CO$_2$), and that abatement using energy forestry (short rotation forestry and coppice) would be even more cost-effective. Indeed, in the case of highly-productive stands the value of timber produced can exceed the cost of establishment and management so that no net social costs arise. In the current assessment, the creation of new native broadleaved woodlands managed for biodiversity objectives is estimated to provide carbon abatement at £41 per tonne of CO_2. However this figure is pessimistic, as it does not include an evaluation of the co-benefits that such woodland creation would provide. Nor does this figure take account of GHG emissions or removal under the land use prior to the creation of the new woodland, which in some cases may be large and add further to the overall benefit of woodland creation. The scale of land-use change envisaged here would clearly require an integrated approach involving full consideration of GHG balances and of all the ancillary benefits of woodland creation.

Impacts and adaptation of forests, woodlands and urban trees

Climate change is already having impacts in UK forestry. These include effects on productivity, tree condition, leaf emergence, woodland soil function, woodland fauna and flora, forest hydrology and, probably, also the incidence of insect pest and tree disease outbreaks. However, there is uncertainty over the likely severity and extent of these impacts in the future.

Increased climatic warmth, the lengthening of growing seasons, and rising atmospheric CO_2 concentrations may improve tree growth rates so long as water is not limiting and pest and disease outbreaks do not have significant impacts. There is good experimental evidence that increases of CO_2 concentration can lead to increased growth of trees although considerable variability of response has been observed determined by genotype, tree age, air pollution and nutrient availability. In the UK, there is only limited evidence that any increases in tree growth or overall forest productivity can be attributed to longer, warmer growing seasons and rising atmospheric CO_2 concentrations. Recent studies attribute increased forest productivity across much of continental Europe mainly to changes in forest management and nitrogen availability.

An increased frequency and severity of summer drought is likely to represent the most immediate threat to UK woodlands from the changing climate. There is a very high likelihood that climate change will have serious impacts on drought-sensitive tree species on shallow freely-draining soils. Over the near future (<40 years) the range of species currently considered suitable for use in woodland creation in the UK is likely to remain the same. Exceptions to this general situation are likely to be found in south and east England and more widely in the latter half of the century. The planning of which species and species mixtures to

plant on particular sites will be the challenge for forest managers. Over the longer term however (>40 years), especially if high GHG emissions scenarios are taken into account, an extended range of species will have to be considered, Further research is needed to establish which tree species will be most suitable for specific requirements.

Because tree crops take many years to mature, planning horizons in forestry are inherently long. In order to enhance adaptability and resilience of new planting programmes both the current climate and the relatively uncertain projections of the future climate must be taken into account. Models indicate that under a High GHG emissions scenario, significant impacts of climate change on the suitability of species currently used for forestry will become evident by the middle of the century. Under a Low emissions scenario similar impacts will be seen towards the end of the century. Predictions of 'suitability' for future UK climate projections have been obtained using a knowledge-based model (Ecological Site Classification) which assesses the influence of temperature, soil moisture, wind risk and soil nutrient availability upon tree performance. This shows that typical conifer stands currently in the ground are likely to reach maturity before serious impacts of climate change are apparent. Importantly, however, the need for adaptation has to be confronted at the time of planting both when restocking and when creating new woodlands. By the end of the century, the climate will have become unsuitable for some native and non-native tree species. The change of climate from suitable to unsuitable for many species will be particularly marked in southern England where even those species which will remain generally suitable for use in the wider UK will struggle on some sites. This is exemplified by the case of oak by the 2050s under a High emissions scenario (see maps opposite). For such species the extent of regeneration from seed and successful establishment may also decline so that they become susceptible to competition from introduced species – by then better suited to the changed climate. The growth of other species may improve, as for example, in the case of Sitka spruce in the west and north west under a Low emissions scenario by the 2050s (see maps opposite). The regional variability of tree responses to climate change that is predicted by current models of perfomance or ecological suitability is crucial because it means that bespoke adaptation strategies are required rather than a generic 'one size fits all' approach. For example, of the 28 species assessed using the Ecological Site Classification model, 20 were predicted to increase in suitability in Central Scotland by the 2080s under a High emissions scenario. In contrast, all but two of those species are predicted to show a decline in productivity in southeast England and only one conifer species is predicted to be anything but 'unsuitable'. Further research is therefore needed to establish which

The 'suitability' (defined as productivity relative to maximum productivity achievable by that species under current climatic conditions) for (a) pedunculate oak, and (b) Sitka spruce under Baseline (1961–90, left) and UKCIP02 Low emissions (centre) and High emissions (right) climate change scenarios. The results are based upon Ecological Site Classification (see text for further explanation).

Dark green = very suitable (>70% of current maximum productivity); light green = suitable (50–70% of maximum productivity); orange = marginal (40–50% of maximum productivity); blue = poor (30–40% of max productivity); red = unsuitable (<30% of current maximim productivity).

Pedunculate oak

Baseline   2050 Low emissions scenario   2050 High emissions scenario

Sitka spruce

Baseline   2050 Low emissions scenario   2050 High emissions scenario

tree species are most suitable for specific requirements and, in particular, to determine which infrequently planted (minor) or untried species are candidates for 'adaptive planting' (see Chapter 6, Table 6.5, page 108). The extent of new planting and changes in species choice must, however, be appropriate and sensitive to the potential implications for biodiversity, agriculture, water harvesting, housing and infrastructure development, alongside the

other associated costs and benefits.

There are likely to be significant changes to the composition, structure and character of the ground flora and other species of priority for biodiversity and conservation, particularly under High emissions scenarios and over longer timeframes. Current species descriptions of native woodland communities are unlikely to remain valid so the changing climate raises difficult questions for conservation of woodland biodiversity. In replanting, the preference for use of native tree species and local provenances under all circumstances will need to be reconsidered. Diverse semi-natural woodlands are likely to be able to adapt through natural processes, particularly since the majority of native tree species will persist across the UK – albeit with changes in their distribution and growth rate. However for this adaptive potential to be realised, management intervention (i.e. conventional good practice for woodland management) will be necessary in most woodlands to create a diverse structure and promote natural regeneration.

Pests and diseases of forest trees, both those that are already present in the UK and those that may be introduced, currently represent a major threat to woodlands, by themselves, and in interaction with the direct effects of climate change. There have been a number of serious pest and pathogen outbreaks in the UK over the past 15 years and the types of attack which we have seen have the potential to compromise the ability of forests to contribute to addressing climate change. The extent to which increased world trade in plants and forest products and climate change contribute to current threats is uncertain, but there is a need to reduce the future risks and to manage the existing outbreaks. It is essential that appropriate and effective interception and monitoring systems are in place to prevent the introduction of pests and pathogens. We also need early warning of impending threats and an effective response when outbreaks do occur; this requires good interaction between scientists and those responsible for outbreak management.

Trees also have an important role in helping society to adapt to climate change, particularly in the urban environment. Tree and woodland cover in and around urban areas will be increasingly important for managing local temperatures and surface water. Large tree canopies are particularly beneficial. Guidelines should be followed by all concerned parties both to ensure that we continue to maintain and plant trees in urban areas, and to overcome perceived risks including subsidence and windthrow. Where soil water stress is likely to be a problem, planting should focus on more drought-tolerant species. It is crucial that we have a thorough understanding of the current pattern of tree cover in urban areas, to target where we need to maintain and increase cover. There is also an important role for planting woodland along urban river corridors to reduce thermal stress to fish and freshwater life. Tree and woodland planting should be targeted to: (a) places where people live (especially the most vulnerable members of society) which currently have low tree cover, and (b) places where people gather (such as town and local centres) which currently have low tree cover.

Wood fuel and wood products substituting for other materials

By substituting for other materials with greater climate impact, wood products and wood fuel have a significant role to play in reducing carbon emissions in the UK. Forest products should comprise a larger share in the supply of biomass energy and of wood products used in construction. The UK Renewable Energy Strategy considered that renewables could contribute 15% of total energy requirements by 2020, with wood fuel making a significant contribution to the electricity, transport and heat-generating sectors. The UK has a significant biomass resource although estimates of the potential contribution from forestry vary considerably. The Renewable Energy Strategy has identified biomass conversion to heat as a least-cost way to increase the share of renewable heat for which there is a target of 12% by 2020. The deployment of forest resources to achieve these renewable energy targets is now a priority. The annual production of 2 million tonnes of wood fuel from English woodlands is based on the Wood fuel Strategy for England, while the Scottish Forestry Strategy commits to delivering 1 $MtCO_2$ abatement per year through renewable energy production by 2020. Over the next five years wood fuel in the UK has the potential to save up to 7.3 Mt of CO_2 emissions per year by substituting for fossil fuel. If 1 million ha of dedicated energy forests were planted on current agricultural land it would be possible to increase this to the equivalent of 14.6 $MtCO_2$ abatement per year over the next decade. Biomass for heat provides one of the most cost-effective and environmentally acceptable ways of decreasing UK GHG emissions.

Of the UK's current 2.5% by volume of liquid transport fuel obtained from biological sources more than 70% is imported, the sustainability of this supply being largely unregulated. Both the UK and EU will address the issue of sustainability in the near future. The changes likely to be made will involve restriction of feedstocks to those showing at least a 35% improvement in GHG balance relative to fossil fuels. In the UK this could favour the development of the woody biomass energy industry and involve the conversion of woody materials to liquid fuels through biological processes and thermochemical routes such as pyrolysis. To date about half the biomass feedstocks for the UK have come from imports, including approximately 1 million tonnes of biomass for co-firing. The co-firing market in the UK grew by c. 150% between 2004 and 2006 and is likely to expand further. This will be driven in part by changes to the Renewable Obligation Certificates (ROCs) which will provide better incentives for home-grown biomass as compared with imported supplies. As a consequence of these measures, bioenergy, including that obtained from wood, is likely to make an increasingly important contribution to UK targets for renewable heat, power and liquid fuels.

Substitution of wood products for materials which release GHG in their manufacture contributes to mitigation of climate change both by storing carbon in our buildings and by reducing fossil fuel consumption. The estimated total quantity of carbon stored in the form of wooden construction products in the UK housing stock in 2009 is 19 Mt. If the wood construction products sector continues to grow as it has in the past ten years there is the potential to store an estimated additional 10 Mt carbon (equivalent to 36.7 $MtCO_2$) in the UK's new and refurbished homes by 2019. This would save a further 20 MtC (73.4 $MtCO_2e$) as a consequence of substitution for more carbon-intensive materials. To date, failure to accept wood products arises in part from conservatism in the construction industry. Outmoded attitudes need to be robustly challenged by drawing on the evidence and promoting the technical properties of wood.

Sustainable development and socio-economic considerations

A complete assessment of the potential for our trees and woodlands to contribute to climate change goals can only be made by examining the social, economic and policy context within which UK forestry operates. The extent to which the potential for additional emissions abatement through tree planting is realised, for example, will be determined in large part by economic forces and society's attitudes rather than by scientific and technical issues alone. Private forest owners will require financial incentives to manage land for carbon sequestration, except where it is a joint venture associated with other types of forestry. Furthermore, trees and woodlands across the UK contribute to a wide range of policy objectives (including, for example, recreation provision and biodiversity protection), and woodlands need to be planned so that these objectives are achieved together with emissions abatement. Clearly, there are demands on land for other purposes – notably food production and urban development – which affect the economic potential for land to be allocated to forestry. Policies and practices in agriculture, planning and development and other urban and rural activities will affect the capacity of woodlands to deliver climate change mitigation and adaptation objectives, whilst the returns available through markets as well as Government support for competing land uses will influence how much new forest planting occurs. Policy incentives need to be re-designed so that adequate reward is given to the provision of non-market benefits, including those relating to the climate change mitigation and adaptation functions of forests. The knowledge built up in the UK and beyond should be used to facilitate more successful mitigation–adaptation interactions in the forestry/land use sectors in the wider

context of sustainable development and promotion of rural livelihoods.

Trees and forests have a strong role in the way that people make sense of their environment and of how it is changing. This suggests a particularly significant role for those involved in forest management to engage with the public. By this means they will contribute to a broader understanding of the challenges posed by climate change. The scale and urgency of these challenges are such that they require to be driven at the institutional level, and cannot be left to the actions of individuals.

Grasping the opportunity – next steps to realising forestry's contribution

This Assessment provides the evidence base for a much greater involvement of UK forests and forest products in the fight against climate change. However, provision of the evidence to substantiate the potential contribution of forestry is only the first step towards its realisation. There remain large areas of uncertainty. We have identified research priorities at the end of the chapters in this report that are targeted in particular at these uncertainties, but as important will be the processes of communication of the findings to those in decision-making positions in both the public and private sectors. Awareness at this level will enable the development of policies putting trees, woodlands and forestry at the heart of the UK's response to climate change. The key message of the Stern review was 'Act now or pay later'. In view of the fact that the strength of the carbon sink provided by UK forests is weakening so rapidly, the key message from this Assessment is **'Plant Now and Use Sustainably.'**

	Chapter
# INTRODUCTION	
P. Snowdon	1
## Key Findings	

Forest cover in the UK has risen substantially in the past 90 years, although it remains significantly below the European average and current levels of woodland creation are low. The Government has recognised the need for a significant increase in planting to enable the potential of trees and woodlands to mitigate climate change.

Climate change has added major new policy objectives for the forestry sector. Sustainable forest management, and the provision of multiple social and environmental benefits, remain at the heart of forest policy. Effective standards, guidance and management plans will be essential in ensuring that climate change objectives are achieved in balance with other objectives.

Maximising the capacity of forests and woodlands to mitigate and adapt to climate change requires actions that are tailored to local and regional conditions. Devolution of forest policy has enabled such conditions to be recognised more explicitly and for appropriate actions to be taken forward.

National and international legislation has placed ambitious targets for the UK to reduce its emissions of greenhouse gases. Achievement of these targets is dependent upon thorough analysis of the potential for forestry to contribute to mitigation.

Forests are a unique and multi-purpose resource. They deliver a diverse range of benefits for economies, society and the environment. In particular, they play a critical role in the global carbon cycle through their role in the exchange of carbon between the land and the atmosphere.

Human activity is increasing the concentration of carbon dioxide (CO_2) in the atmosphere through emissions from fossil fuel combustion (about 30 Gigatonnes[1] of CO_2 globally in 2007; International Energy Agency, 2009) and from land-use change, primarily tropical deforestation (about 5.8 $GtCO_2$ per year globally in the 1990s; Denman *et al.*, 2007). About half of the emissions from fossil fuel combustion remain in the atmosphere. The rest of the emissions are absorbed by oceans and the land surface. It is estimated that the world's forests and other terrestrial vegetation absorbed about 3.3 $GtCO_2$ per year during the 10 years between 1993 and 2003 (Nabuurs *et al.*, 2007).

There is an urgent need to consider what actions are needed to minimise the impacts of the projected changes in our climate. Forestry is a special case because many decisions that we make now about our trees and woodlands will take effect years and decades into the future. The natural qualities and human uses of forests give them the capacity to mitigate against the threats posed by climate change. Forests absorb carbon from the atmosphere and store it in trees, vegetation and the soil. Human uses of wood from forests provide a renewable source of energy from woodfuel and store carbon in wood products. Importantly, woodfuel and wood products can be used in place of more fossil fuel-intensive fuels and materials (this is termed 'substitution'). At a global level, the average mitigation potential of the forest sector has been estimated at 5.4 $GtCO_2$ per year (Kauppi *et al.*, 2001). Trees and forests also increase the quality and resilience of our living and working environment, for example by providing shade and shelter and by helping to control flooding. In these ways, they help society to cope with the impacts of a changing climate.

It is more important than ever that the contributions of UK forests to climate change mitigation and adaptation are properly understood and that the potential for further

[1] 1 Gigatonne or Gt = 10^9 or billion tonnes – see Glossary.

action is realised. To assess the potential of our trees and forests to tackle climate change, we need answers to some important questions:

- How are trees responding to climate change and what is projected to happen in the future?
- Can more carbon dioxide be absorbed from the atmosphere by planting new woodlands, and by changing the ways we manage existing woodlands?
- What is the potential to use wood as a fuel for heat and power instead of fossil fuels?
- What scope is there to use more timber in place of other more fossil-fuel intensive materials?
- How can we adapt our woodlands to climate change?
- How can trees and woodlands improve our urban and rural environment to help us cope with climate change?

This report aims to provide answers to these and many other questions for the UK. It gathers the existing evidence, sets out the current contribution of forests to climate change mitigation and adaptation and examines their potential to make a greater contribution in the future to tackling what is the greatest challenge for the UK in the 21st century. We hope that the contents will inform policy- and decision-makers, scientists, and all who have an interest in the future of our planet.

1.1 Background

Climate change is the focus of extensive research and analysis in the UK and globally. This has greatly strengthened the scientific case that global warming is taking place and that radical actions will be needed to avert its most severe effects. Recent assessment reports by the Intergovernmental Panel on Climate Change (IPCC) have done much to assemble the best evidence available, assess its quality and indicate where improvements are needed. Its fourth and latest report[2] noted that:

- Warming of the climate system is unequivocal, as is now evident from observations of increases in global average air and ocean temperatures, widespread melting of snow and ice and the rising global average sea level.
- Most of the observed increase in global average temperatures since the mid-20th century is very likely due to the observed increase in GHG concentrations from human activities.
- Continued GHG emissions at or above current rates would cause further warming and induce many changes

in the global climate system during the 21st century that would very likely be larger than those observed during the 20th century.

The Stern Review (Stern, 2007) provided an economic perspective and stressed the importance of taking action now to avert the most serious effects (costs) of climate change in the future. Most recently, a major international review of the latest science has concluded that the speed and impacts of climate change are likely to be more severe than previously reported and that action to reduce emissions and to adapt is most urgently needed (Richardson et al. 2009). In deciding how to respond, we need evidence across all sectors of the economy on what actions might be most effective, the potential that they hold and how much they will cost.

The contribution of trees and forests to meeting the climate change challenge is significant. The IPCC's 2007 report stated that:

> 'Forestry can make a very significant contribution to a low-cost global mitigation portfolio that provides synergies with adaptation and sustainable development' (Nabuurs et al., 2007, p. 543).[3]

It concluded, however, that only a small part of forestry's potential contribution has been realised to date, and that better evidence and data are needed to underpin this contribution. This report goes some way to providing this evidence base in the UK and highlights where gaps in our understanding exist. It is the first time that such an assessment has been carried out for the UK forestry sector.

1.2 Aims and objectives

The aim of this report is to provide an expert up-to-date assessment of the current and potential contribution of trees and forests across the UK, both in the private and government sectors, to addressing climate change. Specific objectives are to:

- review and synthesise existing knowledge on the impacts of climate change on UK trees, woodlands and forests;
- provide a baseline of the current potential of different mitigation and adaptation actions;
- identify gaps and weaknesses to help determine research priorities for the next five years.

---

[2] See the Summary for Policymakers of the Synthesis Report. www.ipcc.ch

[3] The report of Working Group III on Mitigation.

The report is intended to provide a critical step in identifying how forestry in the UK can improve its contribution to climate change mitigation and adaptation. It is aimed at those with research and policy interests as well as anyone wishing to increase their understanding of how our trees and woodlands can help to tackle the challenges of a changing climate.

1.3 Structure of the report

The report comprises 14 chapters grouped into six sections as follows:

- **Section 1** provides the background for the report. It describes the structure of UK forests and reviews the policy framework. It then describes current trends and projections in our climate. The final part of this section explains the science behind the relationship between trees and forests (including soils) and greenhouse gases (GHG).
- **Section 2** assesses evidence of the impacts to date and of the likely impacts of climate change in the coming decades, using climate projections from the UK Climate Impacts Programme.
- **Section 3** focuses on the contribution of forests to mitigating climate change. It reviews the role of forest planting and management in absorbing and storing carbon, and examines the use of woodfuel and wood products in place of fossil fuels and products whose manufacture generates high levels of GHG emissions (i.e. substitution). It also assesses different scenarios for woodland creation and management in order to estimate the potential for forestry to abate GHG, taking account of a range of factors that determine the cost of mitigating carbon.
- **Section 4** reviews the scope to adapt our woodland resource to a changing climate and examines the role of trees, woods and forests in helping society adapt to the impacts of climate change.
- **Section 5** places forestry in a broader context of land use and sustainable development. It provides an economic perspective and considers what might affect people's behaviour in responding to climate change, and the role of institutions in assisting appropriate responses.
- **Section 6** summarises our conclusions and sets out future research needs.

The intention throughout the report is to assess existing knowledge and evidence. Authors have identified research priorities at the end of each chapter and these are reviewed in Chapter 14.

1.4 Forests in the UK: structure, trends and values

Trees are a highly effective natural means of removing CO_2 from the atmosphere. It is therefore important to examine the size and structure of our forests, and how these change over time. The amount of woodland cover and the types of trees will have a major effect on the capacity of forests to absorb and store carbon and to help adaptation to climate change.

There has been a major increase in UK forest cover over the past 90 years. Woodland cover has increased from 5% in the 1920s to approximately 12% in 2008, covering an area of 2.8 million hectares (ha). As shown in Table 1.1, the increase has been most marked in Scotland, where woodlands now cover approximately 17% of the land area.

Conifers make up almost 60% of UK forests, although substantial variations exist at a regional level (Table 1.2). English woodland cover is largely broadleaved, whereas over three quarters of woodlands in Scotland are coniferous. The majority of woodland cover is planted forest. Just over 20% (646 000 ha) is semi-natural woodland, of which 326 000 ha are ancient (dating to 1600[4] or earlier). Conifer species tend to grow, and absorb carbon, at a faster rate than broadleaved species. Approximately one-third of the UK's forests is in public ownership (Table 1.2). The proportion of forest cover owned by the private sector has been increasing in recent years.

Most of the expansion of the forest area has been the result of substantial planting programmes in the decades following the First, and particularly the Second, World Wars. Two-thirds of woodlands in the UK have been planted since 1950, predominantly conifers. Since the late 1980s, the rate of new planting has been in steady decline, falling from around 30 000 ha per year to 7500 ha in 2008 (Figure 1.1), of which 99% is on private land. From the perspective of mitigating climate change through planting more trees, this fall in woodland creation is the opposite of what is needed. At a time when GHG emissions continue to rise, there is a pressing need to maximise carbon sequestration in our forests, alongside the other benefits that woodlands provide. An examination of ways forward is provided in Chapter 8.

The potential of our forests to tackle climate change is also constrained by the current age profile of the UK's woodlands. Our forests are dominated by mature coniferous woodlands, which will reach felling age over the next 15

[4] 1750 in Scotland.

Table 1.1
Total woodland cover in the UK, in thousand hectares (ha) and as a percentage of total land area.

Year	England Area (000 ha)	(%)	Scotland Area (000 ha)	(%)	Wales Area (000 ha)	(%)	Northern Ireland Area (000 ha)	(%)	UK Area (000 ha)	(%)
1924	660	5.1	435	5.6	103	5.0	13	1.0	1211	5.0
1980	948	7.3	920	11.8	241	11.6	67	4.9	2176	9.0
2008	1127	8.7	1342	17.2	285	13.7	87	6.4	2841	11.7

Source: Forestry Commission (2008).

Table 1.2
Woodland type and ownership in the UK.

Ownership	Woodland area (000 ha) Conifer	Broadleaf	Total
Forestry Commission/Forest Service (Northern Ireland)			
England	147	55	202
Scotland	424	28	452
Wales	92	14	106
Northern Ireland	56	5	61
UK	719	102	821
Non-Forestry Commission/Forest Service (Northern Ireland)			
England	219	706	925
Scotland	621	269	890
Wales	65	114	179
Northern Ireland	10	16	26
UK	915	1105	2020
All woodland			
England	366	761	1127
Scotland	1045	297	1342
Wales	157	128	285
Northern Ireland	66	21	87
UK	1634	1207	2841

Source: Forestry Commission (2008).

years (Figure 1.2). This will lead to a temporary trough in the net absorption of CO_2 by UK forests in the medium term as trees are felled and young trees are planted.

Changes to management practices have made the forests across the UK increasingly diverse in age and spatial structure and this will steadily reduce the peaks and troughs in the absorption of CO_2. Ongoing action of this kind will be very important in enabling forests to sustain their role as a major carbon sink and to fulfil that role consistently over time.

The main stocks of carbon within forests are the trees, other vegetation and soils. Important carbon stocks also exist outside of forests, notably in timber and wood products. The total stock of carbon in UK forests (trees only) is estimated to be about 150 million tonnes (Broadmeadow and Matthews, 2003). The average is therefore about 54 tonnes of carbon per hectare. Table 1.3 shows that most of this carbon is held in England and Scotland, in roughly equal quantities.

Actively growing forests in the UK sequester on average 3 tonnes of carbon per hectare per year, although this varies

Figure 1.1
Annual areas of new planting 1975–2009, in the UK.

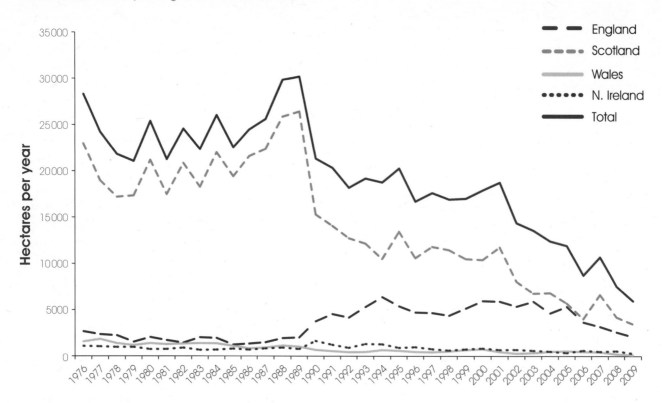

Source: Forestry Commission, 2009.

Figure 1.2
Age structure of forests in the UK.

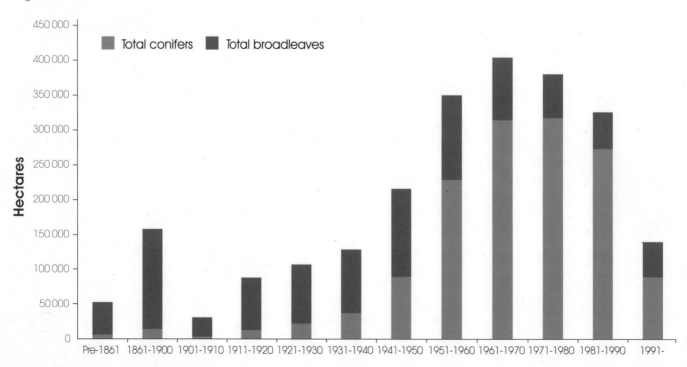

Source: National Inventory of Woodland and Trees 1995–99 (excludes Northern Ireland).

substantially with age, species and growing conditions (Morison *et al.,* 2009; Broadmeadow and Matthews, 2003). The UK GHG Inventory (Centre for Ecology and Hydrology *et al.,* 2008) shows that, in total, UK forests sequester over 4 MtC per year (see Table 1.3). Although the carbon stock in forests is roughly equal in England and in Scotland, the flows of carbon from the atmosphere to forests (i.e. sequestration) is much higher in Scotland due to the predominance of fast-growing conifer species.

Table 1.3
Stock of carbon stored in UK woodlands (trees only).

	Stock of carbon[1] (MtC)	Flows of carbon from atmosphere to forests[2] (MtC per year)
England	63	0.89
Scotland	62	2.66
Wales	18	0.40
Northern Ireland	6	0.17
UK	150	4.12

Note: figures do not sum to totals due to rounding.
[1] Forest Research Woodfuel Resource (www.forestry.gov.uk/woodfuel).
[2] Centre for Ecology and Hydrology *et al.* (2008).

It is estimated that about 640 MtC are stored in UK forest soils, bringing the total carbon stock in all forest carbon pools to over 790 MtC (Morison *et al.,* 2009). Carbon stocks outside of forests (largely timber in construction) are estimated to hold a further 80 MtC (Broadmeadow and Matthews, 2003; Figure 1.3).

Figure 1.3
Carbon pools in UK Forests (MtC).

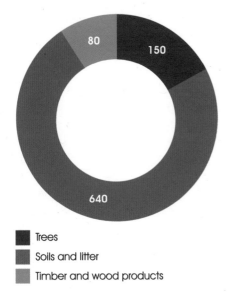

- Trees
- Soils and litter
- Timber and wood products

Forestry contributes to the economy in different ways. Wood production in the UK has been rising in recent years as a growing proportion of coniferous woodlands reach maturity. Production of softwoods, which accounts currently for 95% of the total, is predicted to rise from just over 9 Mt in 2007 to almost 12 Mt per year in the period 2017–21. The value of trees in absorbing carbon has important management implications for timber production, and for future forest management practices. If left standing, trees continue to absorb carbon beyond the age at which they are normally felled for timber and other uses, although this absorption will tail off in the longer term. Extended rotation lengths may, therefore, enhance the carbon sink effect of forests. However, an increase in rotation length affects the production of timber. It would also, after a time, reduce the rate of carbon absorption to low levels and would delay the benefits both from storing carbon in wood products and from the subsequent cycle of carbon absorption in a new tree crop. This issue is examined in Chapters 3 and 6. Analysis is needed to improve understanding on the balance, and potential trade-offs, between the role of forests in producing timber and mitigating climate change.

In 2006, forestry and primary wood processing supported an estimated 43 000 direct jobs and generated approximately £1.6 billion of gross added value to the UK economy (Forestry Commission, 2008). Consumption of wood products in the UK has risen in the past 10 years to over 55 million tonnes per year. The UK derives more than 70% of its timber imports from the countries of the EU27. In addition, about 18% is supplied domestically.

In addition to the timber production benefits from woodland, UK forests deliver a wide range of social and environmental benefits, which exist largely outside formal markets. Willis *et al.* (2003) estimated the non-market benefits of woodlands in Great Britain[5] to be worth approximately £1.25 billion per year (in 2008 prices). The largest values were found to be for forest recreation – over 250 million visits are made each year to UK woodlands (Forestry Commission, 2008) – and for woodland biodiversity, landscapes and carbon sequestration. Subsequent increases in the value ascribed by Government to the price of carbon (Department of Energy and Climate Change, 2009a) mean that carbon sequestration has now become the largest of these values. Trees and woodlands also deliver a range of other ecosystem services ranging from regulating water flows and quality to removing pollutants from the atmosphere. These services are of major value to the economy and society, providing a 'green

[5] In this study data were not produced for Northern Ireland.

infrastructure' on which to base commercial activities, for example in tourism (Hill *et al.,* 2003).

In addition to their value in sequestering and storing carbon, trees and woodlands have become recognised for ameliorating the impacts of climate change. Further work is needed to assess the potential scale of the values pertaining to woodlands and climate change. This will increase awareness of the significant potential for forests in the UK to contribute to climate change mitigation – through sequestration and the use of wood fuel and wood products – and to adaptation to its inevitable impacts. These are the central themes considered in this report.

1.5 Review of forestry policy in the UK: a historical perspective

Recent forest policy in the UK can be traced to the establishment of the Forestry Commission in 1919. The Forestry Commission was empowered through the Forestry Act to purchase land for afforestation and to support and regulate the establishment of woodlands by private owners. The decades following the two World Wars focused on planting new woodlands to ensure a strategic supply of timber. This led to some controversy over the nature of forest plantings and the locations in which they took place, in particular with regard to their impacts on landscapes and the natural environment (Tsouvalis, 2000). In the latter part of the 20th century, a more varied and complex set of policy objectives emerged, based in large part on recognition of the role of woodlands in providing recreation and amenity and as a resource for biodiversity and the natural environment. Forests were recognised for generating multiple outputs, only some of which were valued by markets, and for yielding a flow of environmental benefits and costs which varied according to management and location. Increasingly, forest design and management became more focused on environmental objectives.

The diversity of our forests and their wide geographical spread has subsequently enabled trees and woodlands to contribute to many other objectives, including physical and mental health, urban regeneration, rural development, energy supply and a broader range of ecosystem services. Forests have become viewed as a tool to deliver wide-ranging Government objectives. Most recently, climate change has added another layer to forest policy and poses significant challenges to achieving an appropriate balance between the many demands placed on our forests and woodlands.

1.5.1 Sustainable forest management

The objectives of forest policy across the UK are encapsulated in the principle of sustainable forest management. The Second Ministerial Conference on the Protection of Forests in Europe (MCPFE), Helsinki, 1993 defined sustainable forest management (SFM) as:

> the stewardship and use of forests and forest lands in a way, and at a rate, that maintains their biodiversity, productivity, regeneration capacity, vitality and their potential to fulfil, now and in the future, relevant ecological, economic and social functions, at local, national, and global levels, and that does not cause damage to other ecosystems.

Sustainable forest management is the basis of forestry policy, as demonstrated in the forestry strategies for England, Scotland and Wales, and the Sustainable Development Strategy identifies a similar role for forests in Northern Ireland. Forestry has also played a significant role, particularly since the 1980s, in attempts to develop a more integrated approach to land-use policy in tandem with, *inter alia*, agriculture, tourism, conservation and energy generation. This is evident, for example in the four EU Rural Development Programmes in the UK, which have brought together different activities and (formerly) separate funds into single programmes of support for development.

Compliance with internationally accepted standards of forest management is sought in the UK through both regulatory and voluntary means. On the regulatory side, The UK Forest Standard (UKFS) and procedures for giving grants and licences, require forest owners and managers to comply with wide-ranging environmental and socio-economic standards. The UKFS conforms to a set of pan-European criteria identified by the MCPFE. It defines the criteria and indicators of sustainable forest management for all forests in the UK, both planted and natural. The Forestry Act (1967) requires the granting of a felling licence before woodlands can be harvested. In most cases, permission to fell trees is conditional upon the area being replanted. This is a vital tool in ensuring that forest cover, and therefore its capacity for climate change mitigation and adaptation, is maintained in the long term. Some exceptions to conditional felling apply, for example where there are other over-riding environmental objectives, and where planning permission has been obtained.

Proposals for planting and felling trees and for building forest roads and quarries are also subject to Environmental

Impact regulations. Separate regulations apply in England and Wales, in Scotland and in Northern Ireland. Furthermore, woodland creation and removal proposals are subject to consultation with local authorities and other organisations with statutory powers in relation to land-use before the Forestry Commission gives approval. All such proposals are listed on the Forestry Commission's public register under the relevant country.

Voluntary compliance with international forest management standards is encouraged through internationally-recognised certification processes. All publicly owned forests managed by the Forestry Commission, and a significant proportion of private forests, conform to the UK Woodland Assurance Standard (UKWAS), which sets standards above the legally required minimum. The Forestry Commission's management process is based on a system of forest-wide design plans. These are renewed every 10 years and must comply with the UKWAS, the UKFS and UK grants and licences policy. Forest design plans are subject to public scrutiny. Private owners must submit detailed plans when applying for financial support from the Forestry Commission.

A defining feature of forestry policy in the 2000s has been the devolution of policy to separate administrations in England, Scotland, Wales and Northern Ireland, although some policy areas continue to be administered at a GB level, notably policy on research, international issues and plant health. Forest policy in each of the four countries is now focused on delivering objectives of the respective administrations. This has led to a range of policy objectives across the UK that, while similar in scope, differ in emphasis. For example, with regard to climate change, woodland creation has been given most weight in Scotland – where there is an aspiration to achieve 25% woodland cover by 2050 (Scottish Executive, 2006b). Since the late 1980s, such aspirations have been lacking in England and Wales – Figure 1.1 shows falling planting rates in recent years – but a recent Government white paper (HM Government, 2009b) has urged that new planting is needed to help mitigate climate change (see 1.5.3 below). In England, substantial potential has been identified to increase the supply of woodfuel from existing woodland. Some country-level policies have specific implications for the capacity of forestry to contribute to climate change mitigation and adaptation. These include the restoration of open habitats and an increase in continuous cover forestry (see Chapter 6).

A range of instruments is used to implement forest policy across the UK. Moxey (2009) classifies policy instruments into those that provide information, offer incentives or impose regulatory controls. Examples of each of these are currently in use, including support for research and advisory services, grants and tax breaks, and prescribed forest management activities. Almost one-third of UK woodlands are also under the direct management of the Forestry Commission. Grant payments cover woodland planting and management and other activities, as set out in the respective forestry strategies and policy documents in England, Scotland, Wales and Northern Ireland. Preferential tax treatment for forestry applies largely to productive forests and provides a significant spur for investment in the sector. Most recently, the potential is being explored to expand the use of markets in delivering environmental, non-market goods and services from forests – for example, carbon credits and water regulation – although this remains an emerging area of policy. UK forestry remains excluded from international markets in carbon credits, although voluntary, unregulated forest carbon schemes have developed in the UK private sector (see Chapter 12).

Analysis by Moxey (2009) has explored the extent to which the current range of instruments is suited to policy objectives. Grant payments have evolved significantly since the 1980s to tackle environmental and sustainability objectives and, to some extent, are addressing climate change objectives through, for example, payments towards the costs of investments in wood fuel boilers. It is arguable that major opportunities exist to develop further the suite of instruments so that they incentivise actions to mitigate and adapt to climate change. Such changes may include further alignment of tax breaks to sustainable forest management practices and new incentives for the use of timber as a sustainable material and for its role in carbon storage (see Section 5). Further work is needed to assess the potential from different instruments and where opportunities exist to target support for achieving a low-carbon economy.

1.5.2 International policy on climate change

International policy on climate change is now a major influence on forest policy across the UK. The United Nations Framework Convention on Climate Change (UNFCCC), established at the United Nations Conference on Environment and Development (UNCED) in 1992, is an international environmental treaty signed by the UK and over 180 countries and provides the foundation for the subsequent development of international climate change policy. Its objective is: 'to achieve stabilisation of greenhouse gas concentrations in the atmosphere at a low enough level to prevent dangerous anthropogenic interference with the climate system'.

The Kyoto Protocol, which came into force in 2005, was established under the UNFCCC. Industrialised countries (termed 'Annex 1' countries in the Protocol) that have ratified the Protocol, including the UK, are committed to reduce emissions of carbon dioxide and five other GHG, or to engage in emissions trading. The major distinction between the 'Protocol' and the 'Convention' is that the Convention *encouraged* industrialised countries to stabilise GHG emissions, whereas the Protocol *commits* them to do so. Under the Treaty, countries must meet their targets primarily through national measures. However, the Protocol offers additional means of meeting their targets through three market-based mechanisms that operate across national boundaries. These are:

1. Emissions trading (the European Union Emissions Trading Scheme is the largest in operation)
2. The Clean Development Mechanism (CDM)
3. Joint Implementation (JI).

These 'flexible mechanisms' are intended to allow countries with an emissions reduction or limitation commitment under the Kyoto Protocol (Annex B Party) to minimise their costs by buying and/or selling credits (trading scheme), or by implementing emission-reduction projects in developing countries (CDM) or other Annex B countries (JI). However, the contribution of forestry to mitigating climate change under these mechanisms is limited at present because it has only been used under the CDM to date.

Work is underway to develop a new regulatory framework for international action on climate change following the end of the first commitment period of the Kyoto Protocol in 2012. Negotiators aim to secure a new deal in Copenhagen at the end of 2009. Recent analysis (Eliasch, 2008) has highlighted the importance of including forestry within future emissions trading schemes, principally to reduce forest loss and degradation in developing countries. Deforestation is responsible for approximately 18% of global GHG emissions (Stern, 2007). Securing a deal on Reducing Emissions from Degradation and Deforestation (REDD) will be a major part of the negotiations at Copenhagen.

Complex accounting systems are required for the UNFCCC and the Kyoto Protocol to show whether parties to these agreements are meeting their commitments.

The UNFCCC requires participating countries to submit national reports and annual inventories on implementation of the Convention. Annex I Parties that have ratified the Kyoto Protocol must include supplementary information on their emissions and removals of GHG to demonstrate compliance with the Protocol's commitments. The contribution of forestry is reported, as part of the UK's Greenhouse Gas Inventory, under the inventory and projections for the land-use, land-use change and forestry (LULUCF) sector (Centre for Ecology and Hydrology, 2008).

Commitments have also been made at a European level. Successive resolutions on Forests and Climate Change through the Ministerial Conferences on the Protection of Forests in Europe – most recently in Warsaw in 2008 – have committed signatory countries to play an active role in addressing climate change through forestry's roles in both mitigation and adaptation. These commitments are supported in the EU Forestry Strategy and the EU Forest Action Plan.

1.5.3 UK policy on climate change

The UK has a Climate Change Programme (HM Government, 2006) that aims to reduce carbon dioxide emissions by 20% on 1990 levels, by 2010. It is led by the Department of Energy and Climate Change (DECC) on behalf of the UK and devolved administrations. The UK has taken a proactive approach in developing climate change policy by introducing a Climate Change Act in 2008. The first of its type in the world, it applies to England, Wales and Northern Ireland and sets a legally binding target to reduce total GHG emissions by at least 80% by 2050 – compared with 1990 levels. A Committee on Climate Change was established under the UK Act and is empowered to set successive five-year targets for the UK's GHG account. In response to the Committee's advice (Committee on Climate Change, 2008), the Government has announced the budgets for 2008–12, 2013–17 and 2018–22, leading to a 34% reduction in GHG emissions with respect to 1990 levels, by 2020 (HM Government, 2009a).

A separate Climate Change (Scotland) Act is in force in Scotland. This sets a legally binding emissions reduction target of at least 80% by 2050, with an interim target in 2020 of a 34%, or possibly 42%, reduction from 1990 levels. The Act is supported by the Scottish Government Climate Change Delivery Plan, which describes how the Scottish Government would achieve the interim targets. Adoption of the 42% trajectory is dependent on the targets adopted by the EU and on advice from the Committee on Climate Change. The Delivery Plan lays out woodland expansion targets associated with the GHG reduction trajectories, and also identifies the growing role for biomass in renewable heat production. The Forestry Commission Scotland Climate Change Action Plan lays out a broader set of actions in relation to sequestration of carbon, as well as mitigation and adaptation to climate change. This is further supported by published Scottish Government policies on woodland expansion and woodland removal.

The UK Government has recently published a white paper, the UK Low Carbon Transition Plan. This sets out how the Government will achieve the targets in the climate change budgets, and draws attention to the contribution that forestry can make:

'Woodland creation is a very cost-effective way of fighting climate change over the long term … woodland creation represents 60% of the grant aid administered by the Forestry Commission. But to realise the potential for 2050, we need to see a big increase in woodland creation – and we need to plant sooner rather than later' (HM Government, 2009b).

Five UK forest management scenarios which consider realistic planting and management options to enhance abatement of GHG emissions in the future are analysed in Chapter 8 of this report.

The carbon benefits of wood biomass used to fire or co-fire power generation are recognised by the UK Renewables Obligation (RO), which rewards private sector investment in power generation in order to promote the generation of electricity from renewable sources in the UK. A UK Renewable Energy Strategy (Department of Energy and Climate Change, 2009b) was published in 2009 and sets a target of renewable sources producing 15% of the UK's energy requirements by 2020. Biomass has been identified as having the potential to meet 33% of the renewables target, with woodfuel and forestry making a significant contribution. These are described in Chapter 7.

The Government has also created an Adapting to Climate Change (ACC) Programme to bring together existing adaptation work, and co-ordinate future work. The ACC Programme has two phases. Phase 1, from 2008–11 is laying the groundwork necessary to implement Phase 2, a statutory National Adaptation Programme, as required by the Climate Change Act.

The objectives of Phase 1 of the Programme are to:

- develop a robust and comprehensive evidence base on the impacts and consequences of climate change in the UK;
- raise awareness of the need to take action now and help others to take action;
- measure success and take steps to ensure effective delivery;
- work across Government at the national, regional and local level to embed adaptation into Government policies, programme and systems.

The Programme is focused on England, although some elements will be UK-wide, and will be developed in partnership with the other UK administrations. The Programme is essentially domestic in scope, but consequences of climate change overseas will have an impact on the UK, because of the interconnected nature of our world. The Programme will therefore address those effects where there is a significant domestic risk. By considering the potential of forests to absorb carbon, this study will contribute to the objectives of Phases 1 and 2.

Furthermore, the UK Climate Change Act requires the Government to carry out, every five years, a national Climate Change Risk Assessment (CCRA), of which the first needs to be finished by 2012. An additional Adaptation Economic Analysis (AEA) will consider the costs and benefits of adaptation actions. The results of the CCRA and the AEA will inform a National Adaptation Programme, to be completed by the end of 2012, which will identify priority actions and allocate resources to meet them. The Act also gives ministers the power to require climate risk and adaptation reports from any body delivering a public good, and creates an Adaptation Subcommittee of the Committee on Climate Change, which will scrutinise the Government's adaptation actions. The Forestry Commission has been invited to submit a report to the CCRA (at time of writing). Woodlands have an important part to play in the work required to establish the National Adaptation Programme. They have ecological functions that people enjoy and value, and that can help to protect society from the impacts of climate change. The work reported in this study (in particular Sections 2, 3 and 4) will contribute to the evidence base required to support this risk assessment activity.

1.5.4 Forestry policy on climate change

Climate change is now a major focus of forest policy in

the UK. The forestry strategies for England, Scotland and Wales, and the Sustainable Development Strategy in Northern Ireland, set out the objectives of forest policy on climate change. These emphasise the importance of carbon sequestration through woodland planting and management, the production of woodfuel as a renewable energy source, and the promotion of wood products in place of more fossil fuel-intensive materials. This study will inform the delivery of these objectives. The underpinning science for mitigation objectives is examined in Chapter 6. The strategies also emphasise the importance of adapting forests to the impacts of climate change, and developing the role of trees in helping society cope with a changing climate (e.g. by providing more shade and flood control in urban environments). The evidence underpinning these objectives is examined in Chapters 9 and 10.

At a UK level, the Forestry Commission has identified six priority actions that define the UK approach to forestry and climate change:

1. Protect and manage the forests that we already have
2. Reduce deforestation
3. Restore forest cover
4. Use wood for energy
5. Replace other materials with wood
6. Plan to adapt to a changing climate.

These actions resonate strongly with the conclusions of the IPCC's 4th Assessment Report on forestry and provide the context for much of the content of this report. They are embedded in the respective policies strategies in England, Scotland, Wales and Northern Ireland, and represent the priorities for forestry's contribution to climate change mitigation and adaptation in the coming years and decades. As will be seen in Chapter 8, these priorities are critical to providing a vision – through the identification of scenarios – of how forestry can help to provide a solution to the global challenge of climate change up until 2050 and beyond.

A major task for forest policy on climate change across the UK is to set standards for planning, managing and monitoring woodlands. Standards are important in providing benchmarks against which actions to deliver policy objectives can be assessed. In the forestry sector, it is vitally important that standards apply to the delivery of climate change objectives, while ensuring that the objectives of sustainable forest management are met.

Standards of sustainable forest management are set out

in the UKFS (see 1.5.1 above), which is currently being revised and will include for the first time, a supporting Guideline on Forests and Climate Change. This strategic management guideline sets out requirements and advice on how to achieve the mitigation and adaptation potential of woodlands. This is a significant step in designing policy instruments to address forestry's role in tackling climate change. The contents of this report, including those dealing with the interactions between forests and the atmosphere (Chapter 3); the impacts of climate change on forests (Chapters 4 and 5); management actions for mitigation in different types of forest (Chapter 6) and adapting forests for climate change (Chapters 9 and 10), provide essential evidence to support the guideline.

Standards are also required in the operation of carbon markets. Carbon is widely traded in markets across the world, both in formal markets under the Kyoto Protocol (e.g. the EU Emissions Trading Scheme) and in voluntary markets. Forestry is largely excluded from formal international markets in carbon credits (see 1.5.2 above) – and therefore traded forestry credits are not currently used against international emissions targets – but it has become a significant player in voluntary markets. Forest carbon markets are examined further in Chapter 12. The quality of forestry projects in voluntary markets has been mixed, to date, and the Forestry Commission is currently developing a Code of Good Practice for Forest Carbon Projects to provide confidence in the marketplace that woodlands deliver projected carbon benefits. The Code will assure sustainable forest management (under the UKFS), put in place rigorous protocols for measuring woodland carbon, and establish criteria against which projects can be assessed. The Code will be launched in 2010.

As highlighted above (see 1.5.1), further analysis is needed to assess whether forest policy instruments currently in use, including the use of standards, require further changes to help to achieve the priority actions on climate change set out by the Forestry Commission.

1.6 Forest planning, projections and monitoring

Effective mechanisms for planning, monitoring and projecting the scale of forest resources are essential in assessing the current and future contribution that trees and woodlands can make to addressing climate change.

Planning takes place currently at strategic, forest

management and site levels. The strategic level addresses the broad goals of an organisation and may comprise a number of forest areas. The forest management plan applies to a convenient unit or area for management and on which the requirements of sustainable forest management apply. Site planning is concerned with the operational detail of how proposals will be implemented. These plans also act as reference documents for monitoring and assessing the development of woodlands. Evidence is emerging that forest planning processes are starting to be amended for climate change objectives. This report provides evidence as to why such changes are needed.

The evidence provided by monitoring the amount of carbon in forest ecosystems is very important for developing policies on forestry and climate change, and for reporting forestry's contribution in GHG inventories under international processes for tackling climate change. Monitoring forest carbon is a complex task with significant technical challenges. The size of different stocks, or pools, of forest carbon is quantified above (see 1.4). A more detailed division of these pools is as follows:

1. Trees: both above ground (trunk, branches) and below ground (roots)
2. Wood products
3. Fine and coarse woody debris
4. Soil carbon
5. Other vegetation.

Forest managers in the public and private sectors in the UK routinely monitor the first two of these carbon pools but there is currently little consolidation and reporting of information at a national scale. This is in part due to the different ways that inventory data are managed in different organisations, i.e. the information is recorded in a variety of analogue and digital forms. For example, many private sector managers still rely on paper-based systems, while others have adopted geographic information systems (GIS). The Forestry Commission uses a bespoke GIS called 'Forester' to record all forest management activities and to forecast timber production.

A series of inventories for recording and monitoring forests has been produced in the past 60 years by the Forestry Commission. The latest of these is the National Forest Inventory (NFI), currently under development, which is the successor to the National Inventory of Woodlands and Trees. For production planning purposes, annual timber increment is expressed as Yield Class, which is the maximum mean annual increment of a crop in cubic metres

per hectare. In conjunction with the specific gravity of timber, Yield Class can be used to estimate the amount of carbon in a forest. Yield Classes have been estimated for a range of species and associated management practices.

The NFI is being designed to record levels of carbon stocks in UK forests. This will produce a map, derived from 2003–07 aerial photography, showing the spatial extent, location and nature of all British woodlands over half a hectare in size. It is intended that the map will be updated using 'operational data' to a common baseline of 2008 and biannually thereafter. The scope of this work includes a field survey to estimate timber volumes in both the public and private sectors, and to facilitate more accurate production forecasting. The NFI will provide the most comprehensive assessment of forest resources to date in the UK, and will offer major improvements to the way that forest carbon across the UK is recorded and monitored. It will also become the basis for the forestry component of the UK's GHG Inventory (see 1.5.2 above). The first output of the NFI is the production in 2009 of the digital map of British woodlands. Full results from the NFI will be available by 2014.

The Forestry Commission is also examining how best to integrate assessments of the other carbon pools in woody debris, soil carbon and other vegetation, which are not routinely monitored, within the NFI. Models have been developed in Forest Research – the Forestry Commission's research agency – and elsewhere to estimate the carbon content of woodlands. These models are currently being enhanced to quantify carbon content in all woody biomass (e.g. branches, roots), forest litter and forest soils. In combination with inventory data, they will strengthen the evidence base on the potential contribution of forestry to sequester carbon, store it in wood and deliver substitution benefits through the use of wood fuel and wood products (see Chapter 8).

Other tools have also been developed that assist the monitoring of forest resources in relation to climate change. For example, the Ecological Site Classification (Ray, 2001) has been developed both as a stand-based and spatial tool (on a GIS platform) for matching tree species and native woodland communities to site types. It has been successfully used to inform forest policy, particularly in relation to the impact of future climate scenarios on tree species (see Chapter 5). This research programme is also developing stand-based and spatial climate change impact and adaptation tools, including new ways to assess the risk of biotic impacts of climate change scenarios.

In broad terms, existing and emerging procedures for projecting and monitoring the UK's forests place the UK in a good position to plan and review its forest resource. This should help it to take action on climate change in a sustainable way, depending on the relative costs of forestry investments as a source of mitigation and adaptation. However, it will be important to focus attention on whether the systems in place prove adequate for the new policy and management challenges presented by climate change. Current revisions to the UKFS and the introduction of a supporting guideline on climate change will help. But forest planning faces difficult decisions on how to address the many objectives of forestry. Managers will require ongoing input from the research community as to how woodlands can best deliver against the many demands placed on them. It is the intention of this report to evaluate existing knowledge and to identify gaps in understanding, so that the climate change elements of this management challenge can be met in the future.

References

BROADMEADOW, M. and MATTHEWS, R. (2003). *Forest, carbon and climate change: the UK contribution.* Forestry Commission Information Note 48. Forestry Commission, Edinburgh.

CENTRE FOR ECOLOGY AND HYDROLOGY, UNIVERSITY OF ABERDEEN, FOREST RESEARCH, NATIONAL SOIL RESOURCES INSTITUTE, CENTRE FOR TERRESTRIAL CARBON DYNAMICS, AGRI-FOOD AND BIOSCIENCES INSTITUTE and QUEEN'S UNIVERSITY (2008). *Inventory and projections of UK emissions by sources and removals by sinks due to land use, land use change and forestry.* Annual Report to DEFRA, July.

COMMITTEE ON CLIMATE CHANGE (2008). *Building a low-carbon economy: the UK's contribution to tackling climate change.* The First Report of the Committee on Climate Change, December 2008. The Stationery Office, London.

DENMAN, K.L., G. BRASSEUR, G., CHIDTHAISONG, A., CIAIS, P., COX, P.M., DICKINSON, R.E., HAUGHLUSTAINE, D., HEINZE, C., HOLLAND, E., JACOB, D., LOHMANN, U., RAMACHANDRAN, S., DA SILVA DIAS, P.L., WOFSY, S.C. and ZHANG, X. (2007). Couplings between changes in the climate system and biogeochemistry. In: Solomon S., Qin, D., Manning M., Chen Z., Marquis M., Averyt, K.B., Tignor, M. and Miller H.L. (eds). *Climate Change 2007: the physical science basis.* Contribution of Working Group 1 to the Fourth Assessment Report of the Intergovernmental Panel on Climate Change. Cambridge University Press, Cambridge.

pp. 499–587.

DEPARTMENT OF ENERGY AND CLIMATE CHANGE (2009a). *Carbon valuation in UK policy appraisal: a revised approach.* Department of Energy and Climate Change, London.

DEPARTMENT OF ENERGY AND CLIMATE CHANGE (2009b). *The UK renewable energy strategy.* The Stationery Office, London.

ELIASCH, J. (2008). *Climate change: financing global forests: the Eliasch Review.* The Stationery Office, London.

FORESTRY COMMISSION (2008). *Forestry statistics.* Forestry Commission, Edinburgh

HILL, G., COURTNEY, P., BURTON, R., POTTS, J., SHANNON, P., HANLEY, N., SPASH, C., DEGROOTE, J., MACMILLAN, D. & GELAN, A. (2003): *Forests' role in tourism - phase 2.* Report to the Forestry Commission, Edinburgh

HM GOVERNMENT (2006). *Climate change – the UK Programme 2006.* Cm. 6764. The Stationery Office, London.

HM TREASURY (2009). *Building a low-carbon economy: implementing the Climate Change Act 2008.* HM Treasury, London.

INTERNATIONAL ENERGY AGENCY (2009). *CO_2 Emissions from fuel combustion.* Highlights. Paris. 124 pp.

KAUPPI, P.E., SEDJO, R.A., APPS, M.J., CERRI, C.C., FUJIMORI, T., JANZEN, H., KRANKINA, O.N., MAKUNDI, W., MARLAND, G., MASERA, O., NABUURS, G.J., RAZALI, W. and RAVINDRANATH, N.H. (2001). Technical and economic potential of options to enhance, maintain and manage biological carbon reservoirs and geo-engineering. In: Metz, B., Davidson, O., Swart, R. and Pan, J. (eds) *Climate change 2001: mitigation.* Contribution of Working Group III to the Third Assessment Report of the IPCC. Cambridge University Press, Cambridge. pp. 310–343.

MORISON, J., MATTHEWS, R., PERKS, M., RANDLE, T., VANGUELOVA, E., WHITE, M. and YAMULKI, S. (2009) *The carbon and GHG balance of UK forests: a review.* Forestry Commission, Edinburgh.

MOXEY, A. (2009). *Scoping study to review forestry policy instruments in the UK.* Report to the Forestry Commission, Edinburgh.

NABUURS, G.J., MASERA, O., ANDRASKO, K., BENITEZ-PONCE, P., BOER, R., DUTSCHKE, M., ELSIDDIG, E., FORD-ROBERTSON, J., FRUMHOFF, P., KARJALAINEN, T., KRANKINA, O., KURZ, W.A., MATSUMOTO, M., OYHANTCABAL, W., RAVINDRANATH, N.H., SANZ SANCHEZ, M.J. and ZHANG, X. (2007). Forestry. In: Metz, B., Davidson, O.R., Bosch, P.R., Dave, R. and Meyer, L.A. (eds) *Climate change 2007: mitigation.* Contribution of

Working Group III to the Fourth Assessment Report of the Intergovernmental Panel on Climate Change, Cambridge University Press, Cambridge. pp. 541–584.

RAY, D. (2001). *Ecological site classification decision support system V1.7*. Forestry Commission, Edinburgh.

RICHARDSON, K., STEFFEN, W., SCHELLNHUBER, H., ALCAMO, J., BARKER, T., KAMMEN, D., LEEMANS, R., LIVERMAN, D., MUNASINGHE, M., OSMAN-ELASHA, B., STERN, N. and WAEVER, O. (2009). *Synthesis Report – climate change: global risks, challenges and decisions*. University of Copenhagen, 10–12 March. Online at: www. climatecongress.ku.dk

SCOTTISH EXECUTIVE (2006a). *Changing our ways – Scotland's climate change programme*. Scottish Executive, Edinburgh.

SCOTTISH EXECUTIVE (2006b). *The Scottish forestry strategy*. Forestry Commission Scotland, Edinburgh.

STERN, N. (2007). *The economics of climate change: The Stern Review*. Cambridge University Press, Cambridge.

TSOUVALIS, J. (2000). *A critical geography of Britain's state forests*. Oxford University Press, Oxford.

WILLIS, K.G., GARROD, G., SCARPA, R., POWE, N., LOVETT, A., BATEMAN, I.J., HANLEY, N. and MACMILLAN, D.C. (2003). *The social and environmental benefits of forests in Great Britain. Phase 2 report*. Centre for Research in Environmental Appraisal and Management, University of Newcastle-Upon-Tyne, Newcastle. Report for the Forestry Commission, Edinburgh.

CLIMATE TRENDS AND PROJECTIONS

Chapter

2

C. C. West and J. I. L. Morison

Key Findings

The climate of the UK has already changed recently as a result of global emissions of greenhouse gases through human activity; an increase in mean annual temperatures is particularly evident.

Current Climate Projections (UKCP09) indicate that the climate in the UK will continue to warm through this century and there will be changes in rainfall and its seasonal distribution, which will vary regionally.

Climatic factors, primarily temperature and precipitation, interact with geology, geomorphology and soil characteristics in determining the type and productivity of woodlands and forests. These climate and soil factors determine many of the species and community patterns in our semi-natural woods and they strongly influence the choice of species or provenance for wood production. As climate changes, so the tree and woodland cover will change.

Subsequent chapters of this report assess the scientific understanding of and evidence for the impacts of climate change on forestry and its role in climate change mitigation and adaptation. It is important to set the context for this assessment by describing the latest evidence on how the climate has already changed and is projected to change in the UK over the coming decades under different GHG emissions scenarios. However, it should also be recognised that climate changes elsewhere in the world will affect global patterns of forest productivity, forestry sector activity and the demands of other land uses, which will indirectly influence UK forestry (see Chapters 4 and 5).

2.1 Observed trends in the UK climate

The evidence of recent change in the climate of the UK is compelling with, for example, the instrumental Central England Temperature record showing a substantial increase (Figure 2.1). Jenkins *et al.,* (2008) identified the following climate trends in the UK:

- Central England Temperature has risen by about 1°C since the 1970s, with 2006 being the warmest in the

348-year long record. It is likely that global emissions of man-made GHG have contributed significantly to this rise.
- Sea levels rose by about 1 mm per year during the 20th century but the rate of rise in the 1990s and 2000s has been higher than this.
- Temperatures in Scotland and Northern Ireland have risen by about 0.8°C since 1980, but this rise has not been attributed to specific causes.
- Annual mean precipitation over England and Wales has not changed significantly since records began in 1766. Seasonal rainfall is highly variable, but there has been a slight trend over the last 250 years for decreased rainfall in summer and increased in winter, although with little change in the latter over the past 50 years.
- All regions of the UK have experienced an increase over the past 45 years in the contribution to winter rainfall from heavy precipitation events. In summer, all regions except north east England and northern Scotland show decreases.
- There has been considerable variability in the North Atlantic Oscillation, but with no significant trend over the past few decades.
- Severe windstorms around the UK have become more frequent in the past few decades, although not above that seen in the 1920s.

Figure 2.1

Changes in annual values for Central England Temperatures (green bars) from 1772 to 2008, relative to the average over the 1961–1990 baseline period (about 9.5°C). Decadal variations in temperature are shown in black.

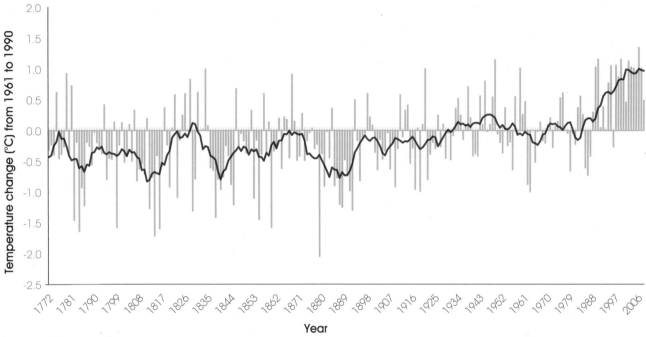

Source: Met Office Hadley Centre.

2.2 Climate Projections for the UK

The UK Climate Projections 2009 (UKCP09) are the latest in a series of government-funded descriptions of future climate in the UK. The principal advance in modelling under the UKCP09 is the ability to run multiple versions of a model in order to explore the modelling uncertainty. Whereas before, one projection based on a single formulation of a model was used, UKCP09 is based on probabilistic data derived from multiple runs of a Hadley Centre model, combined with single runs of other climate models, with weighting from performance of the models over the 20th century. Three IPCC emissions scenarios are used, high, medium and low (respectively, A1F1, A1B and B1). Further details are in Jenkins *et al.,* (2009) and Murphy *et al.,* (2009). Instead of a single outcome, users have a wide range of outcomes with relative measures of the strength of evidence that supports each outcome. This type of information can better inform risk-based decisions, but does require users to explore both the sensitivity of their system to change, and their own attitude to risk. The full range of UKCP09 material, together with scientific reports and guidance on the use of the projections, is accessible online at: http://ukclimateprojections.defra.gov.uk.

A number of issues should be borne in mind when interpreting the findings of UKCP09:

- Projections of climate change take into account uncertainty due to natural variability and due to modelling, i.e. our incomplete understanding of the climate system and its imperfect representation in models. The projections do this by giving the probabilities of a range of possible outcomes, as estimated by a specific methodology.
- Probability in UKCP09 can be seen as the relative degree to which each climate outcome is supported by current evidence, taking into account our understanding of climate science, observations and using expert judgement.
- Probabilistic projections are given at a resolution of 25 km over land, and as averages over administrative regions, river basins and marine regions, for seven overlapping 30-year periods and for three future emissions scenarios.
- Confidence in the projections varies, depending on the geographical scale and the variable under discussion. There is moderate confidence in projections at continental scale. Those at 25 km resolution are indicative to the extent that they reflect large-scale changes modified by local conditions such as mountains and coasts.
- Errors in global climate model projections cannot be compensated by statistical procedures no matter how complex, and will be reflected in uncertainties at all scales.

2.2.1 Projected seasonal and annual changes

The methodology developed for UKCP09 to convert climate model simulations into probabilistic estimates of future change necessitates a number of expert choices and assumptions, with the result that the probabilities we specify are themselves uncertain. We do know that our probabilistic estimates are robust to reasonable variations within these assumptions.

Changes by the 2080s (relative to a 1961–90 baseline) under the 'medium' emissions scenario are given below. Central estimates of change (those at the 50% probability level) are followed, in brackets, by changes which are very likely to be exceeded, and very likely not to be exceeded (10% and 90% probability levels, respectively).

- All areas of the UK warm, more so in summer than in winter. Changes in summer mean temperatures are greatest in parts of southern England (up to 4.2°C; 2.2–6.8°C) and least in the Scottish islands (just over 2.5°C; 1.2–4.1°C). Mean daily maximum temperatures increase everywhere. Increases in the summer average are up to 5.4°C (2.2–9.5°C) in parts of southern England and

2.8°C (1–5°C) in parts of northern Britain. Increases in winter are 1.5°C (0.7–2.7°C) to 2.5°C (1.3–4.4°C) across the country.
- Changes in the warmest day of summer range from +2.4°C (–2.4 to +6.8°C) to +4.8°C (+0.2 to +12.3°C), depending on location, but with no simple geographical pattern.
- Mean daily minimum temperature increases on average in winter by about 2.1°C (0.6–3.7°C) to 3.5°C (1.5–5.9°C) depending on location. In summer it increases by 2.7°C (1.3–4.5°C) to 4.1°C (2.0–7.1°C), with the biggest increases in southern Britain and the smallest in northern Scotland.
- Central estimates of annual precipitation amounts show very little change everywhere at the 50% probability level. Changes range from –16% in some places at the 10% probability level, to +14% in some places at the 90% probability level, with no simple pattern.
- The biggest changes in precipitation in winter, increases up to +33% (+9 to +70%), are seen along the western side of the UK. Decreases of a few percent (–11 to +7%) are seen over parts of the Scottish highlands.
- The biggest changes in precipitation in summer, down to about –40% (–65 to –6%), are seen in parts of the far

Figure 2.2(a)
Mean summer temperature change (June, July, August), relative to 1961–90 means. Data are for 2080s under the 'medium' (SRES A1B) emissions scenario. The change at 50% probability level, called the central estimate, is that which is as likely as not to be exceeded by 2080.

10% = very likely to be exceeded   50% = central estimate   90% = very unlikely to be exceeded

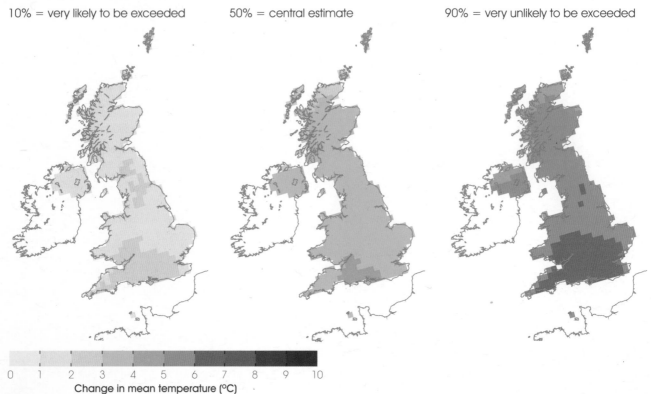

Change in mean temperature (°C)

south of England. Changes close to zero (–8 to +10%) are seen over parts of northern Scotland.

- Changes in the wettest day of the winter range from zero (–12 to +13%) in parts of Scotland to +25% (+7 to +56%) in parts of England.
- Changes in the wettest day of the summer range from –12% (–38 to +9%) in parts of southern England to +12% (–1 to +51%) in parts of Scotland.
- Relative humidity decreases by around –9% (–20 to 0%) in summer in parts of southern England – by less elsewhere. In winter, changes are a few percent or less everywhere.
- Summer-mean cloud amount decreases, by up to –18% (–33 to –2%) in parts of southern England (giving up to an extra +20 Wm^{-2} (–1% to +45 Wm^{-2}) of downward shortwave radiation) but increase by up to +5% (0 to +11%) in parts of northern Scotland. Changes in cloud amount are small (–10 to +10%) in winter. Projected changes in storms are very different in different climate models. Future changes in anticyclonic weather are equally unclear.
- It has not been possible to provide probabilistic projections of changes in snow and wind speed. The Met Office Hadley Centre regional climate model

projects reductions in winter mean snowfall of typically –65% to –80% over mountain areas and –80% to –95% elsewhere. It projects changes in winter mean wind speed of a few percent over the UK, but wind speed projections are very uncertain.
- There is no assessment of how the urban heat island effect may change.
- It is very unlikely that an abrupt change to the Atlantic Meridional Ocean Circulation (Gulf Stream) will occur this century.

Examples of the UK-wide projection for the 2080s for mean summer temperature and mean winter precipitation are given in Figure 2.2, showing the 10%, 50% and 90% probability levels.

2.2.2 Projected changes in daily climate

UKCP09 provides synthetic daily time series of a number of climate variables from a weather generator, for the future 30-year time periods, under the three emissions scenarios.

Analysis of results from the Weather Generator shows that increases in the number of days with high temperatures are

Figure 2.2(b)
Mean winter precipitation change (December, January, February), relative to 1961–90 means. Data are for 2080s under the 'medium' (SRES A1B) emissions scenario. The change at 50% probability level, called the central estimate, is that which is as likely as not to be exceeded by 2080.

10% = very likely to be exceeded   50% = central estimate   90% = very unlikely to be exceeded

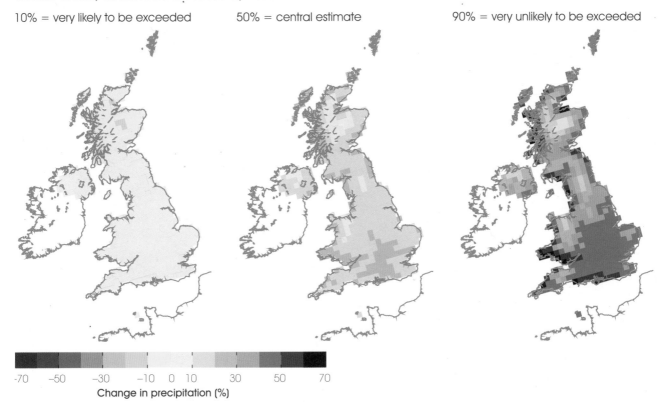

Change in precipitation (%)

found everywhere, particularly in south east England, and reductions in the number of frost days are found, greatest where frost days are currently more frequent. Increases in the number of 10-day dry spells across the UK are found, and are more pronounced in southern England and in Wales.

2.2.3 UKCIP02 and UKCP09

Having stressed the need for users to consider the full range of uncertainty given in UKCP09, it is nonetheless instructive to compare the central estimate (50% probability) of the projected changes with the single projections (for an identical emissions scenario) in the previous projections released in 2002 (UKCIP02). The following conclusions may be drawn.

- In the case of mean temperature, projected changes in UKCP09 are generally somewhat greater than those in UKCIP02.
- The summer reduction in rainfall in UKCP09 is not as great as that projected in UKCIP02.
- The range of increases in rainfall in winter seen in UKCP09 are very broadly similar to those in UKCIP02, although with a different geographical pattern. A few areas are projected to be drier in winter in UKCP09; in UKCIP02 all areas were projected to be wetter.
- Small changes in cloud cover are projected in winter, as in UKCIP02. Projections of summer decreases in cloud are similar to those in UKCIP02.

2.3 Forestry and the changing climate

The projected changes in our climate in the UK are likely to lead to a wide range of direct and indirect effects on trees, woodlands and forests. These include:

- changing growing conditions, including altered cloudiness patterns, extended growing seasons, modified soil moisture seasonal patterns and altered soil nutrient availability;
- modified rates of plant development, growth and wood production;
- changes in the type and frequency of abiotic disturbance such as waterlogging, flooding and storms;
- changes in both types of invertebrate and vertebrate pests and diseases, and in their severity, timing and seasonality;
- changes in distribution and potential range of many species (including invasive species); leading to

- changes in the species composition of woodland communities.

More details on these changes are given in Sections 2 and 3, and the issues about adaptation to them are discussed in Section 4.

The projections of anticipated temperature and rainfall suggest that forests in southern and eastern Britain are likely to experience a greater frequency and severity of summer dry spells, whereas areas in north western Britain will experience a moister and milder climate. Therefore, the climatic limitations to species survival, particularly for southern England, may shift away from factors like frost and cold hardiness, to others such as tolerance of summer drought. There may also be problems associated with earlier flushing in the spring, exposing trees to the risks of late spring frosts and/or delayed onset of dormancy in the autumn resulting in inadequate cold hardening. Such events are likely to be more influential upon species survival than average conditions.

One way to think of these projected changes is that forests in southern Britain will be experiencing climates with some characteristics of west central France by the 2050s and of Mediterranean Europe by the 2080s. Similarly, forests in central Scotland will experience the climate of southern Britain by the 2050s and of central France by the 2080s. This approach is useful for informing adaptation strategies, but such climate analogues are imperfect because many factors, especially latitude, are not included. These projected changes of climate are used as the basis for scientific evaluations of the interactions between UK forests and climate change which are described in this Assessment.

References

JENKINS, G.J., PERRY, M.C. and PRIOR, M.J.O. (2008). *The climate of the United Kingdom and recent trends.* Met Office Hadley Centre, Exeter.

JENKINS, G.J., MURPHY, J.M., SEXTON, D.S., LOWE, J.A., JONES, P. and KILSBY, C.G. (2009). *UK climate projections: briefing report.* Met Office Hadley Centre, Exeter.

MURPHY, J.M., SEXTON, D.M.H., JENKINS, G.J., BOOTH, B.B.B., BROWN, C.C., CLARK, R.T., COLLINS, M., HARRIS, G.R., KENDON, E.J., BETTS, R.A., BROWN, S.J., HUMPHREY, K.A., MCCARTHY, M.P., MCDONALD, R.E., STEPHENS, A., WALLACE, C., WARREN, R., WILBY, R. and WOOD, R.A. (2009). *UK climate projections science report: climate change projections.* Met Office Hadley Centre, Exeter.

THE ROLE OF FORESTS IN THE CAPTURE AND EXCHANGE OF ENERGY AND GREENHOUSE GASES

Chapter

3

P. G. Jarvis, R. J. Clement, J. Grace and K. A. Smith

Key Findings

Climate Change or 'Global Warming' is the rise in temperature of the troposphere (the lowest 17 km of the atmosphere). Greenhouse gases (GHG) absorb particular wavelengths of long-wave radiation and reduce the escape to space of radiation from the surface of earth. The ongoing rise in atmospheric GHG contents resulting from human activity is the major cause of climate change. Forests influence climate in two main ways. First, they absorb solar radiation energy and directly transfer sensible heat into the troposphere. Second, they influence the contents of GHG in the troposphere. UK forests exchange all the naturally occurring GHG (i.e. CO_2, CH_4, O_3, N_2O and water vapour) with the troposphere to a larger or smaller extent.

Because the albedo (solar radiation reflection coefficient) of forest stands is very similar to that of the vegetation that woodlands have replaced in the UK, the increase of forest area which occurred between the 1920s and 1980s has not, in general, changed local solar radiation budgets. Since UK weather and hydrological cycles are, in general, determined by weather patterns formed at large scales over the Atlantic Ocean, Europe and Russia, the contribution of differential energy partitioning by our forested land to our local weather is likely to be small.

The optimum temperature for growth of young Sitka spruce is above current projections at around 18°C, if other variables do not change. But as air temperature rises, the local water vapour pressure deficit (VPD) also tends to increase. Sitka spruce, in particular, and some other conifers, are very sensitive to VPD. On the rare occasions that air temperature currently rises above 20°C in UK forests, the stomatal pores in the needle surfaces of Sitka spruce close, the needles cease to absorb CO_2, and the forest may emit CO_2 for a short period until the VPD declines.

The existing UK forest cover is both a stock of carbon and a system removing CO_2 from the troposphere. If re-stocking follows harvest, without major disturbance, the forest carbon stock remains constant and, in a long-term well-managed forest, all stages in the tree life cycle and forest management cycle are represented equally in the forest. This system has the potential to remove CO_2 continuously from the troposphere and transfer it into storage in the soil and into commercial products that may for example, substitute for fossil fuels or be used for construction. Average annual removal of CO_2, from the atmosphere by closed-canopy Sitka spruce in northern Britain, Yield Class 14–16, is currently about 24 tCO_2 per hectare between years 17 and 40. Taking into account initial losses from 'soil' respiration stimulated by site preparation (see below), a conservative average annual figure over a typical 40-year rotation is about 14 tCO_2, per hectare. This is the average annual rate of CO_2 removal that we can expect from the afforestation of one million hectares of coniferous forest between 1950 and 1990 – unless major disturbance intervenes. Established mixed deciduous oak–ash forest in southern England removes CO_2 from the atmosphere at half to two-thirds of this rate.

Whereas removal of CO_2 from the atmosphere by UK forests currently accounts for only a small fraction of UK GHG emissions, the removals of CO_2 by the forests in Scotland currently account for around 12% of Scotland's GHG emissions.

Soil organic matter (SOM) has accumulated progressively in our northern soils since the ice retreated about 8000 years ago and is, in general, a large reservoir of organic carbon. In northern forests, particularly on the common peaty-gley soils in northern UK, there may be more organic carbon in the SOM than in the trees. This SOM is susceptible to oxidation and consequent emission of CO_2 to the atmosphere, particularly if the SOM is disturbed. Disturbance of the SOM results from either: (1) forces of nature, such as windthrow, or (2) consequences of management practices, in particular site preparation, thinning, harvest and stump removal. Windthrow, site preparation by ploughing and stump removal (on sites in northern Britain and southern Sweden) may cause annual emissions of CO_2 from the SOM of 14–20 tonnes per hectare. It may take 15 years for young trees planted on such disturbed sites to turn the site from a carbon source into a carbon sink.

All crops, forests included, require nutrient resources, particularly nitrogen (N). Emissions of CO_2 are the down-side of site preparation; the up-side is that mineralisation of soil organic matter (SOM) leads to the concurrent release of available nitrogen in the soil. Peaty-gley SOM has a C:N ratio of between 25:1 and 30:1, which is also the range of C:N of spruce needles. As a result of oxidation of the SOM, sufficient nitrogen is made available to provide for growth of the complete forest stand up to canopy closure. Thus, on most forest sites (other than problem sites with, for example, heather-check), there is no requirement for application of nitrogen fertiliser, which carries with it with the associated risk of release of nitrous oxide (N_2O) (a potent GHG). Once the forest canopy is up, resources are to a considerable extent recycled within the forest. The primary requirement for additional resources is to grow the wood, which has a C:N ratio of 400:1 to 700:1, depending on species. In contrast with the situation for conventional agricultural crops, for forest crops sufficient nitrogen is available for growth from the on-going turnover of SOM and the continuous addition of wet and dry deposition of nitrogen in various forms from the atmosphere. Because N_2O is an important GHG, the production of biomass from woody crops without the need for mineral fertilisers with the associated production of N_2O is important in UK climate change mitigation strategies.

There has been controversy over whether current forest CO_2 sinks, particularly those in the tropics, will continue into the future. Mainly as a result of increased photosynthesis caused by elevated CO_2, increases in the troposphere CO_2 lead to appreciable increases in the growth of young trees. A simplistic hypothesis is that as concentrations continue to increase, photosynthesis will saturate with respect to CO_2 and forest respiration will increase in relation to temperature, so that CO_2 emissions will come to exceed CO_2 removals in the foreseeable future. However, three different, independent forest system models (developed respectively in UK, Australia and the USA) that explicitly combine forest carbon and nitrogen cycles (including nitrogen deposition, turnover and emission) with the expected rise in atmospheric temperature and CO_2 concentration, lead to the conclusion that the Norway spruce carbon sink in Sweden and the Sitka spruce carbon sink in Scotland will continue as at present for at least the next 100 years.

Trees and forests interact with the atmosphere through exchanges of energy and greenhouse gases (GHG) such as carbon dioxide (CO_2) in the troposphere – that part of the atmosphere extending from the earth's surface to a height of around 17 km. Current projections (UKCP09) indicate that the climate in the UK will continue to warm through this century, and that there will be changes in rainfall and its seasonal distribution that varies regionally.

Forests will react to these changes but they also have the capacity to influence them at regional, national and global scales in two main ways: first, by absorbing solar radiation energy and directly transferring the energy as sensible heat into the troposphere; second, by exchanging GHG with the troposphere. The processes underlying these interactions between forest and atmosphere are discussed in this chapter.

As the ice retreated from the British Isles after the last glaciations, microorganisms in the rock debris gave way to green plants, and soil formation was initiated; forests followed and carbon (C) and nitrogen (N) accumulated. The most significant aspect of forests from the perspective of climate change is that they comprise both trees and soil. In forests world-wide, there is four times as much carbon in the soils as in the trees. In tropical forests, this ratio is about 1:1, and it increases northwards to 4:1 in the north-temperate forests, up to 8:1 in the boreal forests and over 10:1 in the tundra. While only remnants of those forests remain in Britain, the peaty-gley soils that developed have very largely provided the basis for the woodland creation in northern Britain described in Chapter 1. The carbon in these soils is as, or more, vulnerable to return to the atmosphere as the carbon in the trees. In considering interactions between forests and the atmosphere, we must take into account the role of both the trees and the soils, sometimes together, sometimes separately.

In addition to our historical native broadleaf and Scots pine woodlands, the one million hectares (ha) of conifers, 80% Sitka spruce (Picea sitchensis), recently planted between 1950 and 1990, are a major resource today, available as they mature for a range of end-uses, including structural timber and woodfuel, and for making other contributions to climate change mitigation. The importance of Sitka spruce in commercial conifer plantations has led to it being the focus of much of the detailed characterisation and understanding of the processes underlying conifer tree growth in the UK that is discussed here.

In the first instance in this chapter, we consider a stand or compartment as the spatial scale (i.e. a scale of about 10–20 ha of similar trees at the same point in the management cycle) and we take the period of the rotation as the temporal scale (i.e. around 40 years for conifers). In later sections, for some purposes we consider the scale of a forest of several thousand hectares made up of compartments representing all stages in the management cycle, and ultimately, we consider the forest in a landscape also comprising agriculture and other land uses.

3.1 Greenhouse gases

Greenhouse gases (GHG) are gases that have absorption bands at particular wavelengths that absorb long-wave radiation and thus reduce the escape to space of the radiation emitted from the surface of the Earth. The rise in atmospheric GHG contents attributable to human activity is the major source of anthropogenic climate change. The calculated global warming potential (GWP) of the long-lived GHG depends on the lifetime or turnover time of the gas in the troposphere, and this depends strongly on a number of other, different, reactive atmospheric gases.

The relevant natural GHG in the troposphere are, in order of their significance: water vapour, carbon dioxide (CO_2), methane (CH_4), ozone (O_3) and nitrous oxide (N_2O) (IPCC, 2007). Forests exchange all these GHG with the troposphere to a larger or smaller extent (see 3.6 below).

Water vapour: is the main natural GHG, but its concentration is only indirectly affected by human activity, by our effect on evaporation rates from the land surface which vary with land use, irrigation and drainage. Evaporation from forests may influence the water vapour content of the local atmosphere at any time, but the large-scale tropospheric water content depends on the hydrological cycle, and in the UK case particularly the weather that reaches us from the Atlantic ocean, so that any change in UK forest evaporation would barely affect the water content of the troposphere at the global scale.

Carbon dioxide (CO_2): is the most effective anthropogenic GHG globally. Growing trees take up CO_2 from the local troposphere in photosynthesis, and forests globally mitigate climate warming significantly (Royal Society, 2001).

Methane (CH_4): methane's global warming potential (GWP) is 23 times that of CO_2 on a 100-year timescale. Waterlogged forest soils can be sources of methane, whereas drier forest soils may be sinks. Emission of methane from forest canopies has been postulated, but evidence is lacking.

Ozone (O_3): ozone is both a tropospheric GHG and a strong oxidant with damaging metabolic effects on tree function, particularly on the photosynthetic systems in leaves, as well as on human health (Royal Society, 2008). Compared with the other GHG, ozone is very short-lived, turning over rapidly in the troposphere. In forests, it may be formed continuously from isoprene and terpenes that are produced by tree leaves during photosynthesis. The significance of such forest sources is not well quantified.

Nitrous oxide (N_2O): nitrous oxide's GWP is 296 times that of CO_2 on a 100-year timescale. Both nitrification and denitrification can lead to emissions from forest soils. The application of fertilisers containing nitrogen over crops, pasture and, to a much lesser extent, forest, is a major source of N_2O in the troposphere.

3.2 Forest radiation and energy balance

3.2.1 Solar radiation, albedo and thermal radiation

Life on planet Earth depends on energy from our sun. The sun is a high temperature radiation source, so that solar radiation is essentially short-wave radiation. In addition to solar radiation, tree canopies exchange long-wave (or 'thermal') radiation with the atmosphere in proportion to their temperature (see Box 3.1).

A proportion of the incident solar radiation is reflected by forest canopies back through the atmosphere, the proportion depending on the structure and optical properties of the forest canopy, and the height of the sun above the horizon (i.e. dependent on latitude, time of year and time of day). The average proportion of solar radiation reflected, or *albedo*, defines how much of the sun's energy is not retained in the forest to drive processes. For a closed canopy, the properties of the leaves, their surfaces, arrangement and their spatial distribution,

determine the *albedo*. The midday or mean daily albedo of closed canopy coniferous forest in summer lies in the range 8–12% (i.e. 88–92% of incident solar radiation is

BOX 3.1 Radiation partitioning

The net amount of solar radiation absorbed by a forest compartment (S_n) is:

$$S_n = S(1 - \alpha),$$

where S is the incident solar radiation and α is the solar reflectance (or *albedo*).

The net amount of **thermal radiation, i.e. the radiation emitted by terrestrial bodies such as the atmosphere, soil and biomass**, absorbed by the forest (L_n) is

$$L_n = L_d - L_u,$$

where L_d is the down-welling and L_u the up-welling thermal radiation.

Thus, the **net all-wavelength** absorbed solar and thermal radiation, R_n, is:

$$R_n = S_n + L_n$$

The energy in the **net all-wavelength radiation** is partitioned into different processes:

$$R_n = G + H + \lambda(E+T) + P$$

where G is the heat transfer into the soil; H is the transfer of sensible heat to the atmosphere; E the evaporation of intercepted rain (also called interception loss) and T the transpiration. λ is the latent heat of vaporisation of water and P the amount of energy used in photosynthesis. Note that P is very small compared with the other components of the energy balance, so is usually ignored.

Thus the energy **available (A)** to drive sensible heat transfer and evapo-transpiration is:
$$A = R_n - G = H + \lambda(E+T).$$

Energy partitioning by the forest canopy is expressed by:

- the Bowen ratio $\beta = H/\lambda(E+T)$, and

- the evaporative fraction, $E_f = \lambda(E+T)/R_n$

All the above properties may be expressed as instantaneous values (e.g. midday) or as period-averaged, e.g. over a half-hour, daytime, month, season or year.

absorbed). This range of values of albedo is quite similar to the albedo of heathland that has been replaced by conifers during recent woodland creation. The albedo of closed canopy broadleaf forest when fully leafed is generally much larger, in the range 18–22 %, and comparable with values for agricultural crops. During the leafless phase of the annual cycle and in the absence of snow, in winter the albedo of broadleaf forest drops to lower values.

3.2.2 Leaf dynamics and solar radiation absorption

The growth of a stand of trees, and indeed of forests, depends on the amount of solar radiation that is intercepted and absorbed by the tree crowns that comprise the forest canopy. It is the size, spatial distribution, angular distribution, grouping and overall area of the leaves that determine the amount of solar radiation that is intercepted and absorbed by a forest canopy. Canopies differ widely with respect to these five properties. In general, it is the overall area of leaves, and their distribution within the canopy space, that has the largest effect on CO_2 uptake by a canopy (Wang and Jarvis, 1991).

In Sitka spruce stands, the leaf area index (LAI, see Box 3.2) reaches around 10 at canopy closure and

BOX 3.2 Leaf area index

Leaves absorb solar radiation and absorb CO_2 from the atmosphere. The leaf area index (LAI) is the dimensionless metric used to describe the amount of leaf present in a canopy. It is the summed plan area of leaves per unit area of ground; e.g. a LAI of 8 is 8 m^2 of leaves in a vertical cylinder of 1 m^2 in cross-section through a canopy, measured laid out horizontally without overlap, and expressed per 1 m^2 of ground area.

If leaves are randomly distributed in space through a canopy, with a random distribution of leaf angles, one would expect 95% interception of solar radiation to occur at a LAI of 6. However, if leaves are grouped into crowns, and within the crowns the leaves are grouped into whorls of branches, and on the branches they are further grouped around the shoots, a much larger LAI can be maintained than if the leaves were randomly distributed in space, with random inclination and azimuthal angles (Wang and Jarvis, 1990, 1991).

then drops back to a stable value of around 8 that is maintained through the rotation, as needles are dropped from the two or three lower-most whorls of branches. Viewed looking down from above, the canopy appears as an array of brightly-lit cones with large black holes tapering downwards between them, as a result of multiple reflections and effective absorption of solar radiation (Norman and Jarvis, 1974). A consequence of this structural arrangement is that around two-thirds of the absorption of solar radiation by the trees, once the canopy is closed, occurs in the well-lit upper one-third of the tree crowns. This is maintained throughout the rotation, unless the LAI is temporarily reduced, for example by defoliation or thinning.

In general, canopies of broadleaves have a smaller LAI (4–6) than those of conifers, and the leaves are more horizontal and less grouped than in conifers, but with higher photosynthetic capacity per unit leaf area. The result of a shorter growing season is somewhat lower rate of CO_2 removal from the atmosphere by broadleaves than conifers (Figures 3.3 and 3.6).

3.2.3 Partitioning of the absorbed energy

The radiation energy available to drive exchange processes within a forest, A (see Box 3.1) is the algebraic sum of the incoming and outgoing short-wave and long-wave radiation fluxes. During the daytime, vegetation canopies are generally warmer than the atmosphere because of the absorption of solar radiation, and consequently heat is transferred directly from canopies to the troposphere. In the absence of solar radiation at night however, the net emission of long-wave radiation leads to cooling down of canopies, dew formation and sometimes frost.

The *evaporative fraction* defines the proportion of the absorbed energy that cools the local air but also adds water vapour to it. The Bowen ratio (β), is a measure of the relative proportion of the absorbed radiation energy that is directly transferred into warming the local air.

Energy flux measurements over coniferous forests with dry foliage show that the Bowen ratio is generally larger than 1 and frequently larger than 2; i.e. more than twice as much of the available energy is transferred directly into the troposphere as sensible heat, rather than as transpiration, because transpiration is restricted by the apertures of the pores in the leaves, the stomata. This situation is rather different in the case of broadleaves and agricultural crops, for which β is generally less than unity, i.e. more of the absorbed energy is used to evaporate water by crops and less to heat the ambient air directly.

By comparison with crops and herbaceous vegetation, forests are very well coupled to conditions in the troposphere because of the height of the trees and the aerodynamic roughness of the canopy, both of which enhance turbulent transfer of heat and water vapour (McNaughton and Jarvis, 1983). As a result, evaporation of water from forest canopies is predominantly driven by the water vapour pressure deficit (VPD) of the ambient air, rather than by the available solar energy (A, Box 3.1) (Jarvis and McNaughton, 1986). The VPD is also the primary driver of transpiration from forest canopies, which is consequently more effectively controlled by the aperture of the stomatal pores in the leaves than in shorter agricultural crops for which solar radiation is the primary driver of both evaporation and transpiration (for a summary, see Monteith and Unsworth, 1990, p. 197).

Furthermore, the stomatal pores of Sitka spruce, and many other conifers, are very sensitive to the ambient VPD, closing at large values of the VPD (see 3.3.4 below). Thus, as local temperatures rise and VPD increases, we can expect less transpiration and a larger proportion of the available energy going into *direct* warming of the troposphere as sensible heat. On the other hand, water vapour is also a GHG, and, like CO_2, it impedes the escape of long-wave radiation to space. Thus, closure of stomata, caused by the large VPDs that occur at higher temperatures, results in contrasting reduced addition of water vapour to, and reduced removal of CO_2 from the troposphere (Jarvis, 1994). The balance of effects between 'additional direct heating', reduced 'water vapour emission' and reduced 'CO_2 removal' is not easily calculated and requires a model for solution in particular circumstances.

3.3 Measured growth responses of trees to atmospheric variables

3.3.1 Growth responses to intercepted solar radiation

Growth of a tree stand is proportional to the solar radiation intercepted by the canopy (Monteith, 1977; Jarvis and Leverenz, 1983; Wang *et al.,* 1991) and the relationship is relatively consistent across a range of species (Linder, 1985; Landsberg *et al.,* 1996). The interception of solar radiation depends on the development of canopy leaf area, which depends on a number of environmental resources,

particularly the availability of nutrients during the initial phase of stand development (see 3.4.9 below).

Empirical relationships between the growth in tree and stand dry mass with time and the concurrent intercepted solar radiation have proved to be quite consistent across a range of species, spacing, fertiliser and management treatments and are the basis of more than one widely-used stand growth model (e.g. 3PG). A commonly obtained figure for a number of forest species is 0.85 kg tree dry mass per giga joule (GJ) of intercepted solar radiation (Linder, 1985; Landsberg *et al.,* 1996). This commonality occurs because both interception of radiation and photosynthetic absorption of CO_2 are primarily dependent on the amount of leaf present. An investigation over four years, making use of stands of Sitka spruce in a thinning and fertiliser experiment (± thinning, ± N and ± P) at Tummel forest, central Scotland, gave a maximum of 0.85 kg tree dry mass per GJ intercepted solar radiation (Wang *et al.,* 1991) and this is not atypical for coniferous forest stands. Nonetheless, Cannell *et al.,* (1987), Milne *et al.,* (1992) and Sattin *et al.,* (1997) obtained values of up to 1.4 kg dry mass per GJ intercepted for small plots of short-rotation willows and poplars, near Edinburgh.

3.3.2 Growth responses to atmospheric CO_2 concentration

During the 1990s, a large number of experimental studies were made in Europe to investigate the responses of forest tree species to a doubling of the current atmospheric CO_2 concentration, i.e. to concentrations in the range of around 700–750 ppmv CO_2. Young broadleaves of *Fagus sylvatica, Betula pendula, B. pubescens, Alnus glutinosa, Prunus avium, Quercus robur, Q. petraea, Q. rubra, Populus* hybrids and the conifers *Picea abies, P. sitchensis* and *Pinus sylvestris* between 1 and 10 years old were grown in a range of chambers, most commonly in outdoor open-topped chambers of 3 m diameter and 4 m height (Jarvis *et al.,* 1998). Meta-analysis of the data showed a mean increase in total biomass in response to the doubling in atmospheric CO_2 concentration of 54% over all species (above and below-ground biomass, in stressed and unstressed conditions) over *c.* 5 years of growth: 50% for the broadleaves; 56% for the conifers (Medlyn *et al.,* 1999, 2000). The mean response of Sitka spruce was an increase in dry mass of 32% without fertiliser addition and over 100% with added fertiliser; the response of oak ranged from 5% to over 100% with added irrigation. In general, the effect of growth in elevated CO_2 concentration was large in small, young plants and became magnified as a result of subsequent exponential growth.

While increase in the ambient CO_2 concentration would be expected to increase the concentration of CO_2 at the photosynthetic reaction sites in the chloroplasts within the leaves, and thus to increase photosynthesis and biomass production, the stomata of many species reduce in aperture at higher ambient CO_2 concentrations, thereby reducing the entry of CO_2 into leaves, so that it is not immediately obvious that the projected increase in ambient CO_2 concentration will lead to increase in growth. A meta-analysis of stomatal conductance data from 13 of the long-term, field-based experiments with trees in open-top chambers found an average 21% decrease in stomatal conductance in response to doubling of the ambient CO_2 concentration (Medlyn et al., 2001). However, the stomata of a few species, including Sitka spruce and Scots pine, were not very responsive to the increase in CO_2 concentration, consistent with earlier laboratory-based studies (e.g. Beadle et al., 1979; Morison and Jarvis, 1983).

Trees can be grown to larger sizes and exposed to elevated CO_2 concentrations in individual tree chambers and in stands. For example, Silver birch (Betula pendula), was grown for four years from seedlings to a height of 4.2 m in individual ventilated tree chambers near Edinburgh, in ambient air to which CO_2 was added to give 700 ppmv. Leaf area increased by 60% and net photosynthesis by 68%, relative to the plants in the current ambient CO_2, but biomass growth increased only by 59% because of larger losses of carbon from fine root production and mycorrhizas (Rey and Jarvis, 1997; Wang et al., 1998). In a number of comparable experiments in Europe and in the USA with indigenous species, similar results have been obtained. In central Sweden, for example, individual 10 m tall trees of Norway spruce, growing in plantations with nutrient fertilisation, in air-conditioned chambers, to which CO_2 was added to give a concentration of 700 ppmv, increased in stem growth by 15 to 20% relative to the controls over 3 years (S. Linder et al., pers. comm.). In the USA, free-air exposure of stands of pole-stage Pinus taeda (loblolly pine) to an increase of the CO_2 concentration of 200 ppmv above ambient (using the so-called FACE technology), led to a differential increase in tree height of the dominant trees of 7%, relative to the controls, over 10 years (R. Oren et al., 2009, pers. comm.). Comparable results using FACE methods on other tree stands have been summarised by Karnosky et al., (2005).

There can be no doubt therefore, that young trees and

stands of pole-stage trees grow faster in an atmosphere with increased CO_2 concentration. However, there is limited information on the responses of mature trees to increase in ambient CO_2 concentrations.

3.3.3 Growth responses to ambient temperature and VPD

Examination of independent responses to ambient air temperature and VPD, which are both expected to increase together in the future, requires the use of more sophisticated controlled environment facilities in which only small, young trees up to around 70 cm in height can be grown. Such experiments, in which other environmental variables have been kept constant within narrow limits, have shown that the optimum temperature for growth of young Sitka spruce is a regime of around 20/14°C day/night temperature (Neilson et al., 1972). However, in natural conditions, increase in air temperature is invariably accompanied by an increase in VPD, and other experiments in which VPD was not independently controlled, have shown a lower temperature optimum. Similar experiments have shown a substantial negative growth response to increase in the ambient VPD (Neilson and Jarvis, 1975). Thus, the dominant effect of the projected rise in temperature on Sitka spruce is likely to be mediated through the negative response to VPD in closing the stomata and thereby reducing CO_2 influx and growth.

3.3.4 Stomatal conductance responses to atmospheric VPD

Laboratory investigations of leaf physiology have demonstrated that CO_2 exchanges of Sitka spruce, and other species of both conifers and broadleaves, are very sensitive to the local ambient VPD, because large VPD leads to closure of the stomata, independent of the plant water status, thereby reducing the influx of CO_2 into leaves, as well as the exchanges of other gases and volatile compounds (e.g. Sandford and Jarvis, 1986). A meta-analysis of stomatal conductance data from the long-term, chamber-based experiments across Europe referred to in 3.3.2 above, found that both Sitka spruce and Scots pine showed a substantial decrease in stomatal conductance with increasing VPD but that Quercus robur and other species were rather more sensitive, irrespective of whether grown in ambient or elevated atmospheric CO_2 concentration (Medlyn et al., 2001).

A consequence of this sensitivity of stomata to VPD, observed at stand scale in forests, is that the net carbon influx to stands of Sitka spruce (see 3.4.6 below) may fall to zero or less during days in which the ambient air temperature rises above around 20ºC, in part because of the associated increase in the ambient VPD (Jarvis, 1994). Anecdotal and mensurational forestry evidence that Sitka spruce grows less well in the east of Scotland than in the west, *despite adequate soil water*, is consistent with this negative physiological response to ambient VPD.

3.4 Temporal CO$_2$ dynamics

3.4.1 Stocks and fluxes: some basic concepts and definitions

As a consequence of their photosynthetic activities, forest stands contain significant stocks of carbon in trees and soil, and continuously exchange CO$_2$ with the atmosphere. The main fluxes of CO$_2$ and the two other key GHGs, N$_2$O and CH$_4$, in forest stands are illustrated in Figure 3.1 and appropriate variables are defined in Box 3.3 and Figure 3.2.

3.4.2 Measurements of CO$_2$ fluxes in UK forests

Measurements of the exchanges of CO$_2$ between forests and the troposphere are crucial to identifying, quantifying and understanding: (1) the processes that determine carbon accumulation and growth of forests and (2) the capability of forests to mitigate climate change by removing CO$_2$ from the troposphere. First measurements were made in the early 1970s on a campaign basis using the so-called energy budget or Bowen ratio approach. Our capacity to measure CO$_2$ exchanges (or fluxes) continuously and quantitatively developed in the 1990s, using the so-called eddy-covariance technology, which required very fast (5–10 Hz) CO$_2$-measuring sensors and sonic anemometers.

Although there are today possibly over 100 forest sites where CO$_2$ fluxes are being measured in forests across Europe, only sites in Ireland have a reasonably similar maritime climate and comparable tree species to forests in the UK. The overriding influence of the North Atlantic drives

Figure 3.1
The emissions and removals of the main trace gas fluxes in a forest stand. The interrelationships among the principal carbon fluxes are defined in Box 3.3 and by Figure 3.2.

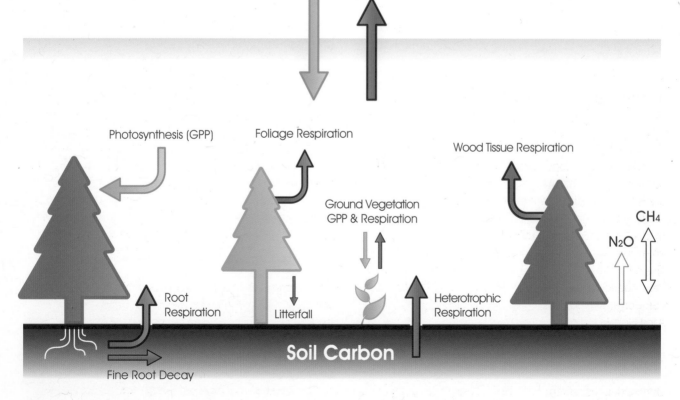

Net ecosystem carbon exchange

Photosynthesis (GPP) Foliage Respiration Wood Tissue Respiration Ground Vegetation GPP & Respiration CH$_4$ N$_2$O Root Respiration Litterfall Heterotrophic Respiration **Soil Carbon** Fine Root Decay

(Note: O$_3$ is not shown in this figure because of indirect formation through VOC production and direct deposition).

BOX 3.3 Some definitions

GPP – 'Gross Primary Production', i.e. removal of CO_2 from the atmosphere by photosynthesis in leaves, is driven by the fraction of visible radiation absorbed by leaf pigments and is a function of LAI, leaf age, position, N content and acclimation to the solar radiation flux

R_A – autotrophic respiration of living cells in twigs, branches, wood, bark, roots

NPP – 'Net Primary Production' (GPP – R_A)

R_H – heterotrophic respiration in the soil

R_R – autotrophic root-system respiration

R_S – soil respiration (= $R_H + R_R$)

R_E – ecosystem respiration (= $R_A + R_H$)

NEP – Net Ecosystem Production (= GPP – R_A – R_H), see Figure 3.2.

NEP is a measure of the total CO_2 influx (or gain) less the total CO_2 emission (or loss) and thus it is *the measure of net removal of CO_2 from the atmosphere* by a forest stand. At the scale of an extensive forest (e.g. 5000 ha or so, with compartments at various stages of management, or a landscape comprising a range of additional land uses, the net CO_2 exchange, as seen from a tall tower or an aircraft, will likely be less than the NEP as a result of natural, accidental or managerial '*disturbance*' (D) of various kinds.

Thus, NBP – Net biome production integrates CO_2 exchanges across the landscape (NEP – D).

Figure 3.2

Diagrammatic representation of the relationships among the carbon fluxes defined in Box 3.3.

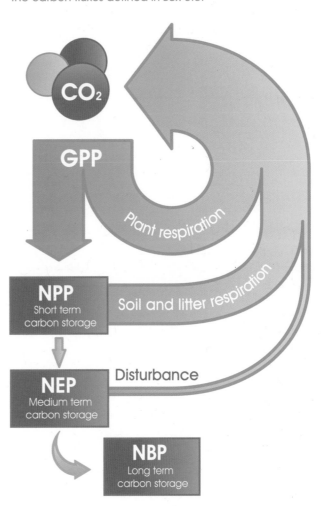

the climate and determines that the forest trees, races and genotypes used in the British Isles are often distinctive. Here, we focus initially on continuous or semi-continuous measurements made since 1997 in the UK, using eddy-covariance technology. Short-term measurements and relevant measurements from nearby Europe supplement the long-term data. There are two on-going, long-term, forest sites in the UK where CO_2 fluxes have been measured continuously, second by second, since 1997: an evergreen, coniferous forest site, with 80% of the tree species Sitka spruce (*Picea sitchensis*) in the Griffin Forest near Aberfeldy, Perthshire, and a deciduous, broadleaf, oak-dominated woodland in The Straits Enclosure, near Alice Holt, Hampshire.

Eddy-flux measurements in coniferous forests

Continuous measurements of CO_2 fluxes have been made

in a young stand (Yield Class 14–16, planted 1980) of Sitka spruce at the Griffin Forest in Perthshire since 1997 (Clement *et al.*, 2003). Figure 3.3 shows the annual time-course of *daily* CO_2 fluxes averaged over five years. The seasonal day-to-day variability in the flux is clear, despite the averaging of five years data, particularly its consistent variability in the autumn. The superimposed cumulative curve gives the 5-year-average total annual removal of CO_2 from the atmosphere at the end of the year (i.e. the NEP) which is 24 tonnes CO_2 per hectare.

Sitka spruce in Scotland is at least as effective as many other spruce forest sites in Europe in removing CO_2 from the atmosphere (Figure 3.4). However, in Ireland, where the national average yield class for Sitka spruce is Yield Class 18, an average annual removal from the atmosphere (NEP) of 33 tCO_2 per hectare has been measured in a stand of Yield Class 24 (Black *et al.*, 2009). Average seasonal

Figure 3.3
The vertical columns show the daily (24 h) net CO_2 fluxes for every day of the year, averaged over 5 years. The columns above the zero line show the net 24 h removals from the atmosphere and gains by the forest; the columns below the zero line show net daily emissions by the forest and additions to the atmosphere. The solid line shows the 5-year average of accumulated removals from the atmosphere. (Sitka spruce at the Griffin Forest, Perthshire 1997–2001.)

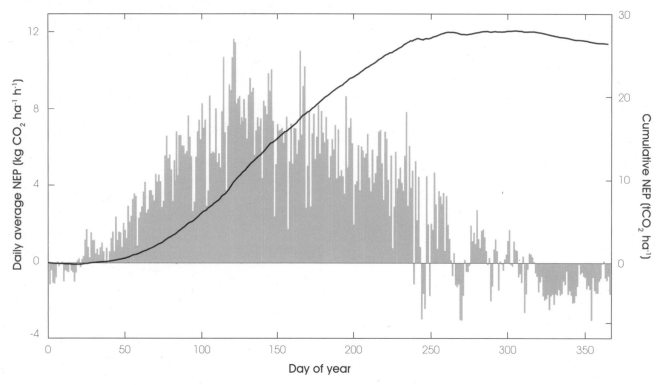

removals of CO_2 from the atmosphere each hour of the day show the daytime absorption and night-time emissions (Figure 3.5). The period of night-time emissions is long in the winter but the magnitude is small because of the low temperatures. As the season progresses and temperature increases, the emissions increase in magnitude but the period of emissions decreases with the increase in day length. The period and magnitude of the daytime removals of CO_2 from the atmosphere very clearly increase through the spring to a maximum in mid-summer, and decrease in the autumn. These data integrated over 12 months and averaged over 5 years give an average annual *removal* from the atmosphere of 24.2±1.4 tonnes CO_2 per ha per year (tCO_2 ha^{-1} $year^{-1}$). The measurements continue to this day; however, the sequence was interrupted at the Griffin Forest in 2004 to investigate the effect of thinning on the CO_2 fluxes.

The Sitka spruce measurements at the Griffin Forest have been supplemented by intermittent measurements using the same methodology in an age series (0–30 years) of comparable stands in the Harwood Forest, Northumberland, during 2000–2003 (Grace *et al.*, 2003; Magnani *et al.*, 2007). Very close synchrony and similarity

in magnitude of the carbon fluxes were obtained between contemporaneous measurements at Harwood in the 30-year-old stand and concurrent measurements at Griffin in the 24-year-old stand, around 150 km apart.

Figure 3.4
Cumulated monthly sums of NEP over the 12 months of 1997 at five spruce forests sites in Europe. The solid black line is for Sitka spruce at the Griffin Forest, Perthshire, Scotland. The other lines represent stands of predominantly *Picea abies* in Germany and Belgium (modified and updated from Bernhofer *et al.*, 2003).

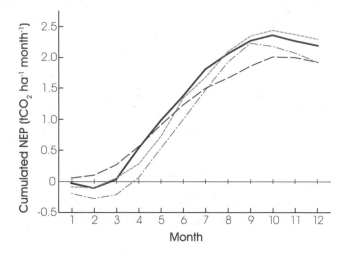

Figure 3.5
The diurnal course of NEP fluxes averaged over two-monthly periods throughout the course of the year. The data above the zero line show the net 24 h removals from the atmosphere and gains by the forest; the data below the zero line show the net daily emissions by the forest and additions to the atmosphere. Sitka spruce at Griffin Forest, averaged over the 5 years 1997 to 2001 inclusive. Note the impact of time of year on day-length and the relative magnitudes of the night-time and day-time fluxes.

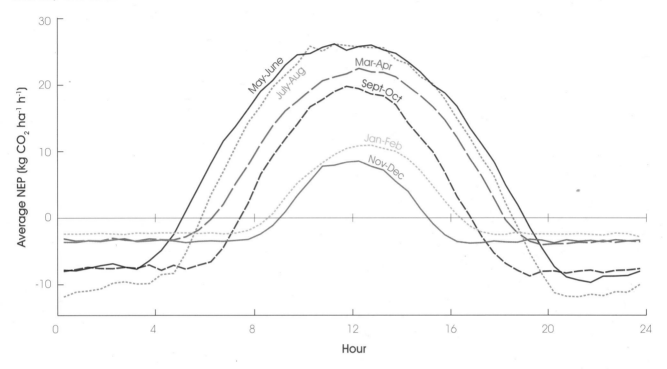

Eddy-flux measurements in deciduous broadleaved forests

Continuous measurements of CO_2 flux, starting in 1998, have been made in a mature stand of mixed broadleaves with a predominant overstorey of oak (*Quercus robur*), with some ash (*Fraxinus excelsior*) at the Straits Enclosure, near Alice Holt (Broadmeadow *et al.,* pers. comm.). The equipment and software used are similar to that used for Sitka spruce at the Griffin and Harwood Forests. The stand was planted in the 1930s and was brought into management as a research site in around 1990. The stand was thinned in 1991, 1994 and 1999, to 495 overstorey trees per hectare. Tree height is around 23 m and the LAI around 6 around the flux tower. When in full leaf, the overstorey absorbs between 60% and 75% of the incident solar radiation. Comparable measurements are being made in stands of broadleaves, predominantly beech (*Fagus sylvatica*) and oak (*Q. robur*) in Denmark and Belgium, respectively, and the results are not dissimilar.

Figure 3.6 shows CO_2 flux data for The Straits broadleaves over the period (1999–2007) The annual average net removal of CO_2 from the atmosphere for the eight-year period is 15.1 tCO_2 ha^{-1} year^{-1}, with a range of 13.1–19.8.

Comparison between Sitka spruce forest in the north and oak forest in the south show that removals of CO_2 from the atmosphere by the oak forest are 62% of those achieved by Sitka spruce. The measurements are continuing at the present time on both sites.

Short-term eddy-flux measurements

Eddy-flux measurements made on a short-term basis at sites of different aged stands of Sitka spruce on *deep-peat* (0.5–5.0 m depth) in the west and north of Scotland, found initial annual *emissions* of 7–15 tCO_2 ha^{-1} year^{-1} from newly ploughed and drained peatlands. The range of emission rates depended on the temperature, depth, water-logging and degree of disturbance of the peat. These emissions turned progressively to annual *removals* from the atmosphere as, first, the ground vegetation returned and, second, as the trees slowly grew, eventually reaching net annual removals of up to *c.* 18 tCO_2 ha^{-1} year^{-1} on sites supporting YC 10. At the same time annual emissions from the peat continued at rates of up to 4 tCO_2 ha^{-1} year^{-1} (Hargreaves *et al.,* 2003). Similar rates of soil CO_2 emissions from an age-series of afforested, drained (but with high water table), blanket peat sites have been

Figure 3.6

The vertical columns show the daily (24 h) net CO_2 fluxes for every day of the year, averaged over 8 years. The columns above the zero line show the net 24 h removals from the atmosphere and gains by the forest; the columns below the zero line show net daily emissions by the forest and additions to the atmosphere. The solid line shows the 8-year average accumulated removals from the atmosphere. (Mixed oak deciduous woodland at the Straits Enclosure, Alice Holt, Hampshire, 1999–2006.)

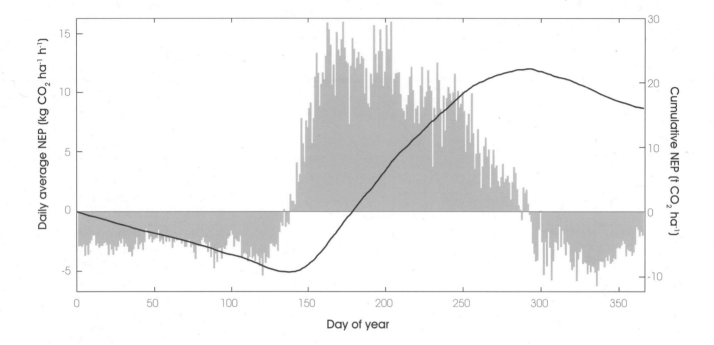

reported from Co. Galway in Ireland (Byrne and Farrell, 2005). In conclusion, the CO_2 fluxes from deep peat sites are in general similar but smaller than fluxes from peaty-gleys, as at Harwood and Griffin, declining pro-rata with the general yield class. Afforestation of deep peat lands is no longer acceptable because of their conservation value for wildlife and water resources; their CO_2 effluxes also support arguments for minimising their disturbance.

The earliest measurements of forest CO_2 fluxes in the UK were made using the so-called Bowen ratio methodology, in Sitka spruce at Fetteresso Forest, Kincardine, in 1970–2 and in Scots pine at Thetford Forest, Norfolk, in 1975–6. The measurements at Fetteresso found NEP close to zero on days on which the air temperature exceeded 17°C and negative when temperatures exceeded 20°C (Jarvis, 1994), largely because of closure of the stomata in response to the increase in atmospheric VPD, but also in part because of the increase in emissions in response to temperature (Jarvis, with a modelling appendix by Y.P. Wang, 1986). Very similar results were obtained with Douglas fir (*Pseudotsuga menziesii*) on Vancouver Island, by Price and Black (1990, 1991). The measurements at Thetford in 1976, a drought year with summer temperatures frequently in the range 25–30°C,

with concurrent exceptionally large VPDs, showed severe reductions in both CO_2 and transpiration fluxes in these extreme conditions (Jarvis *et al.,* 2007) – a possible portent of things to come.

3.4.3 Rotation average CO_2 removal by Sitka spruce forests

Measurements made on the age series of sites at the Harwood Forest have shown that site preparation by ploughing and subsequent planting, led to appreciable decomposition of the exposed soil organic matter and to significant emissions of CO_2 through the rotation (Zerva *et al.,* 2005; Zerva and Mencuccini, 2005b; Figure 3.7), as earlier noted by Cannell *et al.,* (1993). At the same time, growth of the young trees progressively removed more CO_2 from the atmosphere. For the young Sitka spruce it took 12 years for the emissions from the soil (a peaty-gley) to be compensated by removal of CO_2 from the atmosphere by the growing trees. Subsequent to that break-even point, removals from the atmosphere exceeded emissions from the soil, resulting in a *net removal* of CO_2 from the atmosphere by the forest, trees and soil together (i.e. an increasing NEP) (Figure 3.7; Jarvis and Linder, 2007). Canopy closure was reached after another 4 or

5 years, i.e. around 17 years after planting. During this period, the leaf area of the trees further increased to the maximum LAI of 8, which absorbed around 95% of the incoming solar radiation, and the rate of atmospheric CO_2 removals rose to equal the rates measured at both Harwood and Griffin Forests, i.e. an annual *net* removal of CO_2 from the atmosphere of around 24 tCO_2 ha^{-1} $year^{-1}$ (Clement *et al.*, 2003).

Figure 3.7
Net exchanges of CO_2 following site preparation and planting. Emissions of CO_2 from the soil dominate until c. year 11 (the break-even point) followed by increasing net removals of CO_2 from the troposphere by the growing young trees, reaching a plateau at canopy closure c. year 17.

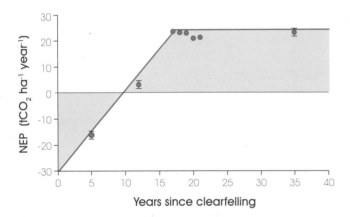

The average CO_2 removal over a rotation depends on the period and magnitude of emissions prior to the break-even point and canopy closure. Integrating over a 40-year rotation, these data indicate an average annual removal over a 40-year rotation of 56% of the closed canopy rate, i.e. 13.5 tCO_2 ha^{-1} $year^{-1}$ (Magnani *et al.*, 2007). The above quantity of 13.5 tCO_2 ha^{-1} $year^{-1}$ would be the appropriate average annual amount for an entire 'mature' forest, *in which all age classes were represented more or less equally throughout the forest.*

However, for historical reasons, the ups and downs of markets, or windthrow, for example, the age distribution of compartments is generally not uniform, and alternative methods are required to estimate the CO_2 dynamics at the larger spatial scales of extensive forest comprising compartments of clumped age and alternative species (see 3.5 below). Where, for example the younger age classes predominate, as in the most recent woodland creation, a more realistic average rate of removal of CO_2 from the troposphere is likely to be around 40% of the

post-canopy-closure rate, i.e. an annual rate of around 10 tCO_2 ha^{-1} $year^{-1}$, whereas in older forests the factor may be around 60% or more of the closed canopy rate.

3.4.4 Nitrogen requirements for effective carbon sequestration

Trees require resources to grow leaves and fine roots for the acquisition of carbon from the atmosphere and nutrients from the soil, respectively. Nitrogen is in general the most significant nutrient required for growth of Sitka spruce on peaty-gley soils. Artificial nitrogenous fertiliser has sometimes been applied in the past to speed-up canopy closure, particularly on nutrient-poor sites and especially on compartments experiencing heather-check as a result of inhibition of the tree root mycorrhizas, but application has sometimes been more general. Urea was, for example, applied universally at the Griffin Forest when the trees were 16–17 years old; but the canopy was already closing fast, and the effect on canopy and stem-wood growth was small and transient. Furthermore, there is some evidence that application of fertiliser nitrogen has an inhibitory effect on decomposition of soil organic matter, and consequently on the mineralisation of the indigenous nitrogen (Fog, 1988; Ågren *et al.*, 2001; Hyvönen *et al.*, 2007).

In addition, there are significant amounts of nitrogen in the rainfall (wet deposition) and the air (dry deposition). Although these amounts have been declining since the 1960s, typical annual wet deposition over Scotland ranges from 5–10 kg N ha^{-1}. The dry deposition probably adds an additional 30% but is less routinely measured and less accurately definable, in part because of the possibility of some direct uptake from the air into leaves within forest canopies (Mencuccini *et al.*, in manuscript). Magnani *et al.*, (2007) found a very significant relationship between NEP and wet deposition of nitrogen across a number of sites in northern Europe, including sites in northern UK, but their relationship was severely criticised as unrealistic because they did not add in estimates for the dry deposition of nitrogen.

Irrespective of the atmospheric deposition, management operations lead to the release of significant amounts of nutrients, particularly nitrogen, from the soil organic matter (SOM). We consider the demand in relation to the supply for the stand life cycle in two periods – the first period to grow the canopy of leaves, the second to grow the wood (see Box 3.4).

Box 3.4 The carbon to nitrogen ratio (C:N) for different components

Leaves	20:1 to 30:1
Fine roots	30:1 to 40:1
Branches and coarse roots	100:1
Stem wood	500:1 (400:1 to 700:1 for different species)
SOM	25:1 to 35:1

Period 1: The N demand to grow the tree to canopy closure (years 0–18):

The nitrogen content of Sitka spruce needles: around 2.0 g N m^{-2} (leaf area).

LAI of closed canopy Sitka spruce: 8

The N required to grow the leaf canopy: 160 kg N ha^{-1}

Estimated approx. total tree N required: around 720 kg N ha^{-1} (= 40 kg N ha^{-1} year^{-1}).

The N supply (years 0–18):

Mineralisation of SOM (Zerva et al., 2005; Zerva and Mencuccini, 2005b): 3.6 tonne C ha^{-1} year^{-1} @ C:N 30:1: 120 kg N ha^{-1} year^{-1}; i.e. approximately 3x the demand (front loaded), not including atmospheric deposition.

Period 2: The N demand to grow the tree trunk (years 18–40):

Trunk carbon content at 40-year harvest (YC 14): 115 tC ha^{-1}.

The N requirement to grow the trunk (C:N 500:1): 230 kg N ha^{-1} (=11 kg N ha^{-1} year^{-1}.

The N supply (years 19–40):

For central Scotland, annual wet deposition is now about 7 kg N ha^{-1}; dry deposition a likely additional 30%, adding up to a total N deposition of around 10 kg ha^{-1}.

In addition, nitrogen continues to be available from oxidation of the SOM, as a result of the original site preparation.

Nitrogen demand to grow the leaf canopy, fine root network and tree structure (years 1–18).

On average more than enough nitrogen is supplied as a result of 'site preparation' to grow the forest canopy and basic tree structures. However, in practice, the annual demand will increase from a small amount in the early years to a much larger amount in the later years of the period, as the trees grow. Thus, supply may considerably exceed demand in the early years, whereas the supply may not be sufficient in the later years so that N-stress may occur (Miller et al., 1979), particularly if a part of the nitrogen mineralised during the early years is not retained but is lost in drainage waters or as atmospheric emissions.

Once the canopy has closed, at around 18 years, new leaves grow at the top of the tree and on the upper whorls of branches, old needles drop from the lower branches and nutrients re-circulate within the stand. The canopy LAI remains at a practically constant value of around 8, requiring additional nitrogen only to replace losses to the atmosphere from leaching and redistribution of leaf litter from the site, and to compensate for any immobilisation in the soil of nitrogen released from the decomposing leaves (Miller et al., 1979; Titus and Malcolm, 1999). Possible immobilisation of nitrogen in the litter could lead to undefined, larger requirements in the second half of the rotation than are indicated in Box 3.4.

Nitrogen demand to grow the wood (years 19–40).

The demand for nitrogen is small by comparison with the growth of the canopy, for example, because the wood has a C:N ratio of around 500:1. The estimated total tree dry mass of YC 14 at 40 years is around 460 tonne ha^{-1}. The total mass of the harvested trunk is close to 50% of the total tree mass, with a carbon content of around 50%, i.e. around 115 tonne C ha^{-1}. With a C:N ratio of around 500:1, the total nitrogen requirement to grow the wood is around 230 kg N ha^{-1}, or on average over 22 years, 10 kg N ha^{-1} year^{-1}.

Nitrogen supply over the period

Nitrogen is supplied continuously as atmospheric deposition, both gaseous (dry deposition) and in precipitation (wet deposition) (Fowler et al., 1989). For central Scotland, the wet deposition is now about 7 kg N ha^{-1} (Magnani et al., 2007) and the dry deposition probably an additional 30%, adding up to a total nitrogen deposition of around 10 kg ha^{-1}, i.e. sufficient to meet the demand

(Cannell *et al.*, 1998). In addition, nitrogen continues to be available from oxidation of the SOM, as a result of the original site preparation, at an annual rate of up to 120 kg N ha^{-1}.

The requirement for nitrogen is usefully separated in both space (establishment of canopy, fine roots vs wood production) and time (the initial 18 years to canopy closure, and the major wood production period of the rotation. Taking the growth pattern into account, during woodland creation, the nitrogen supply to the trees from soil and atmosphere is more than sufficient to grow a substantial crop of wood, *without addition of nitrogen fertilisers*. How the second rotation fares, however, will depend on how the site is managed.

3.4.5 Disturbance

On undisturbed ground, the soil carbon content maintains a relatively stable state of losses balanced by gains. However, disturbance of the soil has major impacts on both carbon emissions and drainage losses, and overall tends to lead to reductions in carbon stocks (Reynolds, 2007). Catastrophic disturbance such as windthrow that leads to uprooting of trees can, for example, result in immediate annual CO_2 *emissions* from the soil and debris at a rate of at least 20 tCO_2 ha^{-1} year^{-1} (A. Grelle, pers. comm.) and this may continue over several years during reclamation of the site. Such large emissions of CO_2 further compound the economic loss of timber and loss of opportunity for CO_2 removal.

There can be substantial losses of carbon from the soil during the first rotation as a result of site preparation (Figure 3.8) but during the second rotation, the soil carbon content may recover to the extent that the original soil carbon content is restored after 30 years of growth (Zerva *et al.*, 2008). A chronosequence of five stands of Norway spruce (*Picea abies*) in Sweden (18–91 years old) planted onto former agricultural land showed an overall increase in the soil carbon stock reaching 191 tC ha^{-1}, despite initial losses from the mineral soil horizons (Cerli *et al.*, 2006). On that basis, it is to be expected that carbon will continue to accumulate, *provided that the soil does not experience further major disturbance*. Across a landscape, disturbance can be very variable as a result of different intensities of management intervention and the vagaries of catastrophic disturbance from natural events, particularly windthrow. On that assumption, the process of carbon accumulation in successive clearfell rotations on peaty-gley soil can be modelled, taking the stand-scale data

Figure 3.8

Changes in soil carbon at a Sitka spruce site in first and second rotation (Zerva *et al.*, 2005). UN is unforested grassland/moorland; CF is clearfell, and remaining bars denote age of the trees at different times in the rotation. The 40-year stand was in the first rotation; 12-, 20- and 30-year stands were in the second rotation.

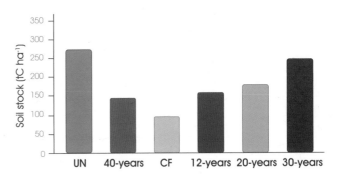

from the Harwood age-series (Magnani *et al.*, 2007), and assuming different degrees of disturbance at harvest (Figure 3.9). In the top line, we see the potential of this type of forest to accumulate carbon in the case of no soil disturbance at the end of each rotation. The broken lines show the possibilities with three levels of soil disturbance: mild, moderate and severe, corresponding to an annual loss of carbon as CO_2 of 30, 60 and 90 tC ha^{-1}. Detailed measurement and modelling of the dynamics of carbon in forest soils is challenging, particularly as substantial carbon is sequestered in humic fractions of soil and peat that have a very slow turnover, as compared with the biomass of humus layers. If these fractions change with disturbance (management or windthrow), it will take a longer time (several decades), to restore the stocks. *Because the soil carbon is a large component of the carbon balance of a forest, more attention needs to be paid to establishing a complete carbon balance of soils under different silvicultural practices*.

Management disturbance also has the capacity to stimulate emission of appreciable amounts of CO_2 to the atmosphere. As we have seen, site preparation by ploughing, leads to major CO_2 emissions throughout the rotation. It seems likely that emissions resulting from mounding are less, if only because a much smaller area of ground is turned over, but that remains to be shown.

A commercial thinning causes only a minor perturbation to the net CO_2 uptake lasting no more than two years (Clement, Moncrieff and Jarvis, unpublished results); first because *both* gains and losses of carbon by the trees are reduced concurrently, and second because thinning

Figure 3.9
Likely changes in the carbon stored in a Sitka spruce plantations on a peaty gley over six rotations. See the text for explanation. The lines show increasing degrees of disturbance at harvest from the top line down.

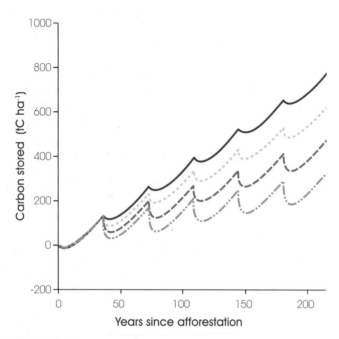

leads to penetration of solar radiation to greater depths in the canopy, with the result that leaves on the lower whorls of the remaining trees rapidly develop capacity to compensate for loss of leaf area above. Brash mats clearly protect the soil surface during thinning, but their effects on soil CO_2 emissions are not known. However, rapid recovery of NEP to its former level after thinning, suggests that there was little stimulation of soil emissions resulting from thinning at the Griffin forest site (data not shown).

Cropping of forest stands occurs over sequential rotations. The carbon fluxes and stocks through two such rotations of Sitka spruce plantation in Harwood, northern England have been analysed (Figure 3.10a,b). The first rotation was established on a heather moorland, with deep ploughing, so the soil carbon stock declined through the first 40-year rotation as a result of enhanced soil carbon oxidation and mineralisation (Figure 3.10b). Thereafter, during the second rotation there was a compensating recovery of soil carbon stocks but this was dependent on minimising the disturbance involved in replacing the first crop and on leaving the below and above-ground residues (roots, stumps, branches and needles) on site. At the whole stand scale the soil carbon losses during the early years of the first rotation (Figure 3.10b) were sufficient to cause

significant negative net carbon gain (NEP, see Figure 3.10a). However, once the canopy closed after around 17 years, CO_2 fixation rates became adequate to yield positive NEP, which reached a plateau of around 24 tCO_2 ha^{-1} year^{-1} at approximately 20 years. CO_2 fluxes were also negative in the early years of the second rotation but they became positive more rapidly as a result of the reduction in loss of soil carbon compared with the first rotation.

Research during the 1970s and 1980s focussed on management effects on nutrient budgets and it became clear that, in general, nutrients should be conserved on site for the following rotation by ensuring that small wood, twigs and leaves in particular, were retained on site (Titus and Malcolm, 1991,1992,1999). Because significant amounts of nutrients are also in the bark, it was also proposed that logs should be barked on site and the bark redistributed, but that proved to be too onerous. Brash is now being put forward as a source of biofuel, and its baling up and removal is being advocated. Nutrients aside (and that issue should be revisited), there is a clear case for ensuring that removal of the coarse brash does not lead to consequent reduction of the soil *carbon* stock.

At time of writing, stump removal is also being advocated on the grounds that stumps are an unused component of the crop that can be exploited as a resource for energy production, and thereby enhance the economic return to the owner. Removal of stumps may also remove pests and certainly can result in a well-prepared level surface for replanting. However, stump removal is tantamount to ploughing a substantial area of the site to a depth of 1 m, and measurements in Sweden have shown consequent annual CO_2 emissions of 25 tCO_2 ha^{-1} year^{-1} (A. Grelle pers. comm., 2008). Since the influence of ploughing in stimulating CO_2 emissions has been found to continue throughout the first rotation, it is likely that the influence of stump removal in provoking CO_2 emissions will continue through the following rotation. Should the enhanced emissions resulting from stump removal continue for no more than 10 years, the gain to the atmosphere from substituting stump-derived chips for fossil fuel will be negated.

Management that leads to ongoing reduction in the soil carbon stock seriously reduces the value of forests as a renewable source for material and energy substitution. As far as the troposphere is concerned, there is little difference between running down the soil carbon and mining coal to burn!

Figure 3.10
A diagram compiled to show measured and inferred carbon fluxes (a) and stock (b), over the first (left) and second (right) rotation of Sitka spruce stands of Yield Class 14–16 m^3 ha^{-1} $year^{-1}$ established on heather moorland in northern Britain. The stocks and fluxes, both presented as CO_2 equivalents, are based on measurements made at the Griffin and Harwood Forests in stands of different ages as part of the EUROFLUX and CARBOAGE programmes. (Further details are given by Zerva and Mencuccini 2005b, Zerva et al., 2005, Ball et al., 2007 and Clement et al., 2003. Conversion factors used and other underlying information are given by Jarvis and Linder, 2007.)

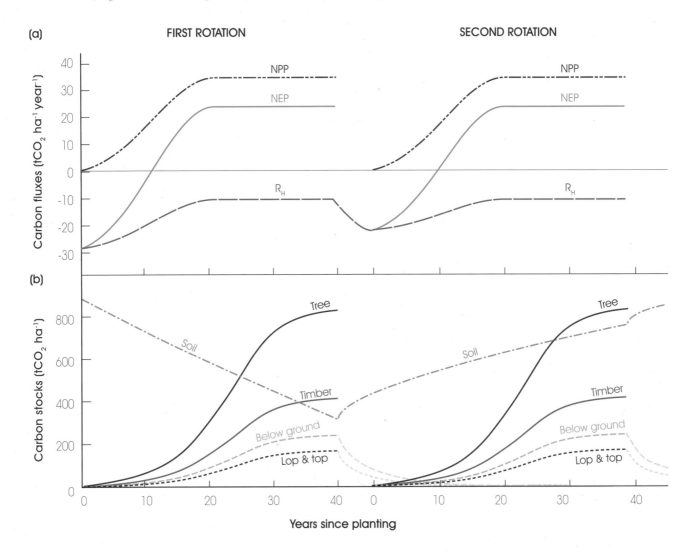

3.5 Spatial carbon dioxide dynamics

3.5.1 Long-term changes in carbon stocks at the forest scale

The long-term trends in carbon stocks of forest biomass associated with land use change may be estimated from the species and age structure of the stands. For conifers in particular, there was considerable planting in the UK from the 1950s to 1990 (see Chapter 1, Figure 1.2). As these 40+ year-old stands are harvested, and other land use changes occur, the carbon stocks inevitably change. The national GHG inventory, based upon this age-structure data, shows a diminishing forest carbon sink from a maximum of 15 Mt CO_2 per year in 2004 to an estimated 10 Mt CO_2 in 2010 (Mobbs and Thomson, 2008). Given a total forest area of 2.84 million hectares (see Chapter 1), this implies an average annual carbon sequestration associated with forested land of 5.4 tCO_2 ha^{-1} (or 1.4 tC ha^{-1}). We may compare this with the 'bottom-up' estimates from measurements of carbon fluxes using eddy covariance, as presented in 3.4.3 above. There we showed for Sitka spruce in the UK that the flux averaged over a stand cycle is reduced to about half of that observed when measurements are made on medium-aged forests, and this has also been shown for a number of other species (Magnani *et al.*, 2007). The reduction results from losses of CO_2 at harvesting and planting, and periods of low uptake in the early stages of the management cycle (Magnani *et al.*, 2007). As shown in 3.4.3 above, for Sitka spruce, the *full rotation flux estimate* of a stand of Yield Class 14, based on the Harwood Forest age series of stands, is around 14 tCO_2 ha^{-1}. For integration over the landscape, however, much more work is needed to establish general relationships between Yield Class and size of the carbon sink for a range of species and age distributions, if we are to use that approach to integrate spatially (Cannell and Dewar, 1995).

The national statistics are based on changes in land use and the consequent stock changes, but have not dealt with changes in the soil carbon during the forest cycle. In most forests there is more carbon as soil organic matter than in the form of biomass, and so a proper understanding of the long-term carbon balance of the forest estate requires knowledge of the behaviour of this carbon pool over the cycle of planting and harvesting. Classical soil survey methods do not give a good assessment of the changes in carbon stocks, because often the sampling methodology does not extend to the parent material and does not take into account changes in depth and bulk density, focusing on *concentrations* not *stocks*. Moreover, at a national level, the soils are extremely variable and it is difficult to sample them adequately. Hence, two recent large-scale surveys of the carbon concentrations in UK soils, by different organisations, have produced somewhat different conclusions (Bellamy *et al.*, 2005; Countryside Survey, 2009). Forests usually do accumulate carbon as organic matter in the soil so long as they are not disturbed (Smith *et al.*, 1997; Cerli *et al.*, 2006; Smal and Olszewska, 2008; Stevens and van Wesemael, 2008; Gadboury *et al.*, 2009), although there are notable exceptions to this and reviewers of the topic are not in agreement (see Reynolds, 2007).

With a number of instrumented flux towers within forests, it is possible to integrate removals and emissions of trace gases across compartments of different species, age-class and disturbance. It is also possible to address particular issues at compartment or small forest scale. However, to obtain quantitative estimates of trace gas fluxes at the larger landscape scale, and to monitor how management is influencing the atmosphere, different approaches are required, including the use of taller towers and aircraft.

3.5.2 Long-term changes in carbon stocks at the landscape scale

Tall Towers. In the past five years, it has become possible, through a series of EU-funded projects, including 'Carbo-Europe', to estimate GHG balances at landscape scale by measuring the concentrations of gases at heights of 200–300 m on tall towers. The network currently is not sufficient to cover all the European landscape, and in the UK, there is only one such tall tower that has been instrumented with this capability, whereas at least four such towers are desirable so as to be able to sample the atmosphere, avoiding local sources, when the wind is coming from different directions.

Integrated Carbon Observing System (ICOS). A new European project (ICOS) aims to deliver a denser network which should provide a rich data source on the fluxes of greenhouse gases at national scale as well as at European scale. ICOS has recently been placed on the UK Joint Research Councils Infrastructure Road Map as an 'emerging topic', with the expectation that by 2012 it will be possible to fund four or five such sites in the UK, along with corresponding ecosystem sites, where CO_2 fluxes and soil carbon stocks will be routinely measured and reported (see details online at: http://icos-infrastructure.ipsl.jussieu.fr/ [and] http://www.rcuk.ac.uk/cmsweb/downloads/rcuk/publications/lfroadmap08.pdf).

3.5.3 The future carbon sink

The one million hectares of conifers planted between 1950 and 1990 are a major resource today, available as the stands mature for a range of end-uses, including structural timber and other material substitution and fossil fuel substitution. Over a full rotation, these forests remove significant amounts of the predominant anthropogenic GHG, CO_2, from the atmosphere. *With appropriate management*, harvesting and replanting 0.025% (i.e. one-fortieth) of the area annually maintains the standing stock of timber, and continues to remove CO_2 from the atmosphere. Thus, harvesting and replanting sustainably transfers carbon into long-term storage in the soil and into wood products that can be used to substitute for products otherwise derived from fossil fuels. As global CO_2 emissions are increasing rapidly, we need to increase the *area* of actively absorbing and productive forest to access these benefits.

As the climate changes, a key question is whether the forest resource will continue as a significant carbon sink in the UK. Earlier analysis has suggested that the likely increase in temperature will enhance CO_2 emissions to the atmosphere and convert the present UK forest CO_2 sink into a CO_2 source. However, such analysis does not take into account the concurrent increase in atmospheric CO_2 concentration and the likely future availability of both wet and dry nitrogen deposition. When temperature, carbon cycle and nitrogen cycle are included in models, it seems very likely that the current plantations will continue as a significant CO_2 sink (Figure 3.11). *Increasing* the forest area and *minimising* disturbance of the soil will also very likely increase the size of the future UK CO_2 sink; such approaches are imperative to increase carbon sequestration (see also Chapter 8).

Figure 3.11
Rotation mean NPP (above) and yield class (below) predicted by the Edinburgh Forest Model. Each point is the mean at age 30 of a 60-year rotation of Sitka spruce growing in the south of Scotland. The model was run to quasi-equilibrium prior to imposition of increases in temperature (T) of 0.1°C per decade up to 1950 and thereafter at 0.2°C per decade, giving a total of 2.5°C warming; (CO_2) from 290 ppmv in 1900 to 510 ppmv in 2050, to 690 ppmv in 2100; and nitrogen deposition (N) from 5 kg N ha^{-1} year^{-1} in 1940 to 20 kg N ha^{-1} year^{-1} in 1970, and thereafter remaining constant (after Cannell *et al.*, 1998).

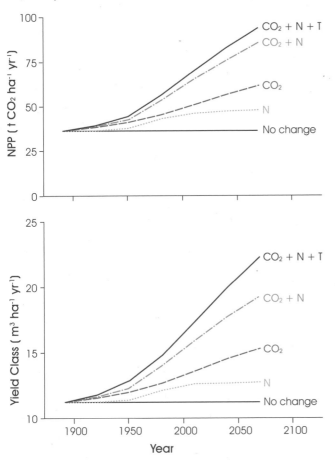

3.6 Interaction of forests with other greenhouse gases

3.6.1 Impact of climate change

In addition to their role in the exchange of CO_2, forests are the source of volatile organic compounds (VOCs) that contribute to the photochemical production of aerosols, and of tropospheric ozone (O_3) (see below); both processes have implications for air quality, and O_3 is a GHG as well as a gas that can damage biological systems. Climate change can also be expected to affect the fluxes of other GHG. Forest soils are sources and sinks of methane (CH_4), and sources of nitrous oxide (N_2O); both of these gases have much higher global warming potentials (GWPs) than CO_2, and their fluxes are strongly affected by temperature and soil moisture status. However, their fluxes usually have a fairly small effect on the overall GWP of a forest system, compared with that arising from CO_2 exchange, but in some circumstances they can make a significant contribution.

3.6.2 Impact of soil physical and nutrient conditions

Land drainage and depth to water table are significant influences on trace GHG fluxes from forest soils, particularly peaty ones. The increased aeration of the surface layers resulting from drawdown of the water table releases more nitrogen by mineralisation that can then act as a substrate for N_2O and NO production. Enhancement of mineral nitrogen by deposition from the atmosphere can have a similar effect. Conversely, improved aeration inhibits the emission, and promotes the microbial oxidation of CH_4 while increased nitrogen availability and soil mechanical disturbance inhibit CH_4 oxidation.

3.6.3 Methane (CH_4)

The relationship between emissions of CH_4 measured at the Harwood forest and soil conditions is shown in Table 3.1. A clearfelled site with the water table close to the surface emitted 6.8–18 kg CH_4 ha^{-1} year^{-1}, whereas the soil under 20- and 30-year-old stands emitted only 0.2–2.8 kg CH_4 ha^{-1} year^{-1}. These trends are in line with previous findings in Finland by Martikainen *et al.*, (1995) and Nykänen *et al.*, (1998), and in Sweden by Von Arnold *et al.*, (2005).

Although peaty soils may turn from being a net source of CH_4 into a net sink after draining, because of microbial oxidation of atmospheric CH_4 in the soil (e.g. Huttunen et al., 2003), generally the CH_4 oxidation rates tend to be low. At Harwood, a maximum sink of 20.8 µg CH_4 m^{-2} h^{-1} was detected by Zerva and Mencuccini (2005a) from a 40-year-old first rotation site before clearfelling, whereas no detectable consistent sink was found at Harwood by Ball et al., (2005). Earlier data (collated by Smith et al., 2000) showed a range of oxidation rates of 1–9 kg CH_4 ha^{-1} $year^{-1}$ (equivalent to 11.4–103 µg CH_4 m^{-2} h^{-1}) for seven UK forest and woodland sites, with the highest rate being in a 200-year-old deciduous woodland on a sandy lowland mineral soil in East Lothian (Dobbie and Smith, 1996) (Table 3.1). Conversion of forest or woodland to agricultural use reduces the size of the methane sink by two-thirds, on average (Smith et al., 2000) – caused by soil disturbance and/or increased nitrogen availability. After reversion to forest it takes in the order of 100 years for the sink activity to recover fully (Priemé et al., 2007); the reasons for the slowness of the process are still unknown.

Reported impacts of temperature on CH_4 emissions vary. Zerva and Mencuccini (2005a) found an increase in CH_4 efflux with increasing temperature at Harwood, but Ball et al., (2005) did not. The impact of the depth to water table also varies; Ball et al., (2005) found a positive exponential relationship between the efflux and the closeness of the water table to the surface (Figure 3.12), which is consistent with much previous work abroad on wetland soils (e.g. Roulet et al., 1993), but Zerva and Mencuccini (2005a) found a decrease in CH_4 emissions with rising water table, for a newly clearfelled site elsewhere at Harwood. However, they found that the emissions gradually increased over eight months following clearfelling, even although the mean monthly water table depths differed little during this period. Possibly, therefore, the effects of controlling variables such as temperature and/or changes in substrate availability were more important than water table depth in these circumstances.

3.6.4 Nitrous oxide (N_2O)

N_2O emissions from a 30-year-old Sitka spruce stand at the Harwood Forest, Northumberland, measured in the field by Ball et al., (2005) were 4.7 and 1.9 kg N_2O ha^{-1} $year^{-1}$ in 2001 and 2002, respectively. At a neighbouring 20-year-old stand, the emissions were 0.2 kg N_2O ha^{-1} $year^{-1}$ in both years (Table 3.1). The results for the more mature stand were very similar to those of Dutch and Ineson (1990) for a Sitka spruce stand in the Kershope Forest (based on soil core measurements in the laboratory), where soil and vegetation are similar to that at Harwood: 4.1 kg N_2O ha^{-1} $year^{-1}$ and 2.0 kg ha^{-1} $year^{-1}$ in a drier year (Table 3.1). Overall, these UK results fit well with those from other northern European forests

Table 3.1
Greenhouse gas fluxes reported for UK forest sites. Negative values indicate uptake by the soil, positive values indicate emissions.

Gas	Emission (kg ha^{-1} $year^{-1}$)	CO_2 equivalent (kg ha^{-1} $year^{-1}$)	Reference
N_2O			
Standing forest			
Glencorse	0.12–0.28	34–83	Kesik et al., 2005
Harwood	0.2–4.7	59–1390	Ball et al., 2005
Kershope	2.0–4.1	592–1214	Dutch and Ineson, 1990
Clearfelled sites (CF)			
Harwood	0.7–2.0	209–592	Ball et al., 2005
Kershope	12—51	3580–15200	Dutch and Ineson, 1990
CH_4 (emission)			
Harwood CF	6.8–18	156–414	Ball et al., 2005
Harwood (20yr–30yr)	0.2–2.8	4.6–64	Ball et al., 2005
CH_4 (uptake)			
7 UK forest and woodland sites	−1.0 to −9.1	−23 to −209	Smith et al., 2000
	Median: −2.4	−55	
Harwood	−1.8	−41	Zerva and Mencuccini, 2005a

Figure 3.12
Relationship between depth to water table and mean methane emission, the Harwood Forest, 2001–02, from 20-, 30- and 40-year-old Sitka spruce stands (F-20, F-30, F-40), two clearfelled plots (CF-1, CF-2) and adjacent non-forested grassland (NF). (Based mainly on data in Ball et al., 2005).

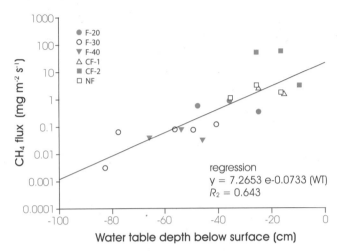

on organic soils, e.g. those of Martikainen et al., (1993) in Finland, and of Von Arnold et al., (2005) in Sweden. Emissions generally show a logarithmic decrease with increasing C:N ratio in the soil (Figure 3.13), although the absolute values in two data analyses shown in Figure 3.13 differ (Klemedtsson et al., 2005; Pilegaard et al., 2006). The values from the Harwood study are included in Figure 3.1; they fit reasonably within the trend, but with a considerable scatter, probably because of variations in soil wetness and aeration (Jungkunst et al., 2004). Fluctuating aeration regimes may be expected to create partially aerobic conditions in which nitrogen mineralisation can take place (producing the substrate necessary for N_2O production), and anaerobic zones or microsites in which denitrification can occur. These contrasting conditions may occur sequentially at a given depth, or in parallel at different points in the profile. Factors which might be responsible for such a difference include temperature and associated transpiration water demands, and pore size distribution, both of which would affect soil water content at a given height above the water table and the propensity for creation of anaerobic microsites.

Zerva and Mencuccini (2005a) reported an N_2O flux over 10 months at a newly clearfelled site at Harwood of 1.7 kg N_2O ha^{-1} (equivalent to 2.0 kg ha^{-1} year^{-1}), and Ball et al., (2005) measured fluxes of 0.7–0.9 kg N_2O ha^{-1} year^{-1}), on another clearfelled area in the same forest. However, Dutch and Ineson (1990) reported denitrification losses

of 10–42 kg N ha^{-1} year^{-1} (the larger part of which was as N_2O) during the first two years after clearfelling. This is an order of magnitude larger than the results of the Harwood studies (see Table 3.1). Huttunen et al., (2003) found average fluxes over three growing seasons after clearfelling on drained peat lands in Finland of 246 and 945 mg N_2O m^{-2} day^{-1}. Assuming a season of five months duration (the longest in their study), and only a small emission (around another 10%) in the colder conditions of the rest of the year, these rates would be equivalent to annual fluxes of around 0.4 and 1.5 kg N_2O ha^{-1}, i.e. very similar to the Harwood results.

Figure 3.13
Relationship between soil C:N ratio and annual N_2O emission, for European forest soils. Data of Klemedtsson et al. (2005) (o); Ball et al. (2005) (●); and Pilegaard et al. (2006), scaled to the same units (▲). Broken line: regression through Klemedtsson et al. (2005) data; solid line: regression through Pilegaard et al. (2006) data.

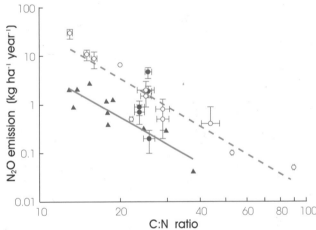

For N_2O, it appears that, in general, relatively high soil water contents (but well short of waterlogged conditions), combined with high soil temperatures and a low C:N ratio, give rise to the highest N_2O emissions. Most soils used hitherto for afforestation in the UK are prone to low soil temperatures, because of their upland locations, and generally have fairly high C:N ratios and rates of atmospheric nitrogen deposition much lower than in Central Europe, all of which serves to explain the modest emissions that have been measured. Nonetheless, if there should be a future trend towards afforestation of lowland soils, especially those that have been in previous agricultural use, with C:N ratios of 10 to 15, the average N_2O emissions would likely be much higher, on the basis of the European data available (Kesik et al., 2005; Pilegaard et al., 2006). Additionally, a future general warming of forest soils may be expected to result in significantly larger

N_2O emissions; complex interactions with soil moisture oxygen status render the response to temperature very non-linear, and large increases with temperature have been observed (Brumme, 1995).

3.6.5 Tropospheric ozone, NO, NO_2 and VOCs

Ozone (O_3) is a major GHG in the troposphere and a strong oxidant. As a GHG, O_3 currently ranks as less effective than CH_4, but more effective than N_2O (IPCC, 2007a). O_3 differs in its characteristics from CO_2, in that it is very labile, continuously being formed and decomposed in the troposphere, so that its effectiveness as a GHG cannot be compared relative to CO_2 in the same way as the long-lived GHG, CH_4 and N_2O.

Formation of O_3 in the troposphere is the result of interactions between volatile organic precursor compounds (VOCs) and NO_x (i.e. NO + NO_2). Precursors are varied and may be derived from industrial sources considerable distances away from the forest. However, O_3 may be generated within forests by sunlight-driven chemical reactions between VOCs produced by trees, provided that NO is also present, and within forests the soil can be a significant source of NO.

Emission of NO from a forest soil at Glencorse, near Edinburgh, has been measured as 1.5 µg NO-N m^{-2} h^{-1}, as compared with emission of N_2O of 11.9 µg N_2O-N m^{-2} h^{-1} (Pilegaard et al., 2006). However, at European forest sites with much higher rates of nitrogen deposition, NO emissions are generally much larger than this (Pilegaard et al., 2006). Another major variable affecting NO emission is the wetness of the soil: the NO:N_2O emission ratio decreases exponentially with increasing water-filled pore space. Once emitted to the air, NO oxidises rapidly to NO_2, creating a mixture of NO and NO_2, i.e. NO_x. In the UK and other industrialised countries in Europe, most tropospheric NO_x originates from combustion processes rather than from the soil, but whatever its origin, in sunlight NO_x reacts with VOCs, including those released by forest trees and soil, to form O_3.

In the UK, the VOCs emitted by plantations of Scots pine (*Pinus sylvestris*), Norway spruce (*Picea abies*), birch (*Betula* spp.) and poplar (*Populus* spp.), are mainly isoprene and monoterpenes (Stewart et al., 2003); carbonyl emissions (mainly of acetone) from coniferous trees are also relatively large (Laurila and Lindfors, 1999). Isoprene is produced in sunlight within the chloroplasts in leaves of both broadleaves and conifers, its production

increasing with the absorbed solar radiation and with air temperature (Baldocchi et al., 1995). During a day of fine weather, production of isoprene follows photosynthesis, increasing during the morning, peaking in the early afternoon and declining in the evening. Sitka spruce is the dominant emitting species, providing approximately 40% of the annual isoprene and monoterpene fluxes, mainly in the forests in Scotland; poplar plantations in eastern England, and other possible future biofuel species, are also strong isoprene emitters (Kesselmeier and Staudt, 1999). Higher temperatures and possibly sunnier summers resulting from climate change are very likely to make our climate more like that currently in regions much further south in Europe, and it is predicted this will give rise to increased biogenic VOC concentrations, and thus to increased tropospheric O_3 (Bell and Ellis, 2005).

Ozone within the forest environment diffuses into leaves via the stomatal pores in the leaf surfaces and its oxidising properties can do major damage to the physiology of both the guard cells that control the aperture of stomatal pores and the photosynthetic system within leaves (Ashmore, 2005), thus reducing the forest carbon sink and further exacerbating climate change. However, the stomata in many species tend to respond to higher than usual CO_2 concentrations by a reduction in pore aperture, so that internal leaf damage by O_3 may be mitigated to some extent by the concurrent rise in CO_2 concentration. Thus with concurrent increase of both CO_2 and O_3 concentrations, damage caused by O_3 may be limited, but at the expense of a lower rate of removal of CO_2 from the atmosphere (Sitch et al., 2005). Sitka spruce and Scots pine, however, are exceptional in that their stomata are not very sensitive to the ambient CO_2 concentration (see 3.3.2 above). Thus Sitka spruce and possibly other conifers with limited stomatal sensitivity to increasing CO_2 concentration may be more at risk to rising tropospheric ozone concentrations than species with marked stomatal closure at higher CO_2 concentrations.

3.7 Summation of greenhouse gas impacts

Overall, the contribution of the trace GHG to the total global warming potential (GWP, see 3.1 above) of most UK forests and forest soils is generally minor compared with that of CO_2. Relatively high N_2O emissions following clearfelling appear to be fairly short-lived, and when averaged over a rotation of perhaps 40 years, are much less significant. Nonetheless, future patterns of land use, with woodland creation on some lowland former agricultural soils being afforested, combined with increased temperatures, could make the trace gas contribution relatively larger. Also, if the fluxes of the short-lived tropospheric O_3, and its GWP, were to be adequately quantified and included, this would further add to the non-CO_2 component of the total GWP.

In the UK context, GHG other than CO_2 may become more important. Disturbance, especially afforestation of gleyed soils can strongly influence GHG flux. For example, lowering of water table can lead to a reduction of CH_4 emissions whereas the application of fertiliser can increase nitrous oxide fluxes. The capability to measure methane and nitrous oxide fluxes on a continuous basis by eddy covariance is only now developing, with the advent of appropriate laser-based instruments that can be mounted on towers alongside the current CO_2 sensors and sonic anemometers. This will enable a major step forward in our understanding of the overall significant GHG fluxes associated with forests throughout the production cycle.

Research priorities

1. Upgrade and ensure the future of the Straits, Griffin and Harwood research sites. Enhance the radiation budget measurement by adding measurements of upward and downward components of both short-wave (solar) and long-wave (thermal) radiation. Increase the focus on energy partitioning and direct transfer of sensible heat to the atmosphere. Add above-canopy, continuous net flux measurements of CH_4 and N_2O, as fast gas measuring sensors become generally available.

2. Increase the number of similar, comparable, long-term GHG measurement sites in our forests to cover geographical locality, species, stand age, yield-class and management operations, such as thinning. Select sites on the basis of production potential and ecological significance.

3. Develop and utilise moveable low-level, flux systems to investigate the impacts at compartment scale on net emissions/removals of CO_2, CH_4 and N_2O consequent on natural events that result in site disturbance, particularly of the soil, such as windthrow, and management operations, such as ploughing, mounding, clearfell and stump removal.

4. Evaluate the role of the GHG gas ozone (O_3) within our forests. Initiate a programme on natural emissions within forests of isoprene, monoterpenes and other VOCs from trees that may lead to generation of the GHG, ozone

(O_3) within forests.

5. Support programmes for using instrumented aircraft to evaluate emissions and removals of trace gases across landscapes to define major sources and sinks of GHG in relation to forestry, agriculture and other land uses, at district and regional scales.

6. Stimulate collaboration with CEH, SEPA and Defra for instrumentation of tall towers (400 m) to measure concentrations of GHG and other trace gases continuously, so as to define daily and seasonal sources and sinks of forested landscapes at regional scales.

References

ÅGREN, G.I., BOSATTA, E. and MAGILL, A.H. (2001). Combining theory and experiment to understand effects of inorganic nitrogen on litter decomposition. *Oecologia* **128**, 94–98.

ASHMORE, M.R. (2005). Assessing the future global impacts of ozone on vegetation. *Plant, Cell and Environment* **28**, 949–964.

BALDOCCHI, D., GUENTHER, A., HARLEY, P., KLINGER, L., ZIMMERMAN, P., LAMB, B. and WESTBERG, H. (1995). The fluxes and air chemistry of isoprene above a deciduous hardwood forest. *Philosophical Transactions of the Royal Society of London, Series A* **350**, 279–296.

BALL, T., SMITH K.A. and MONCRIEFF, J.B. (2005). Effect of stand age on greenhouse gas fluxes from a Sitka spruce [*Picea sitchensis* (Bong.) Carr.] chronosequence on a peaty gley soil. *Global Change Biology* **13**, 2128–2142.

BEADLE, C.L., JARVIS, P.G. and NEILSON, R.E. (1979). Leaf conductance as related to xylem water potential and carbon dioxide concentration in Sitka spruce. *Physiologia Plantarum* **45**, 158–166.

BELL, M.L. and ELLIS, H. (2005). The impact of biogenic VOC emission on tropospheric ozone formation in the mid-Atlantic region of the United States. In: Brebbia, C.A. (ed.) *Air pollution XIII*. WIT Press, Southampton. pp. 89–95.

BELLAMY, P.H., LOVELAND, P.J., BRADLEY, R.I., LARK R.M. and KIRK, G.J.D. (2005). Carbon losses from all soils across England and Wales 1978–2003. *Nature* **437**, 245–248.

BERNHOFER, C., AUBINET, M., CLEMENT, R., GRELLE, A., GRUNWALD, T., IBROM, A., JARVIS, P., REDMAN, C., SCHULZE, E.D. and TENHUNEN J.D. (2003). Spruce forests (Norway and Sitka spruce, including Douglas fir): Carbon and water fluxes and balances, ecological and ecophysiological determinants. In: Valentini, R. (ed.) *Ecological studies: fluxes of carbon water and energy of European forests*. Springer, Berlin. pp. 99–123.

BLACK, K., BYRNE, K.A., MENCUCCINI, M., TOBIN, B., NIEUWENHUIS, M., REIDY, B., BOLGER, T., SAIZ, G., GREEN, C., FARRELL, E.T. and OSBORNE, B. (2009). Carbon stock and stock changes across a Sitka spruce chronosequence on surface water gley soils. *Forestry* **82**, 255–272.

BRUMME, R (1995). Mechanisms of carbon and nutrient release and retention in beech forest gaps. III. Environmental regulation of soil respiration and nitrous oxide emissions along a microclimatic gradient. *Plant and Soil* **169**, 593–600.

BYRNE, K.A. and FARRELL, E.P. (2005).The effect of afforestation on soil carbon dioxide emissions in blanket peatland in Ireland. *Forestry* **78**, 217–227.

CANNELL, M.G.R. and DEWAR, R.C. (1995). The carbon sink provided by plantation forests and their products in Britain. *Forestry* **68**, 35–48.

CANNELL, M.G.R., DEWAR, R.C. and PYATT, D.C. (1993). Conifer plantations on drained peatlands in Britain: a net gain or loss of carbon. *Forestry* **66**, 353–369.

CANNELL, M.G.R., MILNE, R., SHEPPARD, L.J. and UNSWORTH, M.H. (1987). Radiation interception and productivity of willow. *Journal of Applied Ecology* **24**, 261–278.

CANNELL, M.G.R., THORNLEY, J.H.M., MOBBS, D.C. and FRIEND, A. D. (1998). UK conifer forests may be growing faster in response to increased nitrogen deposition, atmospheric CO_2 and temperature. *Forestry* **71** (Suppl.), 277–296.

CERLI, C., CELI, L., JOHANSSON, M.B., KOGEL-KNABNER, I., ROSENQVIST, L. and ZANINI, E. (2006). Soil organic matter changes in a spruce chronosequence on Swedish former agricultural soil. I. Carbon and lignin dynamics. *Soil Science* **171**, 837–849.

CLEMENT, R., MONCRIEFF, J.B. and JARVIS, P.G. (2003). Net carbon productivity of Sitka spruce forest in Scotland. *Scottish Forestry* **57**, 5–10.

DOBBIE, K.E. and SMITH, K.A. (1996). Comparison of CH_4 oxidation rates in woodland, arable and set aside soils. *Soil Biology and Biochemistry* **28**, 1357–1365.

DUTCH, J. and INESON, P. (1990). Denitrification of an upland forest site. *Forestry* **63**, 363–378.

FOG, K. (1988). The effect of added nitrogen on the rate of decomposition of organic matter. *Biological Reviews* **63**, 433–462.

FOWLER, D., CAPE, J.N. and UNSWORTH, M.H. (1989). Deposition of atmospheric pollutants on forests. *Philosophical Transactions of the Royal Society, London, Series B*, **324**, 247–265.

GADBOURY, S., BOUCHER J.F., VILLENEUVE, C., LORD, D. and GAGNON, R. (2009). Estimating the net carbon balance of boreal open woodland afforestation: a case-study in Quebec's closed crown boreal forest. *Forest Ecology and Management* **257**, 483–494.

GRACE, J. *et al.* (2003). *Final Report to the European Commission of the CARBO-AGE Project, a component of CARBOEUROPE.* Online at: www.bgcjena.mpg.de/public/carboeur/

HARGREAVES, K.J., MILNE, R. and CANNELL, M.G.R. (2003). Carbon balance of afforested peatland in Scotland. *Forestry* **76**, 299–316.

HUTTUNEN J.T., NYKÄNEN H., MARTIKAINEN P.J. and NIEMINEN M. (2003). Fluxes of nitrous oxide and methane from drained peatlands following forest clear-felling in Southern Finland. *Plant and Soil* **255**, 457–462.

HYVÖNEN, R., ÅGREN, G.I., LINDER, S., PERSSON, T., COTRUFO, M.F., EKBLAD, A., FREEMAN, M., GRELLE, A., JANSSENS, I.A., JARVIS, P.G., KELLOMÄKI, S., LINDROTH, A., LOUSTAU, D., LUNDMARK, T., NORBY, R.J., OREN, R., PILEGAARD, K., RYAN, M.G., SIGURDSSON, B.D., STRÖMGREN, M., VAN OIJEN, M. and WALLIN, G. (2007). The likely impact of elevated [CO_2], nitrogen deposition, increased temperature and management on carbon sequestration in temperate and boreal forest ecosystems: a literature review. *New Phytologist* **173**, 463–480.

JARVIS, P.G., with an appendix by WANG, Y.P. (1986). Coupling of carbon and water in forest stands. *Tree Physiology* **2**, 347–368.

JARVIS, P.G. (1994). Capture of carbon dioxide by coniferous forest. In: Monteith, J.L., Scott, R.K. and Unsworth, M.H. (eds) *Resource capture by crops*. Nottingham University Press, Loughborough. pp. 351–374.

JARVIS, P.G. (ed.), assisted by Aitken, A., Barton, C.V.M., Lee, H.S.J. and Wilson, S. (1998) *European forests and global change: the likely impacts of rising CO_2 and temperature.* Cambridge University Press, Cambridge.

JARVIS, P.G. and FOWLER, D.G. (2001). Forests and the atmosphere. In: Evans, J. (ed.) *The forests handbook, Vol. 1.* Blackwell Science, Oxford. pp. 229–281.

JARVIS, P.G. and LEVERENZ, J.W. (1983). Productivity of temperate, deciduous and evergreen forests. In: Lange, O.L., Nobel, P.S., Osmond, C.B. and Ziegler, H. (eds). *Encyclopedia of plant physiology*, new series, Physiological plant ecology IV, Vol 12D. Springer-Verlag, Berlin. pp. 233–280.

JARVIS, P.G. and LINDER, S. (2007). Forests remove carbon dioxide from the atmosphere: spruce forest tales! In: Freer-Smith, P.H., Broadmeadow, M.S.J. and Lynch, J.M. (eds) *Forestry and climate change.* CABI, Wallingford. pp. 60–72.

JARVIS, P.G. and MCNAUGHTON, K.G. (1986). Stomatal control of transpiration: scaling up from leaf to region.

Advances in Ecology Research **15**, 1–49.

JARVIS, P.G., STEWART, J.B. and MEIR, P. (2007). Fluxes of carbon dioxide at Thetford Forest. *Hydrology and Earth Systems Science* **11**, 241–255.

JUNGKUNST, H.F., FIEDLER, S. and STAHR, K. (2004). N_2O emissions of a mature Norway spruce (*Picea abies*) stand in the Black Forest (southwest Germany) as differentiated by the soil pattern. *Journal of Geophysical Research* **109**, Art. No. D07302 doi:10.1029/2003JD004344.

KARNOSKY, D.F., TALLIS, M., DARBAH, J. and TAYLOR, G. (2007). Direct effects of elevated carbon dioxide on forest tree productivity. In: Freer-Smith, P.H., Broadmeadow, M.S.J. and Lynch, J.M. (eds) *Forestry and climate change.* CABI, Wallingford. pp. 136–142.

KESSELMEIER, J. and STAUDT, M. (1999). Biogenic volatile organic compounds (VOCs): an overview on emission physiology and ecology. *Journal of Atmospheric Chemistry* **33**, 23–88.

KESIK, M., AMBUS, P., BARITZ, R., BRUGGEMANN, N., BUTTERBACH-BAHL, K., DAMM, M., DUYZER, J., HORVATH, L., KIESE, R., KITZLER, B., LEIP, A., LI, C., PIHLATIE M., PILEGAARD, K., SEUFERT, S., SIMPSON, D., SKIBA, U., SMIATEK, G., VESALA, T AND ZECHMEISTER-BOLTENSTERN, S. (2005). Inventories of N_2O and NO emissions from European forest soils. *Biogeosciences* **2**, 353–375.

KLEMEDTSSON, L., VON ARNOLD, K., WESLIEN, P. and GUNDERSEN, P. (2005). Soil CN ratio as a scalar parameter to predict nitrous oxide emissions. *Global Change Biology* **11**, 1142–1147.

LANDSBERG, J.J., PRINCE, S.D., JARVIS, P.G., MCMURTRIE, R.E., LUXMOORE, R. and MEDLYN, B.E. (1996). Energy conversion and use in forests: an analysis of forest production in terms of radiation utilisation efficiency (ε). In: Gholz, H.L., Nakane, K. and Shimoda, H. (eds) *The use of remote sensing in the modeling of forest productivity at scales from the stand to the globe.* Kluwer Academic Publishers, Dordrecht. pp. 273–298.

LAURILA, T. and LINDFORS, V. (eds) (1999). *Biogenic VOC emissions and photochemistry in the boreal regions of Europe.* Air Pollution Research Report 70. Commission of the European Communities, Luxembourg.

LINDER, S. (1985). Potential and actual production in Australian forest stands. In: Landsberg, J.J. and Parsons, W. (eds) *Research for forest management.* CSIRO, Melbourne. pp. 11–51.

McNAUGHTON, K.G and JARVIS, P.G. (1983). Predicting effects of vegetation changes on transpiration and evaporation. In: Kozlowski, T.T. (ed.) *Water deficits and plant growth*, Vol VII. Academic Press, New York. pp. 1–47.

MAGNANI, F., MENCUCCINI, M., BORGHETTI, M.,

BERBIGIER, P., BERNINGER, F., DELZON, S., GRELLE, A., HARI, P., JARVIS, P.G., KOLARI, P., KOWALSKI, A.S., LANKREIJER, H., LAW, B.E., LINDROTH, A., LOUSTAU, D., MANCA, G., MONCRIEFF, J.B., RAYMENT, M., TEDESCHI, V., VALENTINI, R. and GRACE, J. (2007). The human footprint in the carbon cycle of temperate and boreal forests. *Nature* **447**, 848–850.

MARTIKAINEN, P.J., NYKÄNEN, H., CRILL, P. and SILVOLA, J. (1993). Effect of a lowered water table on nitrous oxide fluxes from northern peatlands. *Nature* **366**, 51–53.

MARTIKAINEN, P.J., NYKÄNEN, H., ALM J. and SILVOLA J. (1995). Change in fluxes of carbon dioxide, methane and nitrous oxide due to forest drainage of mire sites of different trophy. *Plant and Soil* **168**, 571–577.

MEDLYN, B.E., BADECK, F.W., DE PURY, D.G.G., BARTON, C.V.M., BROADMEADOW, M., CEULEMANS, R., DE ANGELIS, P., FORSTREUTER, M., JACH, E., KELLOMÄKI, S., LAITAT, E., MAREK, M., PHILIPPOT, S., REY, A., STRASSEMEYER, J., LAITINEN, K., LIOZON, R., PORTIER, B., ROBERNTZ, P., WANG, K. and JARVIS, P.G. (1999). Effects of elevated $[CO_2]$ on photosynthesis in European forest species: a meta-analysis of model parameters. *Plant, Cell and Environment* **22**, 1475–1495.

MEDLYN, B.E., BADECK, F.W., DE PURY, D.G.G., BARTON, C.V.M., BROADMEADOW, M., CEULEMANS, R., DE ANGELIS, P., FORSTREUTER, M., JACH, E., KELLOMÄKI, S., LAITAT, E., MAREK, M., PHILIPPOT, S., REY, A., STRASSEMEYER, J., LAITINEN, K., LIOZON, R., PORTIER, B., ROBERNTZ, P., WANG, K. and JARVIS, P.G. (2000). *Effects of elevated $[CO_2]$ on biomass increment in European forest species: a meta-analysis of model parameters. Biomass growth and carbon allocation of young trees in response to elevated CO_2 concentrations.* Unpublished Final Report to the European Union of the 'Ecocraft' Project, 1993 to1999.

MEDLYN, B.E., BARTON, C.V.M., BROADMEADOW, M., CEULEMANS, R., DE ANGELIS, P., FORSTREUTER,M., FREEMAN, M., JACKSON, S.B., KELLOMÄKI, S., LAITAT, E., REY, A., ROBERNTZ, P., SIGURDSSON, B.D., WANG, K., CURTIS, P.S. and JARVIS, P.G. (2001). Stomatal conductance of forest species after a long-term exposure to elevated CO_2 concentration: a synthesis. *New Phytologist* **149**, 247–264.

MILLER, H.G. (1979). The nutrient budgets of even-aged forests. In: Ford, E.D., Malcolm, D.C. and Atterson, J. (eds) *The ecology of even-aged forest plantations. Proceedings of Division I. International Union of Forestry Research Organisations,* Edinburgh, September 1978. pp. 221–256.

MILNE, R., SATTIN, M., DEANS, J.D., JARVIS, P.G. and CANNELL, M.G.R. (1992). The biomass production of three poplar clones in relation to intercepted solar radiation.

Forest Ecology and Management **55**, 1–14.

MOBBS, D.C. and THOMSON, A.M. (2008). Inventory and projections of UK emissions by sources and removals by sinks due to land use, land use change and forestry. *Annual Report by CEH to Defra.*

MONTEITH, J.L. (1977). Climate and efficiency of crop production in Britain. *Philosophical Transactions of the Royal Society of London, Series B,* **281**, 277–294.

MONTEITH, J.L. and UNSWORTH, M.H. (1990). *Principles of environmental physics.* 2nd edn. Edward Arnold, London.

MORISON, J.I.L. and JARVIS, P.G. (1983). Direct and indirect effects of light on stomata. I. Scots pine and Sitka spruce. *Plant, Cell and Environment* **6**, 95–101.

NEILSON, R.E. and JARVIS, P.G. (1975). Photosynthesis in Sitka spruce (*Picea sitchensis* (Bong.) Carr.) VI. Response of stomata to temperature. *Journal of Applied Ecology* **12**, 879–889.

NEILSON, R.E., LUDLOW, M.M. and JARVIS, P.G. (1972). Photosynthesis in Sitka spruce (*Picea sitchensis* (Bong.) Carr.) II. Response to temperature. *Journal of Applied Ecology* **9**, 721–745.

NORMAN, J.M. and JARVIS, P.G. (1974). Photosynthesis in Sitka spruce (*Picea sitchensis* (Bong.) Carr.) III. Measurements of canopy structure and interception of radiation. *Journal of Applied Ecology* **11**, 375–398.

NYKÄNEN, H., ALM, J., SILVOLA, J., TOLONEN, K. and MARTIKAINEN, P.J. (1998). Methane fluxes on boreal peatlands of different fertility and the effect of long-term experimental lowering of the water table on flux rates. *Global Biogeochemical Cycles* **12**, 53–69.

PILEGAARD, K., SKIBA, U., AMBUS, P., BEIER, C., BRUGGEMANN, N., BUTTERBACH-BAHL, K., DICK, J., DORSEY, J., DUYZER, J., GALLAGHER, M., GASCHE, R., HORVATH, L., KITZLER, B., LEIP, A., PIHLATIE, M.K., ROSENKRANZ, P., SEUFERT, G., VESELA, T., WESTRATE, H. and ZECHMEISTER-BOLTENSTERN, S. (2006). Factors controlling regional differences in forest soil emission of nitrogen oxides (NO and N_2O). *Biogeosciences* **3**, 651–661.

PONTAILLER, J.Y., BARTON, C.V.M., DURRANT, D. and FORSTREUTER, M. (1998). How can we study CO_2 impacts on trees and forests? In: Jarvis, P.G. (ed.) *European forests and global change: the likely impacts of rising CO_2 and temperature.* Cambridge University Press, Cambridge. pp. 1–28.

PRICE, D.T. and BLACK, T.A. (1990). Effects of short-term variation in weather on diurnal canopy CO_2 flux and evapotranspiration of a juvenile Douglas-fir stand. *Agricultural and Forest Meteorology* **50**, 139–158.

PRICE, D.T. and BLACK, T.A. (1991). Effects of summertime changes in weather and root-zone water storage on canopy CO_2 flux and evapotranspiration of two juvenile Douglas-fir stands. *Agricultural and Forest Meteorology* **53**, 303–323.

PRIEME, A., CHRISTENSEN, S., DOBBIE, K.E. and SMITH, K.A. (1997). Slow increase in rate of methane oxidation in soils with time, following land use change from arable agriculture to woodland. *Soil Biology and Biochemistry* **29**, 1269–1273.

REY, A. and JARVIS, P.G. (1997). Growth response of young birch trees (*Betula pendula* Roth.) after four and a half years of CO_2 exposure. *Annals of Botany* **80**, 809–816.

REYNOLDS, B. (2007). Implications of changing from grazed or semi-natural vegetation to forestry for carbon stores and fluxes in upland organo-mineral soils in the UK. *Hydrology and Earth Systems Science* **11**, 61–76.

ROULET, N.T., ASH, R., QUINTON, W. and MOORE, T. (1993). Methane flux from drained northern peatlands: effect of a persistent water table lowering on flux. *Global Biogeochemical Cycles* **7**, 749–769.

ROYAL SOCIETY (2001). *The role of land carbon sinks in mitigating global climate change.* Policy Document 10/01. Royal Society, London.

ROYAL SOCIETY (2008). *Ground-level ozone in the 21st century: future trends, impacts and policy implications.* Policy document 18/08. Royal Society, London

SANDFORD, A.P. and JARVIS, P.G. (1986). The response of stomata of several coniferous species to leaf-to-air vapour pressure difference. *Tree Physiology* **2**, 89–103.

SATTIN, M., MILNE, R., DEANS, J.D. and JARVIS, P.G. (1997). Radiation interception measurement in poplar: sample size and comparison between tube solarimeters and quantum sensors. *Agricultural and Forest Meteorology* **85**, 209–216.

SITCH, S., COX, P.M., COLLINS, W.J. and HUNTINGFORD, C. (2007). Indirect radiative forcing of climate change through ozone effects on the land-carbon sink. *Nature* **448**, 791–794.

SMAL, H. and OLSZEWSKA, M. (2008). The effect of afforestation with Scots pine (*Pinus silvestris* L.) of sandy post-arable soils on their selected properties. II. Reaction of carbon, nitrogen and phosphorus. *Plant and Soil* **305**, 171–187.

SMITH, K.A., DOBBIE K.E., BALL, B.C., BAKKEN, L.R., SITAULA, B.K., HANSEN, S., BRUMME, R., BORKEN, W., CHRISTENSEN, S., PRIEME, A., FOWLER, D., MACDONALD, J.A., SKIBA, U., KLEMEDTSSON, L., KASIMIR-KLEMEDTSSON, A., DEGORSKA, A. and ORLANSKI, P. (2000). Oxidation of atmospheric methane in Northern European soils, comparison with other ecosystems, and uncertainties in the global terrestrial sink. *Global Change Biology* **6**, 791–803.

SMITH, P., POWLSON, D.S. and GLENDINING, M.J. (1997). Potential for carbon sequestration in European soils: preliminary estimates for five scenarios using results from long-term experiments. *Global Change Biology* **3**, 67–79.

STEVENS, A. and VAN WESEMAEL, B. (2008). Soil organic carbon stock in the Belgian Ardennes as affected by afforestation and deforestation from 1868 to 2005. *Forest Ecology and Management* **256**, 1527–1539.

STEWART, H.E., HEWITT, C.N., BUNCE, R.G.H., STEINBRECHER, R., SMIATEK, G. and SCHOENEMEYER, T. (2003). A highly spatially and temporally resolved inventory for biogenic isoprene and monoterpene emissions: model description and application to Great Britain. *Journal of Geophysical Research* **108**, Art. No. 4644, doi:10.1029/2002JD002694.

TITUS, B.D. and MALCOLM, D.C. (1991). Nutrient changes in peaty gley soils after clearfelling of Sitka spruce stands. *Forestry* **64**, 251–270.

TITUS, B.D. and MALCOLM, D.C. (1992). Nutrient leaching from the litter layer after clear felling of Sitka spruce stands on peaty gley soils. *Forestry* **65**, 389–416.

TITUS, B.D. and MALCOLM, D.C. (1999). The long-term decomposition of Sitka spruce needles in brash. *Forestry* **72**, 207–221.

VON ARNOLD, K., WESLIEN, P., NILSSON, M., SVENSSON, B. and KLEMEDTSSON, L. (2005). Fluxes of CO_2, CH_4 and N_2O from drained coniferous forests on organic soils. *Forest Ecology and Management* **210**, 239–254.

WANG, Y. P. and JARVIS, P.G. (1991). Influence of crown structural properties on PAR absorption, photosynthesis and transpiration in Sitka spruce: application of a model (MAESTRO). *Tree Physiology* **7**, 297–316.

WANG, Y.P., JARVIS, P.G. and BENSON, M.L. (1990). The two-dimensional needle area density distribution within the crowns of *Pinus radiata* trees. *Forest Ecology and Management* **32**, 217–237.

WANG, Y.P., JARVIS, P.G. and TAYLOR, C.M.A. (1991). PAR absorption and its relation to above-ground dry matter production of Sitka spruce. *Journal of Applied Ecology* **28**, 547–560.

WANG, Y.P., REY, A. and JARVIS, P.G. (1998). Carbon balance of young birch trees grown in ambient and elevated atmospheric CO_2 concentrations. *Global Change Biology* **4**, 797–807.

ZERVA, A. and MENCUCCINI M. (2005a). Short-term effects of clearfelling on soil CO_2, CH_4, and N_2O fluxes in a Sitka spruce plantation. *Soil Biology and Biochemistry* **37**, 2025–2036.

ZERVA, A. and MENCUCCINI, M. (2005b). Carbon stock changes in a peaty gley soil profile after afforestation with Sitka spruce (*Picea sitchensis*). *Annals of Forest Science* **62**, 873–880.

ZERVA, A., BALL, T., SMITH, K.A. and MENCUCCINI, M. (2005). Soil carbon dynamics in a Sitka spruce (*Picea sitchensis (Bong.)* Carr.) chronosequence on a peaty gley. *Forest Ecology and Management* **205**, 227–240.

CHANGE

SECTION 2
IMPACTS

Chapter 4 - Observed impacts of climate change
on UK forests to date
Chapter 5 - An assessment of likely future impacts
of climate change on UK forests

OBSERVED IMPACTS OF CLIMATE CHANGE ON UK FORESTS TO DATE

Chapter

4

M. S. J. Broadmeadow, M. D. Morecroft and J. I. L. Morison

Key Findings

There is clear evidence that climate change is having an impact on some aspects of the composition and function of woodland. Leafing has advanced by 2–3 weeks since the 1950s in response to increased temperatures and there is some evidence that this is having a negative impact on woodland flora, particularly vernal species.

Evidence for increases in tree growth rates and forest productivity resulting from lengthening growing seasons, rising atmospheric CO_2 concentrations and climatic warming is limited for the UK. This is consistent with recent studies which attribute the increases reported across much of continental Europe to changes in forest management and nitrogen availability.

In most cases it is not possible to separate the impacts of climate change on forest ecosystems from the effects of forest maturation over the past century, changes in land cover and forest management and the effects of nitrogen deposition.

The likely effects of extreme drought years such as 1976 and 1989/90 provide an insight into some of the well-documented impacts of climate. The effects of drought are not restricted to a single year. Subsequent decline and, ultimately, mortality can occur over a period of more than 10 years.

Current monitoring networks have not adequately addressed climate change. Extensive monitoring networks such as Countryside Survey and the National Forest Inventory may not be suitable for monitoring the impacts of climate change or the effectiveness of adaptation measures. An annual monitoring platform is required to monitor the impacts of climate change, particularly extreme climatic events as they occur, and to provide further evidence for modelling future impacts and developing appropriate adaptation strategies.

Evidence is now compelling that the climate has changed over past decades in the UK, with nine of the ten warmest years on record having occurred since 1990 and a shift in the seasonality of rainfall now emerging.

Although it is difficult to prove that these changes are the direct result of anthropogenic greenhouse gas (GHG) emissions, modelling experiments provide the evidence: only when GHG forcing is included in those experiments can global climate models replicate the climate of the past century adequately (IPCC, 2007). We can therefore link climate-driven observed changes in woodland form and function to the direct effects of anthropogenic climate change; more importantly, we can use these observations, coupled to the fundamental knowledge of tree physiology

outlined in Chapter 3, to project the future impacts of climate change (see Chapter 5).

It is critical that observed impacts of climate change are well documented to enable conclusions on future impacts to be drawn. However, a note of caution should be sounded because the level of climate change witnessed to date (e.g. 0.8°C to 1°C change in mean UK temperatures; see Chapter 2) is small relative to the change that is likely over the course of the 21st century; we therefore can neither predict how woodlands will be impacted nor design or implement adaptation responses to limit the future impacts on the basis of recent responses alone. An equally important point is that trends in response variables may not yet have become apparent as a result of either the limited timecourse of monitoring data or a functional threshold not having been reached; the lack of an observed trend in a response variable does not therefore provide proof that climate change is not having an impact or will not in the shorter or longer term. The corollary of this is that where impacts of climate change are evident, these can provide a powerful indicator of advancing climate change.

This chapter collates evidence of impacts of climate change on woodland ecosystems in the UK to date. It also assesses the impacts of extreme climatic events – whether a result of climate change or not – that are likely to become more common as climate change progresses. Such information represents a powerful resource for modelling studies, to provide an analysis of the likely future impacts of climate change.

4.1 Evidence of climate change impacts to date

A clear temperature trend has only emerged since around 1980 and the magnitude of change observed to date is only 0.8–1.0°C (Chapter 2 and Jenkins et al., 2007). Clear trends in rainfall patterns and other climatic variables are more difficult to detect. These difficulties are compounded by changes in land management (Carey et al., 2008), pollutant deposition (NEGTAP, 2001) and atmospheric composition (IPCC, 2007) that have occurred over the same time period. Furthermore, trees are typically more resilient and slower to respond to environmental pressures than more dynamic elements of ecosystems such as ground flora, fungi and fauna. It therefore follows that unquestionable evidence of impacts of climate change on trees and forests are likely to be limited. However, the impact of recent extreme climatic events has been well documented and when considered with projections for the future and the quantification of uncertainty available through the UK Climate Projections (Chapter 2 and Murphy et al., 2009) they provide an insight into the likely future effects of a changing climate.

A number of formal networks and reporting protocols that are relevant to forestry and woodland have been established to identify impacts of climate change. These

are reviewed here, together with a summary of the information they have provided.

4.1.1 Climate change indicators for the UK

A series of indicators covering aspects of the climate, economy and natural environment were published in 1999 (Cannell et al., 1999) and reviewed in 2003 (Cannell et al., 2003). Two 'impacts indicators' are directly relevant to woodlands (health of beech trees and leafing date of oak), with a further four providing information on the wider natural environment (date of insect appearance/activity, insect abundance, arrival date of the swallow and small bird population size).

4.1.2 Environmental Change Network

The UK Environmental Change Network (ECN) is an ecosystem monitoring and research programme with a range of contrasting sites across the UK established by the UK Government in 1992 and funded by a range of stakeholders that operate the individual sites. Seven of these have forested areas (Alice Holt, Wytham, North Wyke, Hillsborough, Rothamsted, Porton, Cairngorm). The programme has been operating since 1992 and Morecroft et al. (in press) have recently completed a review of trends in the data. The climate of the sites has been monitored and has shown a significant warming trend of 0.9°C over 15 years. There has also been a significant rise in precipitation during this period. There was, however, no consistent evidence of change in species composition of vegetation across sites. Local changes in the composition of functional types of plants within communities were found, but attribution to climate change was generally not possible. Morecroft et al. (2008) studied tree growth at Wytham Woods ECN site and found evidence of low growth rates of sycamore, Acer pseudoplatanus, during dry periods and this was associated with reduced photosynthetic rates. This contrasted with ash, Fraxinus excelsior, which is often identified as sycamore's main native competitor. Morecroft and colleagues concluded that an increase in frequency of summer droughts might reduce the competitiveness of sycamore in future.

4.1.3 UK Phenology Network

The UK Phenology Network (UKPN) is a joint initiative between the Centre for Ecology and Hydrology and the Woodland Trust, using volunteer observers (~20 000) to report on the timing of a range of development processes (phenological indicators) in both spring and autumn

(*Nature's Calendar Survey*). The Network produces a powerful dataset for identifying the progressive impacts of climate change on the natural environment. However, its application to impacts and adaptation studies is currently limited because of its relatively short duration. Other activities of the Network, particularly the collation of historical datasets, address this need.

4.1.4 Countryside Survey

The Countryside Survey is the largest survey of soils and vegetation in the UK, to date, and provides evidence about the state of the UK's countryside today and details of land use and land use change. The most recent survey took place in 2007 and the findings can be compared with those of previous Countryside Surveys from 1998, 1990, 1984 and 1978. There are two parts to the Survey: the Field Survey and the Land Cover Map. The Land Cover Map uses data from satellites to form a digital map of the different types of land and vegetation across the UK. The Field Survey is a very detailed study of a sample of 600 1 km squares, located across England, Scotland, Wales and Northern Ireland. The individual squares are chosen so that they represent all major habitat types in the UK. The field survey includes soil and vegetation assessments.

4.1.5 Forest Health Survey

The Forest Health Survey (later renamed the Forest Condition survey) was established in 1984 (see Innes and Boswell, 1987) to assess, primarily, the crown condition of single species stands of forest trees across Great Britain. Five species (oak, beech, Scots pine, Sitka spruce and Norway spruce) were assessed on an annual basis across up to 350 sites. In addition to crown condition, a range of other metrics were recorded, including stem diameter (increment), level of fruiting (masting), biotic and abiotic damage. In 1993, the network of plots were incorporated within the EC/ICP-Forests Level I survey, providing the opportunity for comparison with wider changes in forest condition across Europe. The initial focus was on the impacts of air pollution, although latterly, climate change became of increasing importance. The survey was suspended in 2006.

4.1.6 National Forest Inventory

Extensive woodland surveys have been carried out across Great Britain in an *ad hoc* way over the past 80 years, with Woodland Censuses in 1924, 1947, 1965 and 1980

and the National Inventory of Woodland and Trees (NIWT) in 2003. These surveys collected data on the extent and nature of the woodland resource and, particularly in the case of NIWT, on its condition. The National Forest Inventory (NFI) is a rolling five-year programme with field assessments covering a minimum of 0.5% of the woodland resource. The field survey began in summer 2009 and the first cycle will report in 2015, with interim reports also to be made available.

4.2 Impacts on the timing of natural events

The study of the timing of natural events (phenology) has a long history, stretching back to 1736 through the 'Marsham Records' from Norfolk that spanned a period of more than 200 years to 1947 (Sparks and Carey, 1995). Because of the longevity of the record (and the availability of the Central England Temperature index), clear relationships can be demonstrated between spring/winter temperature and leafing date, for example. Further analysis of the Marsham record has enabled a temperature sensitivity of leaf flushing in different tree species to be calculated that will be critical for interpreting the likely impacts of climate change on the complex interactions that determine the composition of woodland ecosystems (Table 4.1) (Sparks and Gill, 2002).

Leafing date provides, for the forestry/woodland sector, the most straightforward indicator of the impact of climate change because there is a well-known relationship between the date of occurrence of phenological phases and temperature. There has been an advance in leafing date of oak by about 3 weeks since the 1950s, with leafing now consistently earlier than prior to 1990 (Figure 4.1). In theory, earlier leafing will increase net primary productivity through extending the growing season. Extending the growing season in spring will have a greater effect than in autumn because of the higher incoming solar energy in April (budburst) as compared with October (autumn senescence). However, earlier leafing may have negative impacts on the wider woodland ecosystem through affecting the synchrony between different trophic levels (for example, oak, winter moth caterpillar and blue tit: Buse and Good, 1996) or through reducing the amount of light available for characteristic woodland ground flora specialists such as bluebell, wood anemone and sanicle. This may have contributed to the decline in woodland specialists reported by Kirby *et al.* (2005), see 4.4 below.

Table 4.1
Estimated response of flushing to temperature in the first quarter of the year for 13 woody species. Data taken from the Marsham record (Sparks and Gill, 2002).

Common name	Scientific name	Temperature response (days °C⁻¹)*
Hawthorn	Crataegus monogyna	9.9
Sycamore	Acer pseudoplatanus	6.7
Birch	Betula pendula	5.2
Elm	Ulmus procera	5.7
Mountain ash	Sorbus aucuparia	5.6
Oak	Quercus robur	5.6
Beech	Fagus sylvatica	3.0
Horse chestnut	Aesculus hippocastanum	4.8
Sweet chestnut	Castanea sativa	5.5
Hornbeam	Carpinus betulus	6.1
Ash	Fraxinus excelsior	3.5
Lime	Tilia spp.	5.2
Field maple	Acer campestre	4.4

*Advance in first leafing date in days per degree increase in January–March temperatures.

Figure 4.1
Leafing date of oak in Ashstead, Surrey between 1950 and 2009.

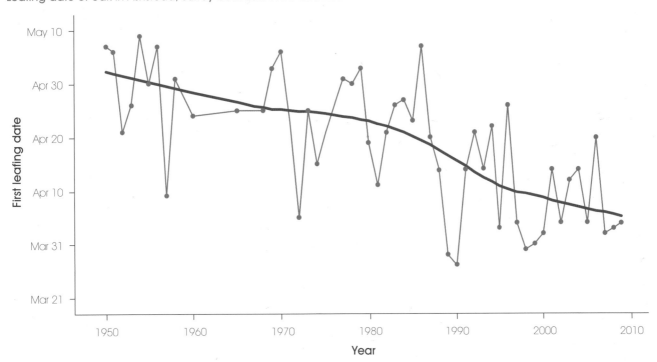

Data courtesy of Jean Coombs and Tim Sparks. See also Sparks and Gill, 2002.

Earlier leafing dates also enhance the risk of frost damage to trees because cold night-time temperatures are stochastic weather events. For example, although the modelled date of leafing advanced significantly between 1950 and 2005 in southern England, there has been no corresponding trend in the date of the last spring frost (less than –2°C, below which tissue damage is likely). There has therefore been an enhanced risk of frost damage over this period, although mean daily temperatures have risen by nearly 1°C. There has been no evidence of increased levels of spring frost damage to date, but the implications of changing vulnerability to frost should be an important consideration in provenance and species choice as part of developing adaptation strategies (see 5.1.1, Chapter 5). For some species, experimental studies have revealed the converse with a lack of winter chilling, leading to delayed budburst (Murray et al., 1989).

A series of other phenological records are available that are relevant to the forestry sector, including appearance dates of woodland butterflies and birds and emergence or flowering dates of woodland flora. As in the case of leafing date of oak, there is clear evidence of enhanced precocity, confirming that climate change is already impacting on woodland ecosystems (Sparks and Gill, 2002).

4.3 Impacts of climate change on tree growth and forest productivity

The climate of the UK has changed considerably over the last century, as outlined in Chapter 2. Mean annual temperature has increased with high temporal consistency in the rate of change between regions of the UK (Jenkins et al., 2007). Although the total annual rainfall has tended to increase in many parts of the UK, summer rainfall has shown a slight decrease, particularly in eastern regions and in southern and central England. These changes have resulted in an increasing frequency of drier and warmer summers. If soil moisture reserves are not depleted during the growing season, such changes would be expected to have increased plant growth, due to higher rates of photosynthetic CO_2 uptake and cellulose accumulation.

An index of seasonal warmth can be calculated from the sum of daily temperatures above a base of 5°C (accumulated temperature). This index has been calculated from monthly temperature records over the past 50 years (Figure 4.2). A sample of nine meteorological stations, three each in England, Scotland and Wales, show that the growing season has become warmer, based on normalised data, over the past 50 years in all regions. There is a direct relationship between accumulated temperature and yield and, if soil moisture deficit is not limiting, trees will grow faster in a warmer climate. Indeed, a number of studies have reported an increased yield from forests in Britain and Europe, which has been attributed to the warming climate, increased nitrogen availability largely because of pollutant deposition (see 3.3.4, Chapter 3) rising CO_2

levels (270 to 390 parts per million; see 5.1.1, Chapter 5) and improved silviculture (Worrell and Malcolm, 1990; Cannell et al., 1998; Cannell, 2002; Magnani et al., 2007). If summers have also been drier, this implies reduced cloud cover, and thus higher solar radiation, which could have caused increased photosynthesis and growth. However, photosynthesis in forest stands can be higher when light is more evenly distributed as under light overcast conditions than in cloudless conditions, so the net consequences of changes in cloud cover are uncertain.

Increased growth in warmer growing seasons is dependent on sufficient moisture. In dry summers, stem increment of beech has been shown to be reduced, particularly on surface water gley soils as the seasonally fluctuating water table tends to restrict rooting depth and thus moisture availability (see Figure 4.3; note dips of radial increment in 1976, 1984, 1989/90 and 1995/6).

Very dry summers have caused serious damage to tree stands, particularly in species that are not well suited to site conditions. In 1975 and 1976 two consecutive dry summers caused more serious damage and die back to beech than to oak (Mountford and Peterken, 2003), in the Denny Inclosure of the New Forest. Very dry summer periods have also caused abiotic damage to Sitka spruce on shallow freely-draining soils in eastern Scotland. Drought causes the xylem to collapse resulting in stem lesions and cracks appearing through the cambium (Green and Ray, 2009). Affected trees have been shown to exhibit shake (stem cracking), rendering the timber poor quality and of no structural use. Examples of this type of damage occurred in spruce forests in eastern Scotland in 2003.

Dendrochronology studies have demonstrated increasing growth rates across much of Europe over the past century (Briffa, 1991; Becker et al., 1994; Spiecker et al., 1996; Kahle et al., 2008). The evidence is compelling, particularly for natural stands, which have shown an increased growth rate in central Europe. However, these studies have focused on conifer species and there has been little examination of broadleaved tree species. Briffa (1991) found no increase in growth (as indicated by diameter increment) for four Scots pine stands in Scotland and one in southern England, contrasting with a significant enhancement across much of central and southern Europe between 1860 and 1975. Kahle et al. (2008) reported a similar lack of response in Great Britain (three sites in Scotland, one in the English Pennines) and northwest Europe. For this reason, the increase in height and diameter increment that was observed across much

Figure 4.2
Accumulated temperature (above a base of 5°C) trend over 50 years for nine meteorological recording stations, three each in England, Scotland and Wales. The trend is shown as the yearly anomaly, i.e. the difference each year from the overall 50-year mean for each station, in order to facilitate comparison between stations.

Wales

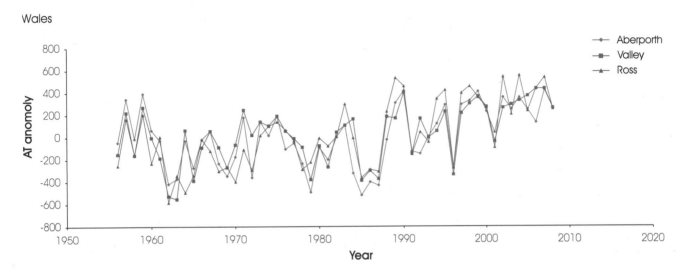

England

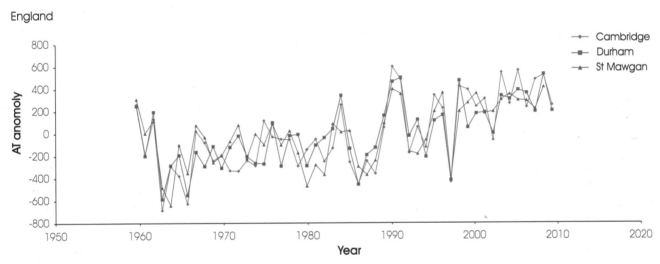

Scotland

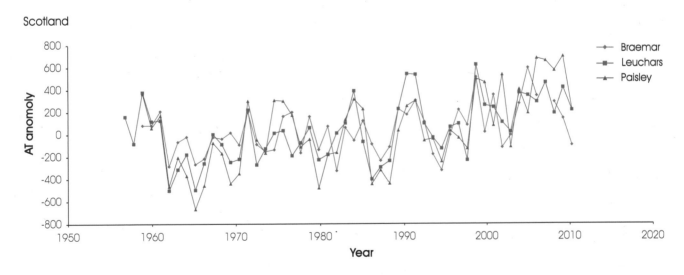

Figure 4.3

Relationship between warm and dry summers and the normalised stem diameter increment of beech (*Fagus sylvatica*) growing on three soil types (c) surface-water gley, (d) podzol and (e) brown earth in southern England. Accumulated temperatures (above a base of 5°C) (a) and summer rainfall totals (b) are shown for a site in Oxford (Wilson *et al.*, 2008, Sanders *et al.*, in press).

(a) Accumulated temperature

(b) Summer rainfall

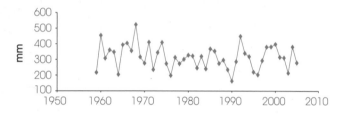

(c) Gley

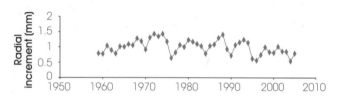

(d) Podzol

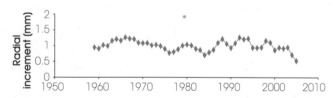

(e) Brown earth

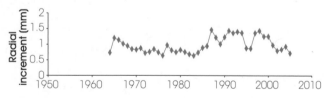

of the remainder of Europe was ascribed to changes in nitrogen availability resulting from changes in atmospheric nitrogen deposition and forest management. No effect of rising carbon dioxide concentration, temperature or rainfall patterns could be identified. Matthews *et al.* (1996) analysed mensuration sample plot collected over the

period 1920–85, reporting an increase in growth rate over this period. However, survey design and planting year had over-riding impacts on the analysis, with observed effects ascribed to improvements in planting practices, site type selection and approaches to management.

More recent analyses of broadleaf growth in southern England have revealed trends and climatic impacts that were not apparent in earlier studies. Sanders *et al.* (2007) reported an increase in tree ring width of old growth (planted 1820) woodland since the mid-1970s (Figure 4.4a). They also report a positive impact of mild spring temperature and above average early summer rainfall and a negative impact of lower than average July–August rainfall. Wilson *et al.* (2008) assessed the response of tree-ring width of beech stands in southern England to summer soil moisture deficit. The strength of the response was dependent on soil type. Growth on the gley soil, in particular, showed a large negative response to the summer droughts of 1976, 1989/90 and 1995 that was maintained for a number of years in each case. On a number of the sites included in the analysis, a decline in growth has been evident since 2000. Sanders *et al.* (in press) report an interaction with frequent masting that has been evident in recent years and earlier periods of reduced growth apparent in the chronology (Figure 4.4b).

4.4 Impacts on woodland flora

A 2001 re-survey of woodland plots first surveyed in 1971 (Kirby *et al.*, 2005) showed some potential evidence of impacts of climate change, as the frequency of 47 out of 332 plant species showed a significant positive relationship with early season temperature change over the period, whereas four species showed negative correlations. Those species which increased in frequency or abundance tended to be associated with locations with higher precipitation and lower July maximum temperatures, but were otherwise hard to categorise. There was also evidence of contrasting relationships between percentage cover and temperature in different species.

The most recent Countryside Survey, carried out in 2007 (Carey *et al.*, 2008), showed that there had been a change in the character of broadleaved woodlands, with tree and shrub species tending to increase in frequency, while ground flora, particularly grasses, tended to decrease (Carey *et al.*, 2008). There were also increases in competitive species and a decline in ruderal (weedy) species. This is indicative of the maturing of woodlands,

Figure 4.4
(a) Annual growth ring increment for oak trees at Alice Holt forest between 1820 (planting) and 2005. *n* = 11; data have been de-trended for age effects with standard dendroclimatological approaches (Sanders *et al.*, 2007). (b) Comparison of annual stem increment of five dominant, 55-year-old beech trees at Covert Wood in Kent, assessed through tree-ring analysis, and the occurrence of repeated heavy 'masting' as determined from various sources (Sanders *et al.*, in press).

(a)

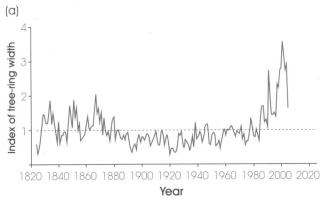

(b)

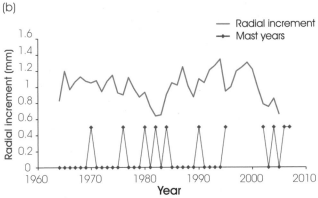

confirming the findings of Kirby *et al.* (2005) and the previous Countryside Survey (Haines-Young *et al.*, 2000). No effects of climate change were evident in CS2007, although the report notes that 'factors such as land-use surrounding the sampled woods, climate change or pollution (sulphur and nitrogen deposition) may also be important. More detailed analysis is required to determine and distinguish the possible roles of the drivers of the changes reported here for the woodlands of Great Britain'.

Where changes in ground flora are apparent, it is important to note that other drivers may be responsible, as outlined by Carey *et al.* (2008) and Hopkins and Kirby (2007). Foremost are changes in levels of management in semi-natural woodland leading to a change in structure. In an analysis of successive woodland surveys, Hopkins and Kirby (2007) report that in 1947, only 51% of the broadleaved resource was classed as high forest,

compared with 97% in 2002. Coppice and 'scrub' made up 21% and 28%, respectively, of the remainder in 1947. The decline in the level of management that this implies was confirmed by Kirby *et al.* (2005) who reported that 68 of the 103 woods visited in 2001 showed no evidence of recent management activity. They also reported both an increase in stem size and basal area of the overstorey tree species. The observation that the 10 sites that were in the track of the 1987 storm that affected southern England tended to show an increase in plant species richness as compared to sites outside the storm track, emphasises the importance of natural disturbance to species composition (see 5.1.5, Chapter 5). The impacts of atmospheric nitrogen deposition are well documented (NEGTAP, 2001) with fast-growing plant species expected to be favoured at the expense of slower growing woodland species (Grime *et al.*, 1988). This trend is evident in the work of Haines-Young *et al.* (2000), Kirby *et al.* (2005) and Carey *et al.* (2008), and is not unsurprising given that 95% of broadleaved woodland in the UK is reported to receive excess nitrogen deposition (Hall *et al.*, 2004). If the impacts of climate change are to be detected at an early stage, those effects will need to be distinguished from changes in flora resulting from the drivers outlined above. This will require that a monitoring system designed for that purpose, such as the Environmental Change Network (see 4.1 above), remains in place to provide a continuous record measured using methods that remain common over time.

4.5 Impacts on woodland fauna

There is clear evidence of changes to emergence and first flight times for a range of butterflies and moths (Burton and Sparks, 2002) and hoverfly species (Morris, 2000). There have also been changes in the arrival times of migrant bird species (Sparks and Gill, 2002). The abundance of moth species has also shown a significant decline over the period 1966–2001 with an average loss of about two species per year (Benham, 2008). Although land use, land management and other factors are implicated, relationships with summer temperature and winter precipitation are evident and help to explain the 60% decline in macrolepidoptera numbers since 1930 (Woiwod, 2003). The BTO Bird Atlas provides an unparalleled resource for the interpretation of changes in bird distributions (www.bto.org/birdatlas and Eaton *et al.*, 2009). For example, there is clear evidence of expanding ranges of some heathland species such as the Dartford warbler (*Sylvia undata*), while there has been a contraction in the range of some upland species such as the twite (*Carduelis flavirostris*) and snow bunting (*Plectrophenax nivalis*). However, changes in the distribution of many species are difficult to interpret because of the complexity of the underlying drivers.

Populations of deer have a significant impact on establishment and regeneration of both plantation forests and semi-natural woodlands. Sparks and Gill (2002) evaluated the relationship between climate and deer survival/recruitment, concluding that populations would be expected to increase in response to climate change. Although data on deer numbers are limited and do not allow a thorough evaluation of trends, deer-cull data for Scotland confirm the more anecdotal evidence that deer populations are increasing, with the cull rising from 35 000 in 1987/88 to 63 000 in 2004/05 (R. Gill and J. Irvine, pers. comm.). Although this trend is consistent with current understanding of the likely effects of climate change, a causal relationship cannot be demonstrated.

4.6 Interactions between climate change and tree health

There is little evidence in the UK that the prevalence and severity of outbreaks of existing forest pests and pathogens have been directly impacted by climate change, to date. There is, however, evidence from elsewhere in the world of changes to the range of destructive outbreaks of forest insect pests. A good example is the southern pine beetle in the southeast United States, for which the effective range has spread north and westwards as minimum winter temperatures have increased (Evans *et al.*, 2002).

Over the past decade, several new pests and diseases have been found in Great Britain, and some have established with potentially serious economic consequences. New pathways for pests to travel and establish are being provided by the ever-growing global trade. Of particular importance is the increasingly rapid trade in live plants and plant products, as well as the timber trade and goods of all kinds shipped in wood packaging material. These pathways provide increasing opportunities for pests to transfer from their native habitats and successfully establish in new areas.

During 2006, several pests made national headlines. On horse chestnut, a significant ornamental tree species, both bleeding canker (caused by the recently discovered bacterium *Pseudomonas syringae* pv *aesculi*) and the

horse chestnut leaf miner (*Cameraria ohridella*) were widely reported. First described in Macedonia in 1985, leaf miner was detected in Great Britain in 2002 in Wimbledon and has spread rapidly over the past seven years. Oak processionary moth (*Thaumetopoea processionea*), native to southern Europe made headlines when it was found in London due to its irritating hairs and adverse reactions it can provoke when in contact with people, as well as its potential to severely damage oaks. Introduced species of the fungal pathogens *Phytophthora* that include *P. ramorum*, the causal agent of the disease known as 'sudden oak death' in California, are a continuing cause for concern because of their potential to adversely affect trees in urban and rural environments. None of these significant pests are native to Britain and have been imported from different parts of the world, often with warmer climates than Britain.

In addition to dealing with the above, the citrus longhorn (*Anoplophora chinensis*) and Asian longhorn (*A. glabripennis*) beetles continue to be intercepted; both of which have the potential to be very damaging. Southern European cicadas now occur in Kent, and the plane lace-bug (*Corythuca ciliata*) was recently found by Defra plant health and seeds inspectors to be breeding and causing damage to plane trees in Bedfordshire. The lists of intercepted and, more significantly, established pests and pathogens grow each year and these changes are accelerating, as global trade expands.

4.6.1 Imported plant material and horticulture

The global trade in live plant material is on an upward trend. The UK is importing more and more plant material from around the world, much of it for use in urban planting. In 2005, tree species imported into the UK had a total value of £65 million, with a further £69 million spent on importing a large variety of outdoor plants. Interest in gardens and gardening is linked to activity in the housing market and this has helped fuel a booming garden-centre trade over the last decade. Of particular concern is the expanding trade in 'instant trees', with giant root balls. Often these are 'salvaged' rather than produced under controlled production conditions. Some of the genera involved are not restricted by quarantine controls and are therefore not subject to any form of inspection, either on arrival or when planted.

The discovery of *Phytophthora ramorum* and *P. kernoviae* in both forest trees and a wide range of ornamental trees and shrubs links the world of gardens and horticulture to the woodland environment. Gardens are a very highly valued aspect of national life and the discovery of *P. kernoviae*, which is not native to Britain, in a national Magnolia collection in Cornwall has highlighted the potential dangers of imported pathogens. An increasing proportion of imported plants are grown in the Far East and then imported into the EU. Once the plants have entered the EU, the scope for health checks and inspections is severely restricted.

4.6.2 Plant health and import controls

The sea represents a barrier to natural colonisation for most pests, and Britain is fortunate in not having many of the major pests prevalent on the European mainland. Under the EU Protected Zone provisions, the UK is able to maintain controls against named pests and pathogens present elsewhere in the EU.

Three levels of risk are identified by import controls: prohibited (plants or plant materials cannot be imported due to a high risk of pest introduction); controlled (materials are inspected before they are permitted into circulation) and unrestricted (material is not checked). The controlled list relates to material that could host pests indigenous to mainland Europe, or elsewhere in the world where climatic conditions are similar and the same hosts are grown. These factors could readily result in pest establishment in this country. The restricted list is very limited and there is a wide range of genera which, because there is no recorded information about pest risk associated with them, are not subject to any regulation. Therefore, although these controls undoubtedly reduce the risk of the harmful introduction of pests, the risk is by no means eliminated.

4.7 Impacts on woodland soils

Forest management and species choice is, in many cases, dependent on soil form and function. Principal concerns over the effects of climate change relate to nutrient sustainability and enrichment (eutrophication), recovery from past acidification, changes in soil carbon levels and the ability to retain moisture. In particular, soil organic carbon content is a key area of concern because of its role in water retention, in binding soil particles together limiting erosion and ability to retain nutrients. Soil carbon stocks also represent a large potential source of GHG emissions to the atmosphere in that soils in Great Britain store more than 10 billion tonnes of carbon – equivalent to 70 times the total GHG emissions from the UK (Milne and Brown, 1997).

Bradley *et al.* (2005) reviewed the likely impacts of climate change on soil form and function, identifying the following effects that are likely to be observed as climate change progresses:

- In isolation, rising temperatures would be expected to increase soil respiration rates and lead to a decline in soil carbon content. Reduced summer rainfall coupled with higher rates of evapotranspiration may lead to drying of moist organic soils for a greater proportion of the year leading to significant soil carbon loss.
- Increased litter inputs as a result of increased tree productivity in response to rising CO_2 levels and climatic warmth may counter the effects of rising temperatures on soil respiration and enhance soil carbon levels.
- Increased tree growth might contribute to nutrient depletion, possible acidification and reduced productivity in second/third rotation crops.
- As a result of changes to soil carbon levels coupled with changes in nitrogen deposition and management practice, the carbon to nitrogen ratio may change with consequent impacts on woodland ground vegetation composition.
- Increased soil disturbance as a result of more frequent/ severe windstorms and consequent windthrow.
- Increased soil erosion on vulnerable sites as a result of heavier rainfall events.
- Dissolved organic carbon losses from forest soils may increase as a result of higher temperatures and heavier winter rainfall leading to impacts on water quality and treatment costs.
- More frequent and widespread forest fires resulting in soil carbon loss with consequent impacts on water-holding capacity and nutrient retention.
- The relationship between mycorrhiza, ground flora and host trees is affected by atmospheric CO_2, levels pollutant deposition and soil moisture content. Given the importance of such relationships in nutrient cycling and conferring resistance against pathogenic soil fungi, this is potentially an area of concern, although poorly documented to date.

The majority of the likely impacts of climate change outlined above result from interactions between vegetation, soils and climate. Detecting a climate change signal for any of these potential impacts in isolation is therefore challenging, particularly given the limited change in climate to date. However, the two aspects of soil form and function outlined below provide some indication of the extent of climate change impacts on the soil environment to date.

4.7.1 Soil carbon

Forest soils typically accumulate carbon during stand development due to woody and other inputs (Poulton, 1996). However, an analysis of soil carbon change from more than 2000 sites across England and Wales between 1978 and 2003 reported significant loss of soil organic carbon from all land covers, including woodland (Bellamy *et al.*, 2005). Mean carbon content of the upper 15 cm declined from 6.8% and 11.0% to 5.46% and 8.74% for broadleaf and conifer woodland soils, respectively. The losses reported for individual land covers were scaled up to a UK level, suggesting that soil carbon losses could amount to up to 13 MtC, or 8%, of total fossil fuel emissions. The authors suggested that climate change might, in part, explain the results. However, Schulze and Freibauer (2005) concluded that rising temperatures can only have had a limited impact. Although the mean carbon content showed a clear decline over the 25-year period, there was close to a 50:50 split between those soils that showed an increase in soil carbon content and those that showed a decline (Figure 4.5). A limitation of the study was that impacts of changes in land management (including harvesting, restocking or changes in woodland type) were not assessed. These changes could explain some of the unexpectedly large losses in soil organic carbon of up to 90%. Powlson (2006) commented that the reported rate of carbon loss was extraordinarily rapid for temperate forest soils in the absence of major management or land use change. The National Soil Research Institute (NSRI) soil survey has been re-evaluated for woodland soils in Wales and revealed no change in soil carbon content between the two sampling points of those soils (Alton *et al.*, 2007).

The Countryside Survey provides an alternative source of information on changes in soil carbon content (Carey *et al.*, 2008), based on a similar methodology sampling to 15 cm at three time-points (1978, 1998 and 2007). Between 1978 and 2007, the carbon content of broadleaf woodland soils showed a significant increase, with no change reported for coniferous woodland soils. A combined analysis also revealed a significant increase. However, between 1998 and 2007 both broadleaf and conifer woodland soils · showed a small but non-significant decline. The authors could not distinguish a climate change signal from the effects of changes in land management or atmospheric pollutant deposition.

The ability to interpret both the NSRI and Countryside Survey datasets (sampled to 15 cm) for woodland soils is limited because many woodland soils have deeper active

Figure 4.5
Distribution of values of percentage soil carbon changes under woodland (a) broadleaves and (b) conifers from the NSRI soil re-survey data sets (1978–2003; Morison *et al.*, 2009). Each bar represents an individual site. For further details and description of methodology, see Bellamy *et al.*, (2005).

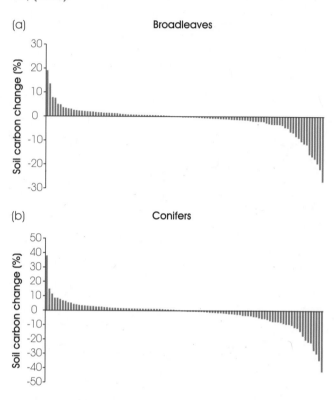

rooting profiles than comparable agricultural soils. The Environmental Change Network provides an alternative source of data, for which the soils have been extensively assessed and characterised to 80 cm depth on a five-yearly basis (1994, 1999 and 2004). Soil organic carbon stocks in the top two (A and B) horizons at the Alice Holt woodland site increased over the three sampling periods at a rate of 1.1 tCO_2 ha^{-1} $year^{-1}$ (Morison *et al.*, 2009: Figure 4.6). A separate study across a chronosequence of oak stands at Alice Holt reported a similar rate of carbon accumulation in woodland soils (0.7–1.1 tCO_2 ha^{-1} $year^{-1}$ (Pitman *et al.*, 2009). The similarity in the reported rates suggests that woodland succession rather than climate change is the primary driver of this increase in soil carbon. An important observation is that when the data were reported by depth rather than by horizon, the increase in soil carbon stocks was not apparent (Figure 4.6). This highlights the inability of sampling the top 15 cm of woodland soils to adequately reflect changes in soil carbon stocks; any changes in soil carbon stocks resulting from the development of the soil profile will not be reported. In part, this may explain some of the conflicting evaluations in soil carbon exchange that have emerged over recent years (see 3.5.1, Chapter 3).

Figure 4.6
Soil carbon stocks measured in (a) soil horizons and (b) soil depths in the ECN oak site at Alice Holt in 1994, 1999 and 2004.

(a)

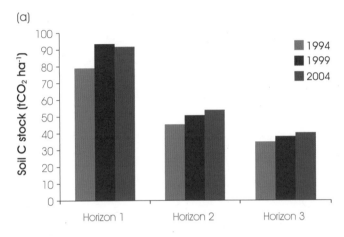

(b)

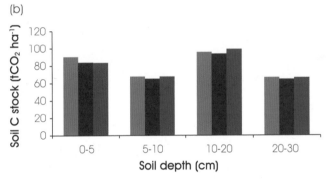

4.7.2 Dissolved organic carbon

Some organic carbon can become dissolved in the soil water (DOC). There is clear evidence of an increase in levels of DOC over the past 20 years (Monteith and Evans, 2000), although the cause of this is unclear. It has been suggested that higher soluble carbohydrate levels resulting from rising carbon dioxide levels in the atmosphere may be responsible. There is also some evidence that rising temperatures leading to increased mineralisation may be implicated, while more frequent and intense wetting/drying cycles have also been shown to result in increased DOC losses (Hentschel et al., 2007).

4.8 Damage to woodland

4.8.1 Forest fires

An increasing trend for outdoor 'grassland' fires has been observed over recent decades (CLG, 2008). This contrasts with the number of forest fires (on Forestry Commission land) which have declined over the period for which data are available (1975–2004: Forestry Commission, 2004). The decline is likely to result primarily from the reduction in the area of thicket stage conifer woodland which is at greatest risk of fire. Cannell and McNally (1997) also noted that climate has generally been the catalyst for forest fires but rarely the actual cause, as most fires are human-induced whether intentional or not. The area of woodland burned increased in the well-documented 'drought' years such as 1976, 1989–90 and 1995, although no increase was evident in 2005. The very large area burned in 1976 was, in part, a result of the ground flora drying out in late summer, leading to a second fire season additional to the normal late-spring fire season where the previous year's litter provides the fuel source. Gazzard (2006) clearly demonstrated the link between soil moisture deficit and number of outdoor fires reported in southern England (Figure 4.7). In most cases, forest fires in the UK are low temperature burns (Mitchell et al., 2007) and do not significantly damage the wider woodland ecosystem, although economic damage to forestry crops is clearly serious. In contrast to the UK, forest fires in more southerly regions of Europe represent a much greater threat with, for example, an area in excess of 800 000 ha burned in 2003. Although this indicates that forest fires are likely to become a greater risk with drier weather, it is unlikely that such large-scale forest fires would result in the UK because of the highly fragmented nature of woodland cover.

Figure 4.7
Incidence of straw/stubble (brown) and grassland/heathland (light brown) fires in southeast England compared with calculated soil moisture deficit for the Alice Holt climatological station. (After Gazzard, 2006.)

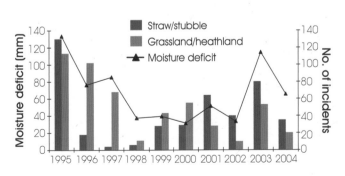

4.8.2 Wind

Wind damage to UK forests has been reviewed by Quine and Gardiner (2002), who concluded that no trend in wind damage has been discernible over the course of the 20th century. However, Lindroth *et al.* (2009) reported a significant increasing trend of forest damage caused by wind storms in continental Europe over the past 60 years. The most significant single storm event in terms of damage to growing stock was the 1987 storm in southern England, during which 3.9 million m³ was lost – a volume equivalent to a little less than half of annual timber production for the UK – thus highlighting the seriousness of the risk.

4.8.3 General condition

Although the Forest Condition Survey was suspended in 2006, it provided a time series reporting mean crown density over nearly 20 years across ~350 plots (Figure 4.8a; Hendry, 2005). Importantly, the methodology was common across the EC ICP-Forests Level I Network of which the Forest Condition Survey was a component. Therefore, a picture of how forest condition in the UK relates to trends across Europe can be drawn (Figure 4.8b; Lorenz *et al.*, 2008).

Across Britain, crown condition of oak showed the greatest decline between 1987 and 2004, in line with observations across Europe that deciduous oak species show the greatest level of 'defoliation' of the six species groups reported. Oak dieback (Gibbs, 1999) was evident over a number of years, while more severe, acute, dieback has been reported more recently on a number of plots. However, abiotic influences (wind and hail damage), together with defoliating insect damage were generally the reason of poor crown condition. Norway spruce also showed a statistically significant decline in crown density over the 20-year period that the survey operated, although there has been no further decline since 1991. Beech has shown large inter-annual fluctuation in crown condition, generally related to the well-documented response of heavy fruiting to weather conditions (Piovesan and Adams, 2001) that has been implicated in the decline of growth in the species (see 4.3 above and Figure 4.4b). Across Europe, beech also showed a marked decline in crown density in 2004, following the severe summer drought conditions experienced over much of Europe in 2003. There has been limited variation in the condition of Scots pine, with no trends evident. Similarly, there has been no trend evident in the condition of Sitka spruce, although the inter-annual variability is large, in part related to green spruce aphid-driven defoliation episodes.

Figure 4.8

(a) Changes in crown density since 1987 for five tree species surveyed annually. The crown density compared with that of an 'ideal' tree with a completely opaque crown is shown for each species. (b) Mean defoliation index in forests in Europe for commonly reported tree species and for the total of all tree species. Samples only include countries with continuous data submission. Sample size for the selected main tree species varies between 1950 and 26 788 trees per species and year. The time series, starting in 1990, is available for a smaller number of countries and is based on between 38 026 and 45 204 trees depending on the year.

(a)
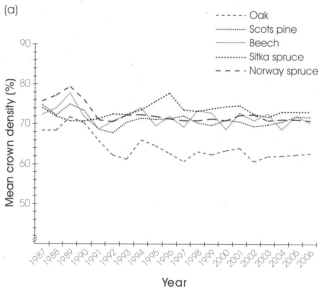

(b)
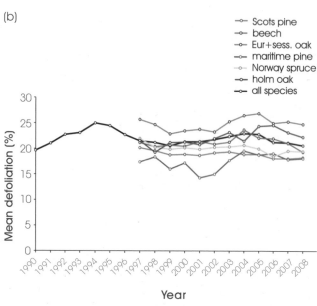

There is therefore no evidence from the Forest Condition Survey of direct impacts of climate change, although an indirect climate change signal could be inferred from the decline in the condition of oak. In reality, the time series of the Forest Condition Survey is too short and the range

Figure 4.9
Comparison between the proportion of severely defoliated trees and soil moisture deficit between 1986 and 2004. Proportion of defoliated trees representative of all five species (oak, beech, Scots pine, Sitka spruce, Norway spruce) in south and east England from the Forest Condition Survey (Hendry, 2005) and Soil moisture deficit calculated for the climatological station at Alice Holt, Surrey.

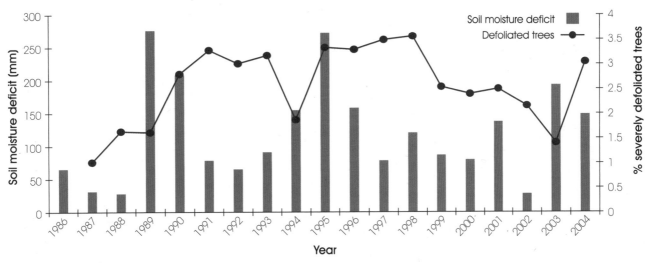

of interacting factors too wide to deduce a clear climate change signal at this stage. However, the impact on crown condition of the well-documented summer droughts of 1989/90, 1995 and 2003 is clear (Figure 4.9), providing insight into the likely impacts on woodland condition as climate change progresses. It is notable that the increase in crown transparency induced by drought the previous year is maintained for 3–4 years after the year in which the drought conditions prevailed.

4.9 Research priorities

- **Monitoring the changes.** There is no established framework for collecting or evaluating information on forest growth and productivity which has contributed to the lack of clear evidence of growth trends of British forests. If an appropriate framework is not put in place, it will therefore be difficult to identify when climate change starts to impact – either negatively or positively – on forest growth. This, in turn will delay the implementation of more extreme adaptation measures and changes in species choice or management systems, should that prove necessary. The development of a framework for monitoring and evaluating changes in increment on an annual basis is therefore an urgent requirement. We suggest that this is developed as a UK climate change indicator for the forestry sector and reported on annually.

- **Understanding the effect of rising temperatures.** There is clear evidence that rising temperatures are having an impact on a range of physiological and ecological functions. In particular, the timing of natural events

shows a strong temperature response. Further research is required to evaluate how the phenological changes observed to date have interacted and impacted on the wider woodland ecosystem.

- **Understanding the impacts of extreme events.** The impacts of extreme events on forest ecosystems needs to be better understood, particularly the effects of prolonged drought on tree physiology and mortality. Responsive monitoring programmes should be planned to ensure that the impacts of extreme climatic events are captured to provide an insight to the likely effects of future climate change. In particular, the early detection and anticipation of forest dieback will be essential to allow early adaptation to limit the impacts of climate change.

References

ALTON, K., BELLAMY, P.A., CLARKE, M.A. and THOMPSON, T.R. (2007). *Soil carbon in Wales – current stocks, trends and opportunity for soil carbon capture.* National Soil Resources Institute, Cranfield University. Report for Environment Agency Wales.

BECKER, M., NIEMINEN, T.M. and GÉRÉMIA, F. (1994). Short-term variations and long-term changes in oak productivity in northeastern France. The role of climate and atmospheric CO_2. *Annals of Forest Science* **51**, 477–492.

BELLAMY, P.H., LOVELAND, P.J., BRADLEY, R.I., LARD, R.M. and KIRK, G.J.D. (2005). Carbon losses from soils across England and Wales 1978–2003. *Nature* **437**, 245–248.

BENHAM, S. (2008). *The Environmental Change Network at*

Alice Holt Research Forest. Forestry Commission Research Note. Forestry Commission, Edinburgh.

BRADLEY, I., MOFFAT, A.J., VANGUELOVA, E., FALLOON, P. and HARRIS, J. (2005). *Impacts of climate change on soil functions*. Defra project, Report SP0538, London.

BRIFFA, K.R. (1991). *Detection of any widespread and unprecedented changes in the growth of European conifers*. Final report to the UK Forestry Commission. Climatic Research Unit, University of East Anglia, Norwich.

BUSE, A. and GOOD, J.E. (1996). Synchronization of larval emergence in winter moth (*Operophtera brumata* L.) and budburst in pedunculate oak (*Quercus robur* L.) under simulated climate change. *Ecological Entomology* **21**, 335–343.

BURTON, J.F. and SPARKS, T.H. (2002). Flying earlier in the year: the phenological responses of butterflies and moths to climate change. *British Wildlife* **13**, 305–311.

CANNELL, M. (2002). Impacts of climate change on forest growth. In: Broadmeadow, M. (ed.) *Climate change: impacts on UK forests*. Forestry Commission Bulletin 125. Forestry Commission, Edinburgh.

CANNELL, M.G.R. and McNALLY, S. (1997). Forestry. In: Palutikof, J.P., Subak, S. and Agnew, M.D. (eds) *Economic impacts of the hot summer and unusually warm year 1995*. University of East Anglia, Norwich.

CANNELL, M.G.R., THORNLEY, J.H.M., MOBBS, D.C. and FRIEND, A.D. (1998). UK conifer forests may be growing faster in response to increased N deposition, atmospheric CO_2 and temperature. *Forestry* **71**, 277–296.

CANNELL, M.G.R., PALUTIKOFF, J.P. and SPARKS, T.H. (1999). *Indicators of climate change in the UK*. DETR, London.

CANNELL, M.G.R., BROWN, T., SPARKS, T.H., MARSH, T., PARR, T.W., GEORGE, D.G., PALUTIKOF, J., LISTER, D. and DOCKERTY, T. (2003). *Review of UK climate change indicators*. Defra contract EPG 1/1/158. Centre for Ecology and Hydrology, Edinburgh.

CAREY, P.D., WALLIS, S., CHAMBERLAIN, P.M., COOPER, A., EMMETT, B.A., MASKELL, L.C., MCCANN, T., MURPHY, J., NORTON, L.R., REYNOLDS, B., SCOTT, W.A., SIMPSON, I.C., SMART, S.M. and ULLYETT, J.M. (2008). *Countryside Survey: UK results from 2007*. NERC/ Centre for Ecology & Hydrology, CEH Project Number: C03259.

CLG (2008). *Fire statistics: United Kingdom, 2006*. Communities and Local Government, London.

EATON, M.A., BALMER, D.E., CONWAY, G.J., GILLING, S., GRICE, P.V., HALL, C., HEARN, R.D., MUSGROVE, A.J., RISELY, K. AND WOTTON, S. (2009). *The state of the UK's birds 2008*. RSPB, BTO, WWT, CCW, NIEA, Joint Nature Conservation Committee, Natural England and Scottish Natural Heritage, Sandy, Bedfordshire.

EVANS, H., STRAW, N. and WATT, A. (2002). Climate change: implications for insect pests. In: Broadmeadow, M. (ed.) *Climate change: impacts on UK forests*. Forestry Commission Bulletin 125. Forestry Commission, Edinburgh. pp. 99–118.

FORESTRY COMMISSION (2004). *Forestry statistics 2004*. Forestry Commission, Edinburgh.

GAZZARD, R. (2006). *Fire and Rescue Services Act 2004, a lost opportunity for forest, heath land and agricultural fire management*. MSc Thesis, University of Surrey.

GIBBS, J.N. (1999). *Dieback of pedunculate oak*. Forestry Commission Information Note 22. Forestry Commission, Edinburgh.

GREEN, S. and RAY, D. (2009). *Potential impacts of drought and disease on forestry in Scotland*. Forestry Commission Research Note. Forestry Commission, Edinburgh.

GRIME, J.P., HODGSON, J.G. and HUNT, R. (1988). *Comparative plant ecology*. Unwin-Hyman, London.

HAINES-YOUNG, R.H., BARR, C.J., BLACK, H.I.J., BRIGGS, D.J., BUNCE, R.G.H., CLARKE, R.T., COOPER, A., DAWSON, F.H., FIRBANK, L.G., FULLER, R.M., FURSE, M.T., GILLESPIE, M.K., HILL, R., HORNUNG, M., HOWARD, D.C., MCCANN, T., MORECROFT, M.D., PETIT, S., SIER, A.R.J., SMART, S.M., SMITH, G.M., STOTT, A.P., STUART, R.C. and WATKINS, J.W. (2000). *Accounting for nature: assessing habitats in the UK countryside*. Department of the Environment, Transport and the Regions, London.

HALL, J., ULLYETT, J., HEYWOOD, L. and BROUGHTON, R. (2004). *Update to: the status of UK critical loads critical loads methods, data & maps*. Centre for Ecology and Hydrology, Huntingdon.

HENDRY, S. (2005). *Forest condition 2004*. Forestry Commission Information Note 75. Forestry Commission, Edinburgh.

HENTSCHEL, K., BORKEN, W. and MATZNER, E. (2007). Leaching losses of inorganic N and DOC following repeated drying and wetting of a spruce forest soil. *Plant and Soil* **300**, 21–34.

HOPKINS, J. and KIRBY, K.J. (2007). Ecological change in British broadleaved woodland since 1947. *Ibis* **149** (Suppl. 2), 29–40.

INNES, J.L. and BOSWELL, R.C. (1987). *Forest health surveys 1987. Part 1: results*. Forestry Commission Bulletin 74. HMSO, London.

IPCC (2007). *Climate change 2007 – the physical science basis. Contribution of Working Group I to the Fourth Assessment Report of the Intergovernmental Panel on Climate Change*. Cambridge University, Cambridge.

JENKINS, G.J., PERRY, M.C. and PRIOR, M.J.O. (2007).

The climate of the United Kingdom and recent trends. Met Office Hadley Centre, Exeter.

KAHLE, H.P., KARJALAINEN, T., SCHUCK, A., ÅGREN, G.I., KELLOMÄKI, S., MELLERT, K.H., PRIETZEL, J., REHFUESS, K.E. and SPIECKER, H. (eds) (2008). *Causes and consequences of forest growth trends in Europe – results of the recognition project*. Brill, Leiden.

KIRBY, K.J., SMART, S.M., BLACK, H.I.J., BUNCE, R.G.H., CORNEY, P.M. and SMITHERS, R.J. (2005). *Long term ecological change in British woodland (1971–2001): A re-survey and analysis of change based on the 103 sites in the Nature Conservancy 'Bunce 1971' woodland survey*. Research Report 653. English Nature, Peterborough.

LINDROTH, A., LAGERGREN, F., GRELLE, A., KLEMEDTSSON, L., LANGVALL, O., WESLIEN, P. and TUULIK, J. (2009). Storms can cause Europe-wide reduction in forest carbon sink. *Global Change Biology* **15**, 346–355.

LORENZ, M., FISCHER, R., BECHER, G., GRANKE, O., SEIDLING, W., FERRETTI, M., SCHAUB, M., CALATAYUD, V., BACARO, G., GEROSA, G., ROCCHINI, D. and SANZ, M. (2008*). Forest condition in Europe: 2007*. Technical Report of ICP Forests. Federal Research Centre for Forestry and Forest Products and University of Hamburg, Hamburg.

MAGNANI, F., MENCUCCINI, M., BORGHETTI, M., BERBIGIER, P., BERNINGER, F., DELZON, S., GRELLE, A., HARI, P., JARVIS, P.G., KOLARI, P., KOWALSKI, A.S., LANKREIJER, H., LAW, B.E. LINDROTH, A., LOUSTAU, D., MANCA, G., MONCRIEFF, J.B., RAYMENT, M., TEDESCHI, V., VALENTINI, R. and GRACE, J. (2007). The human footprint in the carbon cycle of temperate and boreal forests. *Nature* **447**, 848–852.

MATTHEWS, R., METHLEY, J., ALEXANDER, M., JOKIEL, P. and SALISBURY, I. (1996). Site classification and yield prediction for lowland sites in England and Wales. Final report for FC and MAFF joint contract CSA 2119. Forestry Commission, Farnham.

MILNE, R. and BROWN, T.A. (1997). Carbon in the vegetation and soils of Great Britain. *Journal of Environmental Management* **49**, 413–433.

MITCHELL, R.J., MORECROFT, M.D., ACREMAN, M., CRICK, H.Q.P., FROST, M., HARLEY, M., MACLEAN, I.M.D., MOUNTFORD, O., PIPER, J., PONTIER, H., REHFISCH, M.M., ROSS, L.C., SMITHERS, R.J., STOTT, A., WALMSLEY, C.A., WATTS, O. and WILSON, E. (2007*). England Biodiversity Strategy – towards adaptation to climate change*. Final report to Defra for contract CRO327.

MONTEITH, D.T. and EVANS, C.D. (eds) (2000). *The UK acid waters monitoring network: ten year report. Analysis and interpretation of results, 1988–1998*. ENSIS Publishing,

London.

MORECROFT, M.D., STOKES, V.J., TAYLOR, M.E and MORISON, J.I.L. (2008). Effects of climate and management history on the distribution and growth of sycamore (*Acer pseudoplatanus* L.) in a southern British woodland in comparison to native competitors. *Forestry* **81**, 59–74.

MORECROFT, M.D., BEALEY, C.E., BEAUMONT, D.A., BENHAM, S., BROOKS, D.R., BURT, T.P., CRITCHLEY, C.N.R., DICK, J., LITTLEWOOD N.A., MONTEITH, D.T., SCOTT, W.A., SMITH, R.I., WALMSLEY, C. and WATSON, H. (2009). The UK Environmental Change Network: emerging trends in terrestrial biodiversity and the physical environment. *Biological Conservation 142*, 2814–2832

MORISON, J., MATTHEWS, R.W., PERKS, M., RANDLE, T., VANGUELOVA, E., WHITE, M. and YAMULKI, S. (2009). *The carbon and GHG balance of UK forests – a review*. Report for the Forestry Commission. Forest Research, Farnham.

MORRIS, R.K.A. (2000). Shifts in the phenology of hoverflies in Surrey: do these reflect the effects of global warming? *Dipterist's Digest* **7**, 103–108.

MOUNTFORD, E.P. and PETERKEN, G.F. (2003). Long-term change and implications for the management of wood pastures: experience over 40 years from Denny Wood, New Forest. *Forestry* **76**, 19–43.

MURPHY, J.M., SEXTON, D.M.H., JENKINS, G.J., BOORMAN, P.M., BOOTH, B.B.B., BROWN, C.C., CLARK, R.T., COLLINS, M., HARRIS, G.R., KENDON, E.J., BETTS, R.A., BROWN, S.J., HOWARD, T. P., HUMPHREY, K.A., MCCARTHY, M.P., MCDONALD, R.E., STEPHENS, A., WALLACE, C., WARREN, R., WILBY, R. and WOOD, R. A. (2009). *UK Climate Projections Science Report: Climate change projections*. Met Office Hadley Centre, Exeter.

MURRAY, M.B., CANNELL, M.G.R. and SMITH, R.I. (1989). Date of budburst of fifteen tree species in Britain following climatic warming. *Journal of Applied Ecology* **26**, 693–700.

NEGTAP (2001). *Transboundary air pollution: acidification, eutrophication and ground level ozone in the UK*. Report of the National Expert Group on Transboundary Air Pollution. Defra, London.

PIOVESAN, G. and ADAMS, J.M. (2001). Masting behaviour in beech: linking reproduction and climatic variation. *Canadian Journal of Botany* **79**, 1039–1047.

PITMAN, R.M., POOLE, J. and BENHAM, S. (2009). A comparison of Ellenberg values from the ground flora in a chronosequence of lowland oak and Corsican pine stands with forest soil properties. *Journal of Ecology* (in press).

POULTON, P.R. (1996). Geescroft Wilderness, 1883–1995. In: Powlson, D.S., Smith, P. and Smith, J.U. (eds) *Evaluation of soil organic matter models using existing long-term datasets*. NATO ASI Series I, Vol. 38. Springer-Verlag, Berlin. pp. 385–390.

POWLSON, D.S. (2006). *Changes in organic carbon content of soils in the UK*. Report to Defra on interpretation of published analysis of National Soils Inventory (NSI).

QUINE, C. and GARDINER, B. (2002). Climate change impacts: storms. In: Broadmeadow, M. (ed.) *Climate change: impacts on UK forests*. Forestry Commission Bulletin 125. Forestry Commission, Edinburgh. pp. 41–51.

SANDERS, T., BROADMEADOW, M. and PITMAN, R. (2007). Dendrochronology and climate change research. *Forest Research annual report and accounts 2006–7*. The Stationery Office, Norwich.

SCHULZE, E.-D. and FREIBAUER, A. (2005). Environmental science: carbon unlocked from soils. *Nature* **437**, 205–206.

SPARKS, T.H. and CAREY, P.D. (1995). The responses of species to climate over two centuries – an analysis of the Marsham phenological record, 1736–1947. *Journal of Ecology* **83**, 321–329.

SPARKS, T. and GILL, R. (2002). Climate change and the seasonality of woodland flora and fauna. In: Broadmeadow, M. (ed.) *Climate change: impacts on UK forests*. Forestry Commission Bulletin 125. Forestry Commission, Edinburgh. pp. 69–82.

SPIECKER, H., MIELIKAINEN, K., KOHL, M. and SKOVSGAARD, J. (eds) (1996*). Growth trends in European forests*. European Forest Institute Report No. 5. Springer-Verlag, Berlin.

WILSON, S.M., BROADMEADOW, M., SANDERS, T.G. and PITMAN, R. (2008). Effect of summer drought on the increment of beech trees in southern England. *Quarterly Journal of Forestry* **102**, 111–120.

WOIWOD, I.P. (2003). Are common moths in trouble? *Butterfly Conservation News* **82**, 9–11.

WORRELL, R. and MALCOLM, D.C. (1990). Productivity of Sitka spruce in northern Britain, II. Prediction from site factors. *Forestry* **63**, 119–128.

AN ASSESSMENT OF LIKELY FUTURE IMPACTS OF CLIMATE CHANGE ON UK FORESTS

Chapter 5

M. S. J. Broadmeadow, J. F. Webber, D. Ray and P. M. Berry

Key Findings

An increased frequency and severity of summer drought is likely to represent the greatest threat to woodlands from climate change. There is a very high likelihood that there will be serious impacts on drought-sensitive tree species on shallow freely-draining soils, particularly in southern and eastern Britain. These impacts will be widespread in plantations already established. They will necessitate a reassessment of the suitability of species for use in commercial forestry in all regions. The species currently available for use, assuming that appropriate provenances are selected, will remain suitable across much of the UK. However, by the end of the century, impacts in the south and east of England may be sufficiently severe to necessitate the introduction of new species. Planning of which species and species mixtures to plant where, will be the challenge for forest managers.

Models suggest that widespread impacts of climate change on the suitability of most major species currently planted in the UK only become apparent by the middle of the century under the UKCIP02 'High emissions scenario' and towards the end of the century under the UKCIP02 'Low emissions scenario'. Typical conifer rotations currently in the ground are therefore likely to reach maturity before serious impacts are apparent. However, this means that appropriate modification of species choice must be undertaken from now on when restocking and creating new woodlands.

Pests and diseases of forest trees, both those that are currently present in the UK and those that may be introduced, themselves represent a major threat to woodlands. These threats may be increased by interactions with the direct effects of climate change on tree function.

By the end of the century, some native tree species are likely to lose 'climate space', particularly in southern England. The southern limit of the range of many species that retain climate space will have moved close to the UK suggesting that they will struggle on many sites, their regeneration and successful establishment will decline and they could be out-competed by introduced species. The distribution of tree species will inevitably change in response to climate warming. However, if the trees are to stay within appropriate climatic envelopes there will be a requirement for species migration rates to be more than 10 times faster than those achieved in reaching present distributions after the last ice age.

Increased winter waterlogging may render increased areas of upland plantation forest liable to possibly catastrophic wind disturbance, whether or not climate change leads to a change in the wind climate of the UK.

The impacts of climate change are likely to be first seen on establishment, but widespread mortality is initially, unlikely. However, as climate change progresses, the levels of mortality of mature trees will increase as a result of direct and indirect impacts; street, hedgerow and free-standing/isolated trees will show higher levels of mortality because of their higher water demand.

There are likely to be significant changes to the composition, structure and character of woodland ground flora; current species descriptions of native woodland communities are unlikely to remain valid. The range, and ability of the majority of priority woodland species (both flora and fauna) to persist will change as a result of climate change; some will decline further, others will benefit. The current species-based approach to nature conservation will be difficult to maintain in the long term because of the likely degree of change.

Predicting the future can only be an uncertain science, particularly given the interactions between the components – both fauna and flora – of woodland ecosystems. However, the link between climate and tree growth has been well documented, from as far back as 1662 (Evelyn, 1729). The knowledge gained, mostly over the course of the 20th century, on the specific requirements of individual tree species and woodland ecosystems has enabled a range of models of tree growth and suitability to be developed.

A number of such models (see Ray et al., 2002) have been used to identify how forest productivity and ecosystem function may be influenced by projected climate change. Although it is impossible to make such predictions with certainty (and their output should be treated with caution – see Walmsley et al., 2007), these models do represent a useful tool for developing adaptation strategies.

An alternative approach is to interpret the documented impacts of extreme climatic events or past climatic trends in terms of future climate projections. Again, these provide a useful insight into the future, but such analyses should be treated with some caution because the combination of conditions in extreme climate events of the past is unlikely to be replicated in the future.

A third option for informing us of the likely impacts of climate change is to investigate the composition and performance of forest plantations and woodland ecosystems growing in climates more representative of that of the future for a given region (see Broadmeadow et al., 2005). Although 'climate analogues' for the range of climate variables that influence tree growth and ecosystem function rarely exist, the approach has merit in providing information at a broad level on change impacts that might be expected and changes in management practice that would be appropriate.

This chapter therefore considers information on the future impacts of climate change on trees and woodlands from a range of approaches. Evidence from experimental impact studies, spatial analyses linked to climatic variation, and model predictions is evaluated. This then allows a broad assessment of the likely impacts of climate change on UK forestry and woodland ecosystems to be made, and enables a commentary on the robustness of the predictions.

The likely changes in climate for the UK have been outlined in Chapter 2, which takes into account key results from the recent release of the UK Climate Projections (UKCP09: Murphy et al., 2009). However, much of the existing analysis of future climate impacts has used the UKCIP02 projections (published in 2002), and impact studies using UKCP09 are not yet available. Where appropriate, the High and Low scenarios for these earlier projections can give an indication of the uncertainty associated with such impact studies.

5.1 Direct impacts of climate change on forest productivity

5.1.1 Direct impacts of rising carbon dioxide levels

Pre-industrial concentrations of carbon dioxide are sub-optimal for tree photosynthesis (see 3.2.2, Chapter 3). Rising concentrations will therefore have a 'growth-stimulating' effect and there is a large body of supporting experimental evidence (see Broadmeadow and Randle, 2002). Most experiments have been carried out on young trees under controlled environment conditions (Curtis and Wang, 1998). An average growth enhancement for above ground biomass of 51% across 21 studies for a doubling of CO_2 from 350 to 700 parts per million volume (ppmv), corresponds with a 51% increase in maximum photosynthetic rates (ECOCRAFT, 1999: see Broadmeadow and Randle, 2002). Other aspects of tree physiology were also affected, including a small reduction in respiration rates; a larger reduction in stomatal conductance and corresponding water use; reduced leaf nitrogen (and other nutrients) content; larger leaf area index (leaf area per unit ground area); and a tendency to greater below-ground biomass allocation. In particular, the effects of CO_2 concentration on function have implications for tree responses to the changing climate. The reduction in stomatal conductance for some species will enhance water use efficiency and also reduce the damaging effect of ozone by reducing uptake (see below); the larger leaf area may increase risk of wind damage and also reduce light transmission to the forest floor affecting ground flora community structure. There may also be interactions with insect pests as a result of higher soluble carbohydrate levels in the leaf and phloem sap (Broadmeadow and Jackson, 2000; Evans et al., 2002), but also other effects through higher concentrations of plant secondary metabolites (Penuelas et al., 1997), some of which have effects on insect herbivores.

A recent meta-analysis of experimental data concluded that an average 23% increase in net primary productivity would be expected for a doubling of CO_2 from the pre-industrial concentration to 550 ppmv (Norby et al., 2005). However, for practical reasons the majority of experiments have been carried out on young trees. Four free air carbon enrichment (FACE) experiments in mature woodlands of sweetgum (Liquidambar styraciflua), loblolly pine (Pinus taeda) and aspen (Populus tremuloides) show that there is significant photosynthetic downregulation (Schafer et al., 2003; Norby and Luo, 2004; Körner et al., 2005), so that growth rate stimulation is not as large as with young trees. These FACE studies have also demonstrated increased carbon storage in the soil, although it is uncertain what the overall balance will be between CO_2-driven increases and temperature-driven decreases in soil carbon. One FACE system has been operating in the UK, investigating competition between native trees and impacts of elevated CO_2 on soil processes, but this study was also with young trees (see: www.senr.bangor.ac.uk/research/themes/ess/climate_change.php). Karnosky et al. (2007) reviewed the published results from all FACE experiments identifying an overall increase of forest productivity but with considerable variability determined by genotype, tree age, climate, air pollution and nutrient availability. The authors also identified a number of research gaps.

Modelling systems and operational decision support systems that are used to predict the impacts of climate change on forest growth, tree species distribution or woodland community composition (e.g. ESC: see Ray et al., 2002; SPECIES: see Pearson et al., 2002) are based on relationships between past/current climate and plant performance. While these models can accommodate climate change scenarios with relative ease, they do not incorporate the direct and indirect effects of rising carbon dioxide levels outlined above. It is therefore likely that they underestimate the positive effects of future climate and

atmospheric composition change and may provide an unduly negative picture of the impacts. The limitations of the modelling approaches are further explored in 5.2 and 5.3 below.

5.1.2 Impacts of ground-level ozone pollution

Models of climate change and atmospheric chemistry suggest that levels of ozone pollution are likely to increase significantly during this century (NEGTAP, 2001; Royal Society, 2008). Unlike most other pollutants, ozone affects foliage directly, and its harmful effects are primarily caused by ozone molecules entering the leaf through the stomata (see Chapter 3). There is some evidence that ozone can degrade leaf surface waxes increasing water loss. However, ozone principally affects the internal components of leaves through an acceleration of cell senescence and local cell necrosis in the photosynthetically active tissues. The physical disruption of the photosynthetic apparatus often results in reduced levels of chlorophyll, a lower photosynthetic capacity and advanced leaf senescence. The degradation of chlorophyll can be seen as generalised, diffuse 'chlorosis', i.e. yellowing of the foliage. The lower photosynthetic capacity and the continuing costs (in energy terms) of repairing cellular damage often results in reduced growth, even at current ambient exposure levels in the UK. In conifer species, advanced leaf senescence appears as reduced needle retention, with fewer age-classes of needles retained in areas experiencing ozone pollution (NEGTAP, 2001).

Changes in carbon allocation resulting from diversion to ozone damage repair also leads to reduced root biomass with the result that ozone exposure can increase the vulnerability to drought. Furthermore, ozone impairs the functioning of the stomata with the result that stomatal closure in response to drought is compromised, thus compounding any effects of water shortage. Several statistical analyses (outside the UK) have shown links between ozone exposure and forest productivity and condition supporting the findings of research in controlled environment facilities. Dose–response relationships have recently been developed for a limited number of tree species. Further research may enable a quantitative assessment of the cost of ozone pollution to be made.

Emissions control policies for volatile organic compounds (VOCs) and nitrogen oxides, the main precursors to the formation of ozone (see Chapter 3), have led to a recent reduction in peak ozone concentrations. However, against this fall in peak concentrations has been a steady rise in background concentrations. Model predictions are for this trend to continue, with background concentrations rising to approximately double the current value by the end of this century. This increase should be viewed in the context of the impacts on tree health and productivity that current ozone concentrations already have, which may be reducing the productivity of sensitive tree species in some parts of the UK, and in some years, by up to 10%. Ozone pollution may also make trees more susceptible to biotic and abiotic damage, thus compounding the direct effects on growth.

5.1.3 Direct impacts of rising temperature

Most aspects of the direct impacts of climate change (wind, rainfall patterns and rising CO_2 levels) as well as indirect impacts (through changing pest/pathogen risk, for example) are dealt with either in specific sections or as part of the overall impacts on growth and distribution assessed in 5.2.1 and 5.3 below. However, two effects of rising temperature are worthy of further consideration here.

Leafing date and frost risk
Rising temperatures would be expected to lead to a decline in spring frost risk. However, late spring frosts are unpredictable and there is little evidence that the date of the last spring frost has changed in recent decades. In contrast, the advance in leafing is well documented (see 4.2, Chapter 4). If the trend in leafing date was to continue, as seems likely on the basis of models of budburst available in the literature (Sparks and Gill, 2002), but with no or little corresponding change in the date of the last frost, the risk of late spring frost could increase (Figure 5.1). Spring frost damage is an area of concern because of its effect on form and resulting timber quality. Furthermore, repeated frost damage can reduce productivity significantly and, in extreme circumstances, result in mortality, particularly of young trees (Redfern and Hendry, 2002).

Chilling requirement
Some growth stages of some tree species have an obligate chilling requirement. These processes include bud set and the breaking of seed dormancy. The majority of tree seeds in the UK exhibit one of two sorts of dormancy. A few (alder, birch and Scots pine) have seeds that exhibit 'shallow dormancy'. In this type of dormancy a varying proportion of seeds germinate at different temperatures and all seeds respond to a relatively short pre-chill which stimulates faster germination at all temperatures and improves the maximum germination percentage at most

Figure 5.1
Impact of projected climate change (UKCIP02 low and high emissions scenarios) on (a) leafing date of oak (expressed as day of year from 1 January) and (b) proportion of years in which leafing will occur before the date of the last −2°C spring frost, assuming no change in the date of the last frost. See Sparks and Gill (2002) for details of the leafing model.

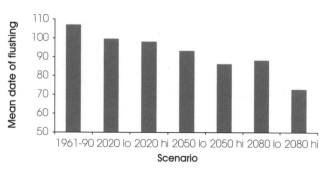

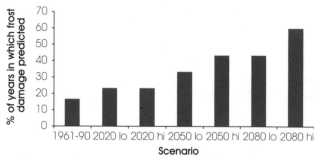

temperatures (Gosling and Broadmeadow, 2006). If climate change brings warmer autumn temperatures, there is a risk that the seeds of these shallow-dormancy species may germinate too soon in autumn and be vulnerable to frost that winter. However, the native tree species that are most likely to be affected by climate change are those with 'deeply dormant' seeds, including juniper, yew, and nearly all broadleaves (e.g. ash, beech, cherry). Freshly shed seeds of these species have a complete metabolic block to germination at any temperature and there is an absolute requirement for a relatively lengthy and unbroken period of cold moist conditions to bring about any germination at all. If climate change brings about winters that are warmer or shorter or both, and these are succeeded by faster rising spring and summer temperatures, then many of these species may not be as well suited to natural regeneration in the projected climate of the future. Variability exists within populations and natural selection will favour those individuals that obtain sufficient chilling. However, variation between populations is likely to mean that more southerly provenances that require less winter chilling may be better adapted to the climate of the future than local/native populations. This potentially has very serious implications for native woodlands and requires further investigation through controlled experimentation and field-based 'reciprocal transplant' trials.

5.2 Assessing future impacts of climate change on tree species suitability and forest productivity using Ecological Site Classification

5.2.1 Modelling approach

The Forestry Commission has developed a knowledge-based model, the Ecological Site Classification (ESC; Pyatt et al., 2001), to map the suitability of tree species to a site. Suitability in this context is defined in terms of growth relative to maximum growth rates achieved in the UK. ESC also has the capability to model the suitability of native woodland ecosystems on the basis of climatic and edaphic conditions. It is important to note in this and subsequent sections the differences between suitability for commercial timber production, the ability to persist in the medium to long term and other ecological functions.

The ESC approach uses six biophysical factors to describe tree species suitability and yield potential for application at individual site level. Two are soil factors – fertility and moisture availability; the remaining four are climatic factors including accumulated temperature (warmth index), moisture deficit (droughtiness index), wind exposure, and continentality. The ESC system has also been developed as a spatial tool for selecting tree species in design plans at the operational scale and for assessing regional suitability of species for different forest types (Ray and Broome, 2003). The regional scale spatial tool has also provided a framework to evaluate the likely impacts of climate change on tree species suitability.

When interpreting the output of the Ecological Site Classification model, particularly in the context of climate change, a number of important caveats (most of which equally apply to the SPECIES model: see 5.3 below) should be considered:

- ESC evaluations are based on mean climate, and extreme climatic events such as windstorms and unseasonal frost (see 4.2, Chapter 4) are not considered;
- Climatic and, in some cases, edaphic factors are represented at coarse resolution (5 km grid), which does

not fully represent the variability in site and micro-climate that can be exploited by trees;
- The impacts of pests and diseases are not considered;
- The beneficial effects of rising atmospheric carbon dioxide levels are not considered;
- Competition between tree species is not considered;
- ESC is an empirical model based on tree performance under British conditions. Climate change projections, particularly the more extreme scenarios, are beyond the knowledge-base of the model and extrapolation has therefore been necessary.

5.2.2 Model results

Climate projections for the UK (UKCIP02: Hulme et al., 2002) have been used to adjust accumulated temperature and moisture deficit, two of the climate factors in the ESC model (Ray et al., 2002; Broadmeadow et al., 2005) with dynamic linking between climatic moisture deficit and summer soil moisture regime. The data and ESC model have provided a low resolution, strategic planning tool to indicate the kinds of changes that may occur in the climatic factors used to specify tree species suitability and site yield potential.

By the 2080s, climate change is projected to have a profound effect on accumulated temperature and moisture deficit in Britain (Figure 5.2). UKCIP02 high emissions projections (consistent with the IPCC A1FI greenhouse gas (GHG) emissions scenario; Nakicenovic and Swart, 2000) show accumulated temperature could increase by as much as 40–50% in south Wales and central Scotland, and by as much as 60% in south east England. For a relatively cool climate, such as the baseline climate of Britain, large increases in accumulated temperature occur due to small increases in the mean daily temperature above 5°C. There are two reasons for this, warmer seasonal temperatures and a longer growing season. Moisture deficits are also likely to increase by up to 40% in eastern and southern Wales and central Scotland, and increases of up to 50% are likely in southeast England. This would cause forest soils in affected areas of Britain to become depleted of moisture for tree growth and, on freely draining and shallow soils, would be serious for species that are sensitive to drought conditions in excess of 180–200 mm (e.g. Sitka spruce, beech, ash). Indeed, on such sensitive sites, the High emissions scenario projections of frequent and extreme moisture deficit would seriously reduce the growth and suitability for most tree species currently grown in Britain. Work by Herbst et al. (2007) showed that evapotranspiration at forest edges is

Figure 5.2
Projected changes in (a) accumulated temperature above a base of 5°C and (b) calculated moisture deficit for future climate projections simulating the UKCIP02 2080s low (IPCC B1) and high (IPCC A1FI) emissions scenarios.

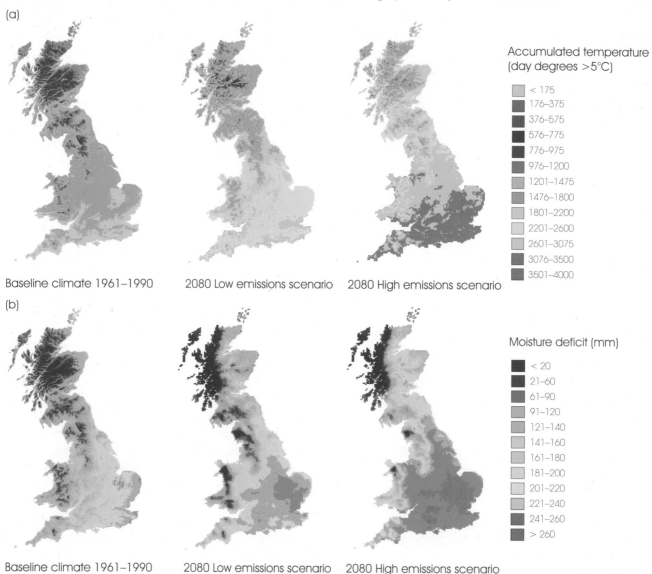

(a)

Baseline climate 1961–1990   2080 Low emissions scenario   2080 High emissions scenario

Accumulated temperature
(day degrees >5°C)

	< 175
	176–375
	376–575
	576–775
	776–975
	976–1200
	1201–1475
	1476–1800
	1801–2200
	2201–2600
	2601–3075
	3076–3500
	3501–4000

(b)

Baseline climate 1961–1990   2080 Low emissions scenario   2080 High emissions scenario

Moisture deficit (mm)

	< 20
	21–60
	61–90
	91–120
	121–140
	141–160
	161–180
	181–200
	201–220
	221–240
	241–260
	> 260

much higher than from the middle of the forest. Where the impacts of climate change are primarily related to water availability, it would be expected that small fragmented woodlands would be more susceptible than large, continuous forest areas.

Outputs from the ESC spatial model show changes in the suitability of different tree species by region. Figure 5.3 shows Britain divided into eight regions. For the area within a region described as woodland (Forestry Commission, 2003), the suitability of five major forestry tree species has been calculated for the baseline climate, and the projected climates of 2050 and 2080 simulated for both Low (IPCC B1) and High (IPCC A1FI) emissions scenarios. It is

important to note that these projections do not differentiate between where a particular species would or would not be planted; they are projected changes in average suitability for an individual species across the current total planted area, much of which in the uplands is marginal land. As a consequence, they do not predict performance of these key species on less marginal land.

Several points emerge from this analysis of predicted changes to suitability relative to the baseline climate scenario. In the case of beech under all but the 2080s High emissions scenario, the area of woodland falling into the 'very suitable' category in east and west Wales, northern England, and eastern and western Scotland is predicted to rise. In contrast, the areas designated 'suitable' for this species decline in eastern England to the extent that, under the 2080s High emissions scenario, there is virtually no area deemed 'suitable' for beech in the region. The model thus predicts the likelihood of a significant regional shift in the occurrence of beech as a productive forest species. In the case of the major commercial timber species, Sitka spruce, there are projected increases in the very suitable category in west Wales, west Scotland and northern England. A decline in suitability of this species, relative to the baseline climate scenario is seen in east Wales and western England. Under most scenarios, with the exception of the 2080s High emissions scenario, Scots pine maintains its suitability in all regions, although it is noteworthy that under this extreme scenario it is predicted to be completely unsuitable for use in western and eastern England. The pronounced decreases in the areas of existing woodland designated as suitable or very suitable for most of the five species in east and west England is attributable to the higher soil moisture deficits projected for the growing seasons in these areas under all future climate scenarios. It should be emphasised here that although the suitability of many of our tree species is likely to decline under these scenarios, (particularly on freely draining shallow soils) the expression of these effects is likely to be seen in the form of reduced timber productivity rather than complete loss of the species from our woodlands. Further, under the projected conditions of climatic stress it is likely that there will be associated biotic damage.

Wetter winters throughout Britain, and particularly in the uplands on imperfectly- and poorly-draining soil types, will cause increased areas, and longer periods, of anaerobiosis. This is very likely to reduce the rooting depth for many shallow-rooting tree species, as well as tree species that are intolerant of a fluctuating water table. The impact of a reduced rooting depth is a reduction in tree stability in forest stands. Whether or not the winter wind climate becomes increasingly stormy, with higher magnitude and/or more frequent damaging events, the reduction in stability through increased water-logging will lead to greater areas of endemic and occasionally catastrophic wind disturbance. Forest management systems must be adapted to reduce the potential for increased damage through using self thinning mixtures, short rotations and perhaps a reversion to native woodland or open habitat on the most affected forest sites.

A summary of the results for an extended range of forestry species is presented in Table 5.1, but on the basis of all

land within a region, rather than being restricted to existing woodland areas as is the case in the analysis presented in Figure 5.3. Because all land within a region is considered in the analysis, the indicative suitabilities given in Table 5.1 are low, reflecting the inherrent unsuitability of much of the land for most tree species in some regions. The important point is to consider the change relative to the baseline scenario. Suitability is presented for three contrasting regions in Great Britain (southeast England, north Wales and Perth and Argyll) to indicate the direction of change of suitability of individual species. Further details are available at www.forestresearch.gov.uk/climatechange.

The analysis presented in Figure 5.3 suggests that in most regions all five of the tree species considered – whether grown for commercial timber production or as productive components of semi-natural woodland – will be challenged by climate change. However, the broader picture presented in Table 5.1 suggests that for many species in some regions, particularly 'minor' or less commonly planted species, the consequences of climate change may be less extreme. Furthermore, for both North Wales and Perth and Argyll, the analysis suggests that an increase in productivity for many species – both conifers and broadleaves – is likely as a result of climate change. Indeed, in Perth and Argyll, the suitability of 20 out of the 28 species assessed is predicted to increase under the 2080s High emissions scenario relative to the baseline. In contrast, the implications for southern England are of real concern and suggest that a different approach to silviculture will be required, including the use of alternative species. Here, the suitability of all but two species (Norway maple and sweet chestnut) is predicted to decline under the 2080s High emissions scenario and of the conifer species, Corsican pine is the only one predicted not to be unsuitable under this more extreme scenario.

Although there will clearly be challenges for British forestry, there will also be real opportunities for the forestry sector to exploit:

- increases in productivity for some existing commercial species in the north and west;
- changes in the range of existing commercial forestry species;
- maintenance of productivity of major commercial species conferring competitive advantage in comparison with many other areas of Europe;
- opportunities to plant new species and provenances.

Figure 5.3
Suitability of five tree species, described as the proportion of the current woodland area that is unsuitable, suitable or very suitable for that species in eight regions of Britain (see map on page 75), and simulated for the baseline climate, and the future climate scenarios 2050 Low (50 l), 2050 High (50 h), 2080 Low (80 l) and 2080 High (80 h) scenarios of UKCIP02 (Hulme *et al.*, 2002).

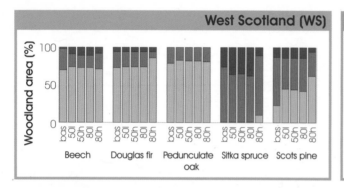

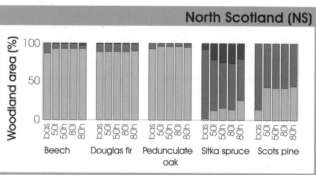

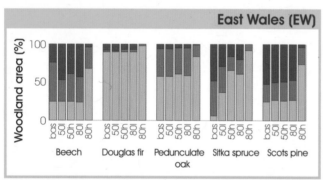

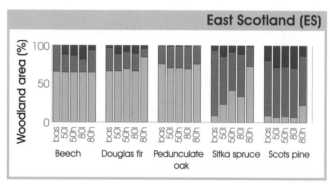

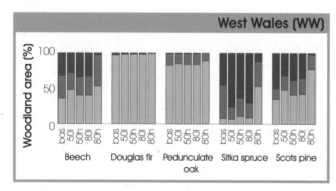

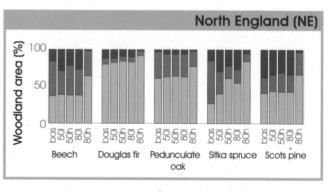

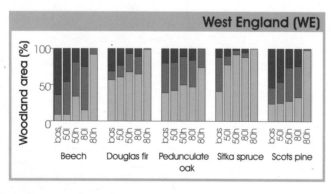

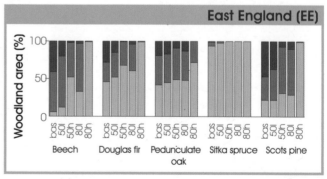

■ Very suitable ■ Suitable ■ Unsuitable

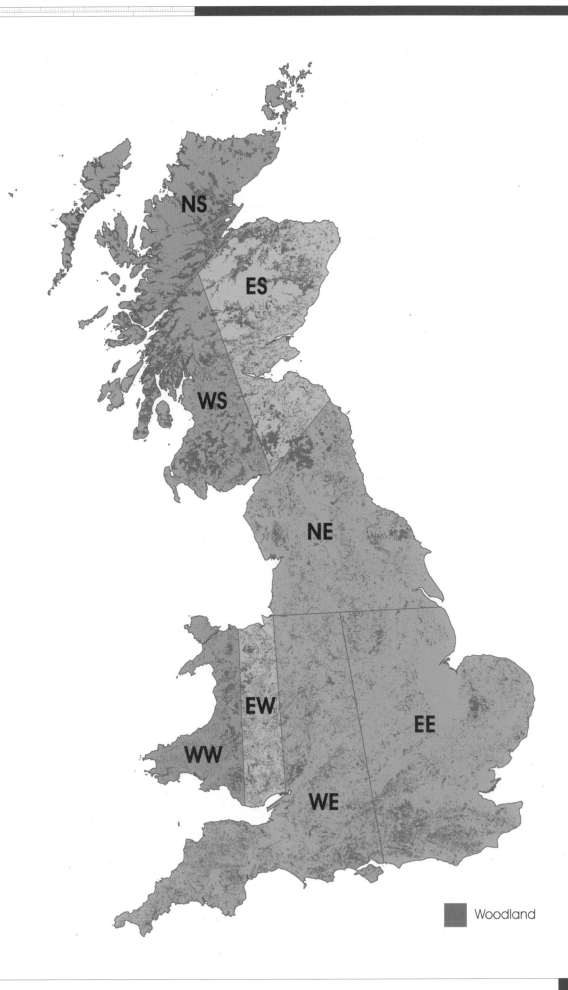

NS

ES

WS

NE

EW

WW

WE

EE

Woodland

Table 5.1
Suitability as predicted by Ecological Site Classification (ESC) for a range of tree species in southeast England, north Wales and Perth and Argyll under four climate change scenarios. See text for guidance on interpretation of results. Values show productivity (YC) relative to the maximum currently achieveable in the UK and is the average for all land within the region.

Key: ■ Very suitable (>70%)   ■ marginal (40–50%)   ■ unsuitable (less than 30%)   ■ Suitable (50–70%)   ■ poor (30–40%)

	Suitability											
	Southeast England				North Wales				Perth and Argyll			
	Baseline	2050 Low	2050 High	2080 High	Baseline	2050 Low	2050 High	2080 High	Baseline	2050 Low	2050 High	2080 High
Broadleaves												
Alder (*Alnus glutinosa*)	0.56	0.62	0.28	0.25	0.44	0.51	0.3	0.24	0.31	0.32	0.23	0.25
Ash (*Fraxinus excelsior*)	0.61	0.64	0.52	0.38	0.4	0.44	0.41	0.46	0.23	0.23	0.22	0.32
Aspen (*Populus tremula*)	0.65	0.69	0.45	0.38	0.55	0.61	0.48	0.45	0.36	0.38	0.33	0.42
Beech (*Fagus sylvatica*)	0.67	0.64	0.46	0.15	0.54	0.57	0.59	0.53	0.32	0.34	0.38	0.51
Downy birch (*Betula pubescens*)	0.39	0.38	0.16	0	0.55	0.6	0.34	0.24	0.46	0.46	0.28	0.3
Norway maple (*Acer platanoides*)	0.64	0.71	0.63	0.66	0.43	0.46	0.45	0.53	0.27	0.27	0.28	0.39
Pedunculate oak (*Quercus robur*)	0.63	0.68	0.56	0.37	0.36	0.42	0.41	0.41	0.21	0.23	0.23	0.33
Poplar (*Populus* spp.)	0.66	0.63	0.45	0.02	0.39	0.46	0.41	0.28	0.18	0.24	0.24	0.32
Rauli (*Nothofagus procera*)	0.38	0.46	0.23	0	0.25	0.34	0.29	0.28	0.09	0.11	0.11	0.17
Roblé beech (*Nothofagus obliqua*)	0.6	0.66	0.35	0.21	0.35	0.41	0.3	0.25	0.17	0.2	0.16	0.23
Silver birch (*Betula pendula*)	0.56	0.63	0.46	0.41	0.44	0.49	0.42	0.46	0.3	0.3	0.27	0.39
Sessile oak (*Quercus petraea*)	0.57	0.59	0.44	0.27	0.45	0.5	0.44	0.41	0.24	0.26	0.27	0.38
Sweet chestnut (*Castanea sativa*)	0.48	0.61	0.57	0.67	0.22	0.31	0.32	0.45	0.09	0.13	0.15	0.26
Sycamore (*Acer pseudoplatanus*)	0.65	0.53	0.35	0.16	0.52	0.54	0.46	0.4	0.31	0.33	0.3	0.38
Wild cherry (*Prunus avium*)	0.75	0.72	0.63	0.25	0.47	0.49	0.48	0.45	0.28	0.28	0.3	0.4
Wych elm (*Ulmus glabra*)	0.61	0.44	0.21	0.00	0.51	0.50	0.43	0.23	0.28	0.29	0.28	0.30
Conifers												
Corsican pine (*Pinus nigra*)	0.73	0.78	0.9	0.6	0.48	0.47	0.6	0.75	0.24	0.28	0.36	0.57
Douglas fir (*Pseudotsuga menziesii*)	0.67	0.65	0.56	0.28	0.48	0.51	0.56	0.5	0.22	0.22	0.34	0.44
European larch (*Larix decidua*)	0.54	0.52	0.27	0	0.49	0.48	0.42	0.35	0.28	0.26	0.25	0.34
Grand fir (*Abies grandis*)	0.30	0.24	0.01	0.00	0.36	0.41	0.29	0.18	0.21	0.22	0.16	0.19
Japanese larch (*Larix kaempferi*)	0.24	0.08	0.08	0	0.59	0.57	0.57	0.2	0.42	0.4	0.4	0.29
Lodgepole pine (*Pinus contorta*)	0.69	0.62	0.55	0.07	0.73	0.67	0.68	0.57	0.56	0.54	0.54	0.67
Noble fir (*Abies procera*)	0.59	0.53	0.25	0.05	0.64	0.65	0.37	0.14	0.39	0.40	0.29	0.28
Norway spruce (*Picea abies*)	0.59	0.52	0.26	0	0.63	0.62	0.54	0.39	0.4	0.4	0.37	0.45
Western red cedar (*Thuja plicata*)	0.41	0.29	0.10	0.00	0.55	0.54	0.42	0.22	0.39	0.39	0.34	0.32
Scots pine (*Pinus sylvestris*)	0.65	0.57	0.49	0.03	0.62	0.57	0.6	0.47	0.43	0.41	0.45	0.58
Sitka spruce (*Picea sitchensis*)	0.36	0.32	0.14	0	0.57	0.59	0.48	0.34	0.46	0.46	0.39	0.43
Western hemlock (*Tsuga heterophylla*)	0.51	0.41	0.32	0.01	0.54	0.55	0.18	0.29	0.39	0.4	0.38	0.38

5.3 Projected changes in tree species distribution

5.3.1 Modelling approach

The availability of species distribution maps and climate and soil data across Europe has allowed the development of models of potential species distribution (modelled on suitable 'climate space' – the geographical area that a given species could occupy on the basis of its climatic requirements alone) that can be used to examine the effects of projected climate change. Foremost among these is the SPECIES model (SPatial Estimator of the Climate Impacts on the Envelope of Species; Pearson *et al.*, 2002). Soil water availability, growing degree-day and temperature indices are used to define current climate space for individual species, with statistical comparison with distribution data enabling model performance to be assessed. Figure 5.4 presents the outputs from this model using climate projections from the Hadley Centre HADCM3 climate model assuming the IPCC SRES A2 GHG emissions profile (equivalent to the Medium–High emissions scenario of UKCIP02). Further details are available at: www.branchproject.org.uk and in Berry *et al.* (2007a). Although the maps presented in Figure 5.4 depict changes in climate space at a 10' resolution, this level of detail is considered by many as inappropriate for interpretation of necessary responses to meet conservation or forestry objectives because of uncertainties in climate and biological responses, as outlined in Walmsley *et al.* (2007). A number of important caveats should also be considered in interpreting the results. The key caveat is that such maps represent projections of where future climate space may be located for each species. They do not attempt to simulate the future distribution of species in response to climate change, and they do not take account of a species' capacity to disperse, or the presence of suitable habitat. Other caveats are identified in Walmsley *et al.* (2007). There are some additional points that also apply for the maps' interpretation for forestry:

- The maps assume that the genotypes (i.e. provenances) of individual species currently present in the UK are as well adapted to the climate of the future as genotypes from more southerly regions.
- Current species distributions are determined by reproductive capacity and/or dispersal. This may underestimate the climatic and/or geographic range over which they can successfully be grown. Beech is a good example, with its commercially viable range extending well beyond its 'ecological distribution' of southern England well into Scotland.
- The current (and therefore projected future) distribution may be the result of competition. Forest management, for example single species stands, may enable a species to endure beyond its ecological climate envelope.

5.3.2 Model results

Putting the caveats outlined above aside, projected changes in species envelopes provide a powerful picture of how species' suitability might change. These should not be used as the sole basis on which to base forest plans and the fine detail of the maps should not be explored. However, as presented in Figure 5.4, they provide a broad assessment of those species that should continue to be well matched to the UK's climate towards the end of the century, those that are likely to be towards their range limits and those that are likely to struggle as a result of climate change.

The results presented in Figure 5.4 provide a 'snapshot' of possible future patterns of species distribution at a single timepoint (the 2080s) based on a single global climate model (GCM: HADCM3) and GHG emissions scenario (IPCC SRES A2). However, as demonstrated in Figure 5.5 for *Quercus robur*, one of the species examined in Figure 5.4, a range of different outcomes can be obtained for different timepoints (the 2020s, 2050s and 2080s), utilising the output of different GCMs (PCA and HADCM3) and assuming different GHG emissions scenarios (IPCC SRES A2 and B1). The significant losses in climate space for the species in the 2080s predicted from the HADCM3 GCM output and IPCC SRES A2 GHG emissions scenario (see Figure 5.4) are not apparent if the output from the PCA GCM is used (assuming the IPCC SRES A2 emissions scenario). The predicted loss of climate space is also much reduced when the IPCC SRES B1 GHG emissions scenario is assumed for the HADCM3 GCM. The implications that arise from uncertainty in GHG emissions scenarios and Global Climate Model output must be recognised and considered alongside the uncertainties inherent in the ecological models.

Of the main species grown for timber production, beech shows the greatest benefit from climate change, with its distribution extending in the north, and with minimal contraction of its range in southern England. However, this outlook should be tempered by the well-documented impact of drought on the species (Peterken

Figure 5.4
Projected changes in climate space for 24 native tree species according to the SPECIES model (Harrison *et al.*, 2001) for the 2080s. Projections assume the IPCC SRES A2 scenario (equivalent to UKCIP02 Medium–High scenario) and are based upon the output from the HADCM3 global climate model. Red depicts loss of climate space, blue gain of climate space and green, the maintenance of climate space. See Berry *et al.* (2007a) for further details.

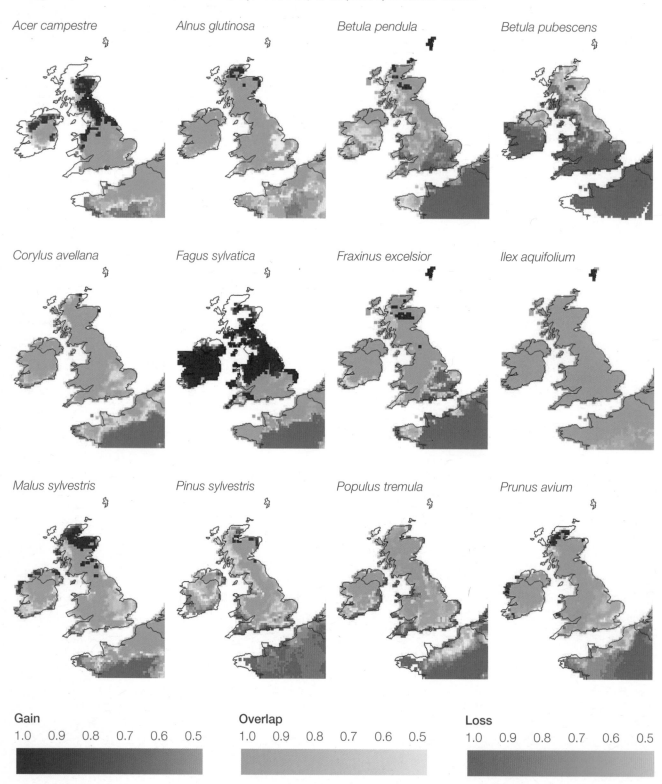

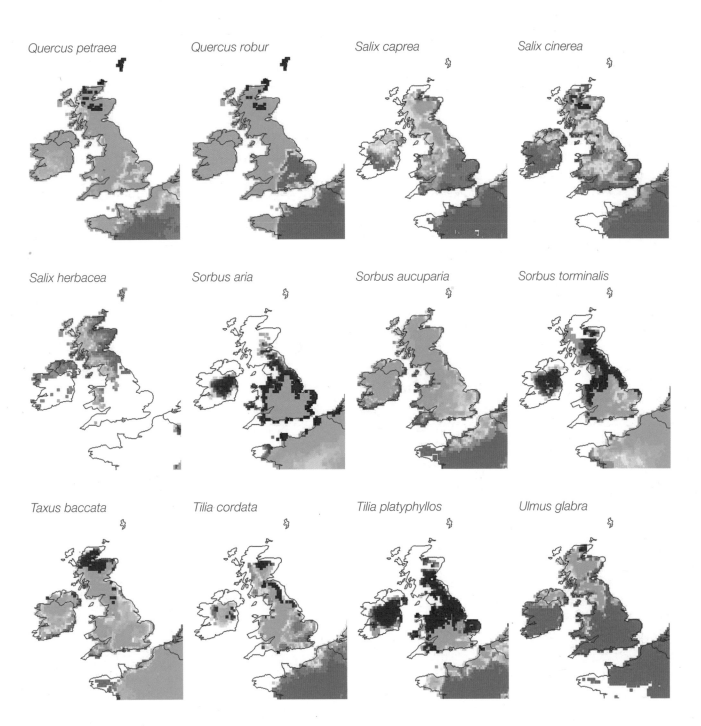

and Mountford, 1996) that has been shown to be linked to soil type (Wilson *et al.*, 2008). It is therefore likely that as a commercially viable species, beech will be limited to good soils in southern and eastern England, but with real promise further north. This northward expansion of its range may have implications for nature conservation policy and management and further consideration of its management outside its 'historical native range' is required, as outlined by Kirby (2009). This future distribution also highlights the point that, in contrast with many sensationalist popular articles, beech is highly

unlikely to disappear from the landscape of southern England. The two native species of oak both show a contraction in their range in south and east England, pedunculate oak (*Quercus robur*) more so than sessile oak (*Q. petraea*). It is noteworthy that the relative changes in *range* of the two species predicted by SPECIES differs from the relative changes in *suitability* projected using ESC, based on their current performance (Table 5.1). The climatic range of ash (*Fraxinus excelsior*) is also projected to move outside parts of southern England, presumably because of limited water availability. However this could

Figure 5.5
Comparison of projected changes of climate space for *Quercus robur* for different timeframes, global climate models
(HADCM3 and PCA) and GHG emissions scenarios (IPCC SRES B1 and A2). Key as for Figure 5.4.

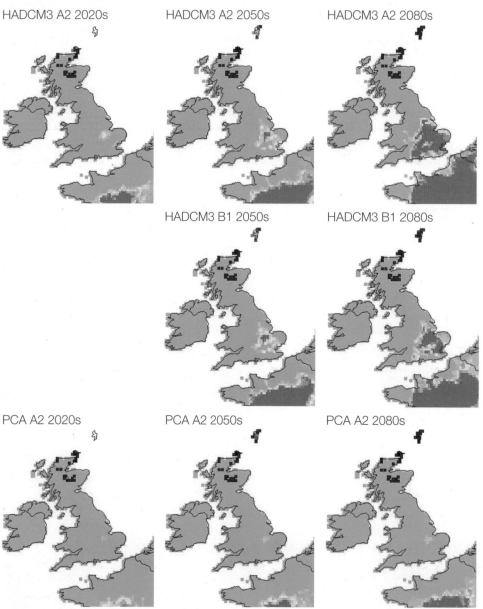

HADCM3 A2 2020s   HADCM3 A2 2050s   HADCM3 A2 2080s

HADCM3 B1 2050s   HADCM3 B1 2080s

PCA A2 2020s   PCA A2 2050s   PCA A2 2080s

be complicated by the European distribution of ash being
limited by competition with other species (principally
F. angustifolia), rather than climate; silver birch (*Betula
pendula*) shows a similar but greater decline in range,
which is not surprising given its well documented sensitivity
to moisture deficit (Peterken and Mountford, 1996). An
important point to note is that most of the 'minor' native
species of British woodland maintain climate space in
these model projections, with the exception of relict boreal
species such as dwarf willow (*Salix herbacea*), downy birch
(*Betula pubescens*) and wych or Scots elm (*Ulmus glabra*).
Apart from beech, the other potential timber species to
gain climate space is European lime (*Tilia platyphyllos*),

in contrast with small-leafed lime (*Tilia cordata*) for which
climate space contracts in southern England.

An additional consideration for nature conservation policy
and the development of adaptation strategies is whether
species will be able to disperse and move through the
landscape as their climate space moves – assuming
that suitable habitat is available. For tree species, this
is highly unlikely as the typical change in climate space
depicted in Figure 5.4 will require that species move at
a rate of 2–5 km per annum, in line with the 'anecdotal
rule of thumb' of a one degree temperature difference
being equivalent to a 100 km shift in latitude. This rate of

migration is well in excess of the 500–1000 m per year that tree species moved after the last ice age to reach present distributions (Huntley and Birks, 1983) a rate that is now considered an over-estimate (Birks and Willis, 2008). For this reason, Aitken et al. (2008) suggest that some tree species, particularly those with fragmented ranges, small populations, low fecundity or suffering decline, should be candidates for facilitated migration.

The overall conclusion from this modelling work is that the majority of native species maintain climate space with the possible exception of in coastal areas where moisture deficits are exacerbated by high wind speeds. The HADCM3 A2 scenario includes relatively severe summer moisture deficits and the projections based on this model are for a number of native species to lose climate space. Climate models with less extreme water deficits than those assumed in the HADCM3 A2 scenario, but which are consistent with UKCP09, indicate that the majority of native species will retain existing climate space in the UK, although these analyses suggest that the edge of the climate envelope for most of those species will move to northern France (data not shown, see: www.branchproject. org.uk). However, most native species in southern England will be at the edge of their climate space by the end of the century and will therefore be challenged. They will not disappear from the landscape, even in southern England where the impacts of climate change are likely to be most severe. The model outputs also confirm the risk to timber production outlined in the ESC analysis.

5.4 Impacts of climate change on woodland habitats and priority species identified in the UK Biodiversity Action Plan

The UK Biodiversity Action Plan (UKBAP) was published in 1994 (Department of the Environment, 1994) as a response to the United Nations Convention on Biological Diversity (United Nations, 1992). It identifies and prioritises actions to protect threatened, and nationally important species and habitats. Of the 400 species listed in the original Plan, around 130 were associated with woodlands, often as their main habitat. The list has now been extended to cover 1150 species and 65 habitats. Of the 65 priority habitats which have UK-wide plans, eight are for native woodlands.

Mitchell et al. (2007) assessed the risk of future climate

change to woodland habitats in England, concluding that summer drought rather than temperature represented the greatest risk to broadleaved mixed and yew woodland, particularly in southeast England. However, change in species composition rather than loss of woodland would be the main impact. Rising sea level was unlikely to represent a major threat, with less than 1% of woodland and only 154 ha of ancient and semi-natural woodland falling within the Environment Agency's tidal flood risk map. An earlier evaluation by Hossell et al. (2000) rated the vulnerability of this broad habitat as low–medium, although Mitchell et al. note that revised climate change scenarios (UKCIP02: Hulme et al., 2002) published in 2002 probably mean that Hossell et al. underestimated the risk. However, the most recent climate projections (UKCP09: Murphy et al., 2009) suggest a less extreme reduction in summer rainfall in southeast England, so the earlier evaluation of Hossell et al. may be appropriate. This changing risk assessment over time highlights the difficulty in making firm predictions on the future of woodland and, particularly, in developing appropriate adaptation strategies. For the coniferous woodland habitat, Mitchell et al. (2007) concluded that although the suitability of different species for different sites would be affected by climate change, the persistence of the habitat would be dependent on management decisions rather than climatic conditions. The potential impacts of climate change during this century on priority woodland habitats in England, Scotland and Wales are summarised below. Further details are provided by Mitchell et al. (2007) and Ray (2008a,b).

5.4.1 Lowland mixed deciduous woodland

This priority habitat covers a broad range of species and woodland types on heavier soils. As identified above, the majority of native tree species will persist, although there will clearly be differential effects of climate change resulting in changing competitive advantage and a shift in species composition. In particular, sycamore and beech may tend to increase at the expense of oak and ash (Ray, 2008b). Bramble, nettle and other species comprising 'rank' vegetation may become more dominant on the heavier, rich soils that are characteristic of the habitat, at the expense of forbs and grasses. Although it is impossible to be precise over its future, lowland mixed deciduous woodland is likely to continue as a functional system across the UK, even if the tree species and ground flora communities within them may change. The greatest threat to the habitat, in common with others, probably comes from the potential impacts of pest and disease outbreaks, particularly where species diversity is limited.

5.4.2 Lowland beech and yew woodland

The impacts of drought on beech are well documented (e.g. Peterken and Mountford, 1996) and a number of studies (e.g. Harrison et al., 2001; Berry et al., 2002a b; Broadmeadow and Ray, 2005; Berry and Paterson, 2009) have suggested that the species will struggle for climate space, or as a commercially viable species. However, the natural distribution of the species coupled to more recent modelling results suggest that the species will persist across southern England. Soil type will be a key determinant of which sites it continues to thrive on (Wilson et al., 2008), as will other site conditions including aspect and topography. Roberts and Rosier (2006) highlight the ability of beech to access water at greater depth on some chalk soils as a result of the nature of the chalk matrix. Stribley and Ashmore (2002) also note the interaction between ozone levels and climate on the condition of beech. Harrison et al. (2001) and more recent work (Berry et al., 2007b) indicates that yew is unlikely to lose climate space. Although the individual species are likely to persist as climate change progresses, beech is likely to be challenged on some sites, particularly those with free-draining thin soils. This is likely to allow ash and, in time, oak to colonise and the nature of the habitat is likely to change, and will probably also extend beyond its native range (Kirby, 2009). There is already an increased acceptance of the species beyond its native range (Wesche, 2003; Kirby, 2009), although continued expansion has the potential to impact negatively on other priority habitats, particularly upland oakwoods, if it is not appropriately managed on those sites.

5.4.3 Lowland wood pasture and parkland

The persistence of the habitat will be determined by past and future management decisions, including species choice, and this is reflected in Hossell et al.'s assessment of vulnerability as 'low', as compared to medium for other priority habitats in England. Broadmeadow (2000) also noted that the impacts of climate change are likely to become evident first on young trees and isolated trees because of their higher water requirement.

5.4.4 Upland mixed ashwoods

There is little evidence to suggest that upland ashwoods will be negatively impacted by climate change with the main species (ash, hazel, oak, birch) highly likely to persist in the wetter north and west where the habitat occurs. However, Mitchell et al. (2007) note that species composition may change, suggesting that the northward movement of small-leafed lime may be a possibility. Ray (2008a) highlights the particular importance of the habitat in Wales (25% of the semi-natural woodland area) and concludes that the habitat, in both Wales and Scotland, may show slower changes in species composition than other habitats because of the ability of ash to regenerate in dense shade.

5.4.5 Upland oakwood

The distribution of tree species in upland oakwoods is unlikely to be affected by projected climate change this century. However, the habitat, which is restricted to high rainfall areas, is highly valued for its fern, bryophyte and lichen flora which are sensitive to changes in rainfall and humidity. ESC NVC analysis suggests that W11 would replace W10 as the most suitable of the oak dominated woodland types in the less wet upland oakwood range (Broadmeadow and Ray, 2005). Of particular note is the very restricted range of W11 under the 2050s high emissions scenario, particularly in Wales, confirming concerns over the possible impact on the community composition of upland oakwoods. Ray (2008a) also notes the possibility that more frequent disturbance events coupled to the changing climate may allow other species such as birch, hazel and rowan to colonise Atlantic oakwoods. Of particular concern in this context is the potential for beech to colonise upland oakwoods with the deeper shade cast by the species impacting negatively on the ground and epiphytic flora.

5.4.6 Wet woodland

Mitchell et al. (2007) suggest that the persistence of wet woodland will depend on local factors rather than regional climate change. Ray (2008a,b) highlighted the role that wet and riparian woodland might play in flood risk management and the maintenance of freshwater temperature to protect fisheries. The extent of these communities may therefore be increased as a management response to climate change. In southern England, the increasing frequency and severity of summer drought represents a threat to these communities, but such threats may be countered by increased winter rainfall maintaining ground water supplies through the early part of the summer. If individual sites do become drier, ash may colonise at the expense of alder (Alnus glutinosa), which is currently the dominant species of many sites. Ray (2008a,b) suggests that water supply is unlikely to limit the habitat in Scotland and Wales, and that wet woodland might expand into upland mires and flushes as a result of changes in climate and land management. The greater fluctuations in water levels that would be expected to result from climate change could enhance the risk of significant dieback of alder through *Phytophthora* infection (Lonsdale and Gibbs, 2002).

5.4.7 Native pinewoods

The persistence of the principal tree species, Scots pine, is unlikely to be affected by climate change. However, the composition of the ground flora is likely to be affected with plant communities associated with the drier sub-communities of the west and central Highlands favoured (Ray, 2008a). There may also be colonisation by other tree and shrub species (e.g. oak, birch, rowan) and the appearance of other flora not generally associated with native pinewoods. Disturbance through fire may represent an enhanced risk, particularly in woodlands used extensively for recreation. Where grazing levels permit, there may be a colonisation of scrub (juniper and montane willows), and ultimately pine, above the current tree-line. This is consistent with the average rise of 80 m (maximum 200 m) observed in the tree-line across sites in Sweden over the past century (Kullman and Oberg, 2009) and more widely (Grace *et al.*, 2002).

5.4.8 Upland birchwoods

The persistence of birch in upland birchwood habitats is unlikely to be challenged by climate change as currently projected. However, Ray (2008a) noted the recent planting and limited tree species diversity that might render these woodlands vulnerable to pest/disease impacts or the effects of extreme climatic events. On drier sites, silver birch may become more competitive at the expense of downy birch.

5.4.9 Priority species

Of the 1150 priority species identified in the UK Biodiversity Plan, 243 have been classified as woodland species, although not confined solely to woodlands. The impacts of projected climate change (using the HADCM3 model and IPCC A2 GHG emissions scenario) on the distribution of a number of woodland priority species (for England) have been modelled using the SPECIES model (see 5.3 above). The same caveats outlined in 5.2.1 apply to the changes in distribution given in Table 5.2, as well as the fact that many of these species had poor European distribution data for model training and thus the projections are subject to a high level of uncertainty.

From the limited range of priority species modelled, higher plants and lichens show a large loss of climate space in contrast to most invertebrate species which show a large expansion in climate range across England, Scotland and Wales. The terrestrial vertebrates show a range of responses, varying from the current very limited distribution and simulated climate space projected to expand to cover the majority of Great Britain in the case of the barbastelle bat, to an almost complete loss of climate space for the black grouse. Very few of the species show a small change to their climate space (and modelled potential distribution) as a result of climate change.

If climate change progresses as projections suggest and if these changes affect priority species distributions as models of climate space indicate, nature conservation policy will need to be revisited as a matter of urgency. Many of the species given as examples in Table 5.2 will no longer be at the edge of their climatic range and pressures on existing populations would be expected to decline. In contrast, other priority species (e.g. the lichen *Biatoridium monasteriense*) will lose all 'climate space' as a result of climate change and efforts to maintain the species on all but a handful of sites are likely to fail. The fact that climate space (and inferred distribution) in the UK is likely to be unaffected for very few species and that the scale of these impacts will vary with time, suggests that a habitat-focussed approach to nature conservation will better address climate change than a species-based approach.

5.5 Impact of changing frequency of wind storms

There is high uncertainty in the wind speed element of future climate change scenarios, with different global climate models showing disparity in both the magnitude and direction of change (see Chapter 2). The methodology behind the recently published UKCP09 projections does not allow projections of changes in wind speed (Murphy *et al.*, 2009). However, the impact on windthrow risk of the changes in mean climate as depicted in the UKCIP02 climate change scenarios (Hulme *et al.*, 2002) have been assessed (Ray, pers. comm.). The projected changes in winter wind speed (up to ~10% increase) resulted in minimal change to the DAMS (Detailed Aspect Method of Scoring) wind hazard classification system developed by the Forestry Commission for UK conditions. However, Quine and Gardiner (2002) outline the importance of the distribution of wind speeds, concluding that a small change in mean wind speed could result in a very large increase

Table 5.2
Projected change in climate space for English woodland priority species as modelled by the SPECIES model (Berry *et al.*, 2007a) using the HADCM3 climate model and the IPCC SRES A2 GHG emissions scenario.

Species	Common name	Projected change in distribution		
		England	Scotland	Wales
Vascular plants				
Arabis glabra	Tower mustard	– –	(L) – – –	(A) +
Lichens, fungi and bryophytes				
Schismatomma graphidioides	Lichen	– –	+	+
Biatoridium monasteriense	Lichen	– – –	– – –	– – –
Catapyrenium psoromoides	Tree Catapyrenium (lichen)	– – –		
Terrestrial invertebrates				
Andrena ferox	Mining bee	–	–	
Lucanus cervus	Stag beetle	+	+ + +	+ + +
Limoniscus violaceus	Violet click beetle	+	+ +	+ + +
Argynnis adippe	High brown fritillary (butterfly)	+ +	(A) + + +	(A) + + +
Boloria euphrosyne	Pearl-bordered fritillary (butterfly)	–	– –	
Erynnis tages	Dingy skipper (butterfly)	+	+	+ +
Leptidea sinapis	Wood white (butterfly)	+ + +	+ + +	+ + +
Melitaea athalia	Heath fritillary (butterfly)	+ + +		+
Terrestrial vertebrates				
Triturus cristatus	Great crested newt (amphibian)	– – –	– –	–
Caprimulgus europaeus	Nightjar (bird)	(L) + + +	(L) +	(L) + +
Emberiza schoeniclus	Reed bunting (bird)	– – –		– –
Lullula arborea	Wood lark (bird)	+ + +	+	+ + +
Muscicapa striata	Spotted flycatcher (bird)	–		– –
Passer montanus	Tree sparrow (bird)	+	–	+
Pyrrhula pyrrhula	Bullfinch (bird)	– – –		–
Streptopelia turtur	Turtle dove (bird)	+ +	+	+ +
Tetrao tetrix	Black grouse (bird)	– – –	– – –	– – –
Turdus philomelos	Song thrush (bird)	– – –		–
Barbastella barbastellus	Barbastelle bat (mammal)	+ +	(A) + +	(A) + + +
Lepus europaeus	Brown hare (mammal)	– –	+ – –	
Muscardinus avellanarius	Dormouse (mammal)	– – +		–
Myotis bechsteinii	Bechstein's bat (mammal)	– + +	+	+
Rhinolophus hipposideros	Lesser horseshoe bat (mammal)	+ + +	+ + +	+ +
Sciurus vulgaris	Red squirrel (mammal)	– –		+

Key: (A) currently absent; (L) currently very localised distribution; –/– –/– – – contraction in climate space; +/+ +/+ + + expansion in climate space.

in wind risk if the frequency or speed of extreme gusts increases. Changes in wind direction that were not included in the UKCIP02 scenarios could also result in changed wind risk; particularly if the storm track across the UK became more southerly, as suggested by Hulme et al. (2002). Improved data availability including global climate model (GCM) pressure field outputs may allow the risk associated with a change in wind direction to be further evaluated.

Climate change may also lead to an indirect effect on wind risk, even if the wind climate itself remains unchanged. The projected increase in winter rainfall is likely to result in more frequent and long-lasting periods of water-logging, which may reduce tree stability. The increase in leaf area that has been reported in many experiments indicate that increased carbon dioxide levels could also enhance the risk of wind damage, as would an extension to the duration of the leafed period for deciduous trees (Broadmeadow and Ray, 2005). Finally, changes to root:shoot allocation, structural properties of wood or tree form/taper could also affect wind risk. It could therefore be argued that amending the ForestGales decision support system to reflect the potential indirect effects of climate change is more of a priority than further evaluation of highly uncertain changes in wind speed. However, if the output from different GCMs converge to give greater confidence in wind speed and storm track projections, a re-evaluation of the impacts of changes in windstorms would be a high priority, given the potentially catastrophic impacts exemplified by recent major storms in the UK and Europe. When new information on a wind speed projections becomes available, it will be important to evaluate vulnerability of woodlands and risks to infrastructure from falling trees. In both cases, contingency planning should be examined.

5.6 Impacts of pests and diseases under future climate change scenarios

Predicting changes to the impact of specific insect pests and tree diseases on woodlands is difficult because of the fine balance between pest/pathogen, host tree and natural enemies. However, it is possible to make two generalisations – stressed trees are more susceptible to insect pests and diseases, and the majority of insect pests that currently affect UK forestry are likely to benefit from climate change as a result of increased activity and reduced winter mortality (Straw, 1995). The impact of facultative pathogens may worsen, while some insect pests that are present at low levels, or currently not considered important, may become more prevalent. Examples of the latter include defoliating moths and bark beetles. In addition, the 'effective' range of existing pests or pathogens may change, including a northwards expansion of those with a southern distribution and the likely appearance of some from continental Europe. Firm predictions cannot be made, although expert judgement of forest pathologists and entomologists allows some assessment to be made of changes in the prevalence of certain diseases and insect pests, based on their current distribution and associated climatic conditions, known biology and epidemiology. Considerable caution should be exercised in extrapolating this analysis to a future climate. For some pests and diseases, likely trends cannot be predicted even on the basis of expert judgement; in this category, and of particular concern, is *Phytophthora ramorum*, the agent responsible for sudden oak death (see 4.1.6, Chapter 4). The higher level of uncertainty associated with the biology of fungi compared to insect pests is reflected in the less specific predictions of future trends in the incidence of fungal diseases and disorders.

5.6.1 Forest insect pests

Climate change will influence the distribution, abundance and performance of forest insect pests, with their impact dependent, in part, on their feeding habits, life cycle characteristics and the relationship between individual climatic variables and population dynamics (Masters et al., 1998). As a general guide, rising temperature will have its greatest impact on the development rate of insect populations, leading to faster insect development, a larger number of generations (voltinism) in a year, as well as range extensions. Insect groups most likely to be affected are multi-voltine aphids, semi-voltine bark beetles, sawflies, weevils and wood-boring lepidoptera. Apart from being linked to drought stress, thereby making trees more vulnerable to the effects of insect pest outbreaks, changes in precipitation patterns are likely to be associated with changes in resin flow of conifers and the nutritional quality of foliage (Wainhouse, 2005). Insect groups most likely to be affected include bark beetles, aphids and longhorn and buprestid beetles. More general forest damage resulting from windstorms or forest fires is also likely to promote outbreaks of bark and ambrosia beetles (Långström, 1984; Fernandez, 2006). A more detailed risk assessment of possible interactions between climate change and insect pest outbreaks is given in Table 5.3 (after Wainhouse, 2008), with a commentary on individual insect groups given below, based on the review of Wainhouse (2008).

Table 5.3
Risk of increased damage by UK forest insect pests as a result of projected climate change. Risk is rated as low, moderate, high or very high, based on an assessment of life history characteristics, population dynamic, historical patterns of damage and likely changes in planting. Where species have a wide host range, risk is assessed for the main host only.

Insect pest	Host	Risk	Main risk factors	Geographical area
Bark beetles, weevils and related species				
Spruce bark beetle (*Dendroctonus micans*)	Spruce	High	Range extension, reduced generation time, drought stress of host trees	Northern Britain
	Pine	Low		
Pine weevil beetle (*Hylobius abietis*)	Spruce	High	Reduced generation time, forest management	Large forest areas managed by clearfell and re-planting throughout the UK
	Pine	High		
Pine shoot beetle (*Tomicus piniperda*)	Pine	High	Windblow, sister broods, heat stress through drought or defoliation	Southern Britain
Larch bark beetle (*Ips cembrae*)	Larch	Moderate	Windblow, sister broods, heat stress, range extension, defoliation	Northwest Scotland, northern England
Oak pinhole borer (*Platypus cylindrus*)	Oak	High	Link to oak decline, reduced generation time, increased importance of oak	Southern Britain
Oak buprestid (*Agrilus pannonicus*)	Oak	High	Link to oak decline, reduced generation time, increased importance of oak	Southern Britain
Aphids and scale insects				
Green spruce aphid (*Elatobium abietinum*)	Spruce	Very high	Reduced generation time, increased winter survival, drought stress of host trees	Most spruce growing areas
Beech scale (*Cryptococcus fagisuga*)	Beech	Low	Increase in young plantations	
Defoliators				
Pine looper moth (*Bupalus piniaria*)	Pine	Moderate	Low rainfall sites	Most pine growing areas
Pine beauty moth (*Panolis flammea*)	Pine	Low	Lodgepole pine	Northern Britain
Winter moth (*Operophtera brumata*)	Spruce	Moderate	Phenological synchrony on oak. Host range extension	
	Oak	Moderate		
Oak processionary moth (*Thaumetopoea processionea*)	Oak	High	Range extension	Southern Britain
Gypsy moth (*Lymantria dispar*)	Oak	Low	Range extension	Southern Britain
Larch budmoth (*Zeiraphera diniana*)	Spruce	Low	Phenological synchrony with host tree	
	Pine	Low		
	Larch	Low		
Lesser pine sawfly (*Neodiprion sertifer*)	Pine	Low	Possible increased risk on dry nutrient-poor sites	
European spruce sawfly (*Gilpinia hercyniae*)	Spruce	Moderate	Reduced generation time, range extension	
Web-spinning larch sawfly (*Cephalcia lariciphila*)	Larch	Low	Range extension, overlap with *Ips cembrae*	Northwest Scotland, northern England

Aphids, scale insects and adelgids

Insects in these groups can reduce growth and lead to cosmetic damage to high value trees through gall production (Bevan, 1987). They typically have a close and sometimes highly specialised relationship with the host tree which can, in turn, significantly influence the timing of the life cycle and population dynamics.

Many are relatively small, sedentary, insects that are often exposed on the surface of the plant. As a result, they are vulnerable to the effects of heavy rainfall, unseasonal cold temperatures and to natural predators. Projected climate change would therefore be expected to have both direct and indirect impacts on populations (Evans et al., 2002; Straw, 1995), resulting in a general increase in the damage caused by aphids and related insect pests. More specifically:

- Higher temperatures will increase the reproductive rate and those species that have multiple generations or can remain active throughout the winter will benefit the most;
- Drought stress of host trees through changes in rainfall patterns and increases in evapotranspiration will favour many species of insect.

Bark beetles, weevils and related species

Bark beetles are among the most important forest insect pests because they attack mature trees and can introduce harmful pathogens (e.g. Redfern et al., 1987; Gibbs and Inman, 1991). The abundance of species and their population dynamics is primarily determined by environmental influences on the availability of suitable breeding substrate. As a consequence, climate change would generally be expected to lead to an increase in the level of damage caused by bark beetles and related insects through:

- more widespread and frequent forest fires;
- increased summer drought stress leading to greater tree mortality and a larger proportion of stressed or compromised living trees that would be vulnerable to attack;
- changes to the wind climate, although projections are highly uncertain; if there is more frequent and widespread windblow this will increase the prevalence of bark beetles.

Defoliators

Defoliators are a highly diverse group of insects for which the population dynamics and abundance are driven by complex interactions between climate, site factors, host tree suitability and predators. Most have a single generation a year, but sawflies have the potential for two or more a year (Knerer, 1993). The most damaging defoliators have 'eruptive' population dynamics. Broadleaved tree species are generally affected early in the growing season and are able to re-flush. Significant damage and reductions in productivity generally only occur as a result of repeated outbreaks in successive years (Tortrix and winter moth, Operophtera brumata; Gradwell, 1974). Conifers are particularly vulnerable to defoliators that can feed on more than one age-class of needle (Watt and Leather, 1988) and those that feed late in the growing season and affect flushing the following year. Again, as a general rule, projected climate change is likely to lead to an increase in the prevalence and severity of defoliation by insect pests through:

- an increase in the number of generations per year for sawflies and other multi-voltine insects;
- drought-stress increasing susceptibility to defoliating pests;
- range extensions of some species, particularly those with a southerly distribution.

Changes in management

Changes in forest management practice, in part as a direct response to the impacts of climate change – or to its mitigation – may affect the fine balance between host trees and insect pests. Such changes include increases (or decreases) in rotation length, altered stocking densities and different age structures across large forests. Changes to stand structure through conversion to continuous cover systems of management may also have an impact. Finally, increased rates of woodland planting, larger areas of young woodlands and new species or provenances may also affect the frequency and severity of insect pest outbreaks.

The threat of insect pests that have not yet been introduced to the UK is poorly understood, but the risk is potentially very significant. Examples of devastating introductions in other parts of the world, for example the emerald ash borer in North America (see: www. emeraldashborer.info), highlight the potential impacts of such interactions. Climate change will make the climate of the UK suitable for an increasing range of forest insect pests, and there is an urgent need to review the global literature on these potential threats to enable appropriate surveillance and contingency plans for forest management to be drawn up.

5.6.2 Tree pathogens

Although there has been some analysis of the impact of climate change on insect populations and the associated damage in forests, in general there has much less consideration of the effect on pathogens. Typically, the key factors in the development of plant disease epidemics are temperature and moisture, and it is well known that rainfall patterns affect the frequency and severity of certain diseases from year to year. Inevitably therefore, climate change will alter the activity of tree pathogens, both through direct and indirect effects. Changes in temperature, precipitation, soil moisture and relative humidity will all have a direct influence on the infection success of pathogens. Indirect impacts will also result as trees suffer episodes of climatic extremes which cause water stress or undermine resistance mechanisms, thus making them more susceptible to latent or opportunist pathogens. If there are more frequent storm events this will increase levels of physical injury on trees, and thereby create wounds that allow pathogen entry. A summary of the risks posed by major pathogens in the context of a changing climate is given in Table 5.4, together with an assessment of current and future risks associated with bacterial pathogens in Table 5.5.

Latent pathogens

As a particular group of organisms that include significant tree pathogens, latent pathogens or 'endophytes' have long been predicted to pose a heightened threat under conditions of climate change. These organisms are highly specialised, able to infect trees and remain asymptomatic for years until environmental factors – often drought stress – trigger the development of disease. Examples include *Biscogniauxia* species which cause damaging strip cankers on beech and oak; these are already observed to be more common on beech in areas of low rainfall and high temperatures (Hendry *et al.*, 1998). Sooty bark disease of sycamore (*Cryptostroma corticale*) is another temperature dependent latent pathogen which only becomes active after hot dry summers (Dickenson and Wheeler, 1981), particularly when the mean monthly temperature of more than one summer month equals or exceeds 23°C (Young, 1978). However, latent pathogens are not just limited to broadleaved trees. The disease

known as Diplodia blight (*Diplodia pinea*) is one of the most common and widely distributed pathogens of conifers worldwide (Burgess *et al.*, 2004). Although considered a southern fungus, its impact becomes visible in the north during drought periods as it is released from its quiescent stage in hosts under water stress (Stanosz *et al.*, 2001). Severe drought in 2003 encouraged *D. pinea* to become epidemic in central Europe, and reports of damage caused by this pathogen have become much more frequent in Britain over the past 10 years (Brown and MacAskill, 2005).

Foliar pathogens

The most immediate and visible changes in tree health in response to climate change could result from increased activity of foliar pathogens. In fungi, sporulation and infection are strongly linked to changes in temperature and precipitation (Peterson, 1967) and, by their nature, foliar pathogens are directly exposed to fluctuations in air temperatures and moisture. Consequently, these agents which often require free moisture for host infection, sporulation and spore dissemination are likely to become more damaging, particularly in western parts of the UK where increased spring rainfall is indicated in climate projections.

Typical examples expected to worsen on broadleaved tree species include *Marssonina* and *Melamspora* species, while some of the warm temperature *Melamspora* species which are currently damaging in central Europe are predicted to become problematic in southern Britain as they extend their range (Lonsdale and Gibbs, 2002). Warmer weather will probably also favour attacks of powdery mildew on foliage (e.g. oak mildew, *Erysiphe alphitoides*), making outbreaks more intense and longer lasting. Moreover, if heavy infestations occur over successive years the vigour and productivity of affected trees is likely to be reduced. Conifer needles and shoots are also susceptible to various foliar pathogens. The most striking example is *Dothistroma* needle blight, which has escalated in incidence markedly over the past decade in Britain (Brown and Webber, 2008) and one of the drivers of this increase is thought to be climate change (Woods *et al.*, 2005; Archibald and Brown, 2007). The shift towards an increased frequency of prolonged periods of rainfall in eastern England combined with temperatures greater than 18–20°C in spring and early summer since the mid- to late-1990s appears to have favoured the spread and intensification of *D. septosporum* and this could be replicated elsewhere.

Table 5.4
Analysis of risk posed by major pathogens present in Britain in relation to climate change (1–5 = low to high risk).

Pathogen	Disease symptoms	Affected genera/species	Likelihood of increased activity	Potential impact	Level of risk
Foliar pathogens					
Marssonina spp.	Leaf spots, shoot blights, branch and stem cankers on young trees	Poplar, birch and willow	Low–moderate	Moderate[3]	2
Erisyphe spp.	Mildew causing leaf and shoot blight	Oak	Moderate–high	Moderate	3
Venturia spp.	Death of leaves and shoots	Poplar and willow	Low	Moderate	2
Melampsora spp.	Death of leaves, premature leaf fall	Poplar and willow	High	Moderate–high[4]	4
Dothistroma septosporum	Needle death, premature defoliation and tree mortality	Pines, especially Corsican, lodgepole and now Scots pine	High	High	5
Diplodea pinea	Shoot blight, top dieback, and cankers on stem and branches	Pines, particularly black pine	Moderate–high[1]	Moderate	4
Phytophthora spp., e.g. *P. ramorum, P. kernoviae*	Leaf and shoot blights and stem cankers	Broadleaf species	Moderate[2]	High	4
Root rots					
Heterobasidion annosum	Decay and mortality (particularly of young trees)	All conifers	Moderate–high	High	4
Armillaria spp.	Decay, tree decline and mortality	Wide range of conifers and broadleaves	Moderate–high	Moderate[5]	4
Collybia fusipes	Decay, tree decline and mortality	Predominantly oak	Moderate	Low–moderate[6]	3
Phytophthora spp., e.g. *P. cinnamomi, P. alni, P. cambivora*	Root death, bleeding canker and tree mortality	Wide range of broadleaves	High	High[7]	5
Stem cankers					
Bacterial diseases	Bleeding stem canker, shoot tip die back, gummosis	Broadleaf species, e.g. alder, ash, cherry, horse chestnut and oak	High	Moderate–high	5
Stress related or latent pathogens					
Biscogniauxia spp.	Strip cankers and dieback	Beech and oak	Moderate–high	High[8]	3
Botryosphaeria stevensii	Cankers and dieback	Ash and oak	Moderate	Low[1]	2
Cryptostroma corticale	Bark death and dieback	Sycamore	High	Low-moderate	2
Nectria coccinea	Bark death	Beech	Moderate	Moderate[9]	3
Phomopsis spp.	Bark cracking and stem lesions	Spruce and larch	Moderate	Low	2
Other established diseases					
Phacidium coniferarum	Bark killing and cankers	Conifers	Low	Low	1
Ophiostoma novo-ulmi	Vascular wilt (Dutch elm disease)	Elm	Moderate	High	2
Ophiostoma and *Ceratocystis*	Bluestain of wood and bark death	Pines and other conifers	High	Moderate–high[10]	4

[1] Impact could be especially high in nurseries, but also recognised as an endophyte which can be very damaging to stressed trees. [2] Some of the aerial Phytophthoras infect best at moderate temperatures (18–22°C) with high humidity, but can persist over hot summers via resting spores. [3] Moderate but localised impact, dependent on species involved. [4] Likely to have a high impact on clonal polar/willow biomass crops but this will be localised. [5] Main impact may still be on ornamental rather than commercial plantation/woodland species, although increased incidence of Armillaria attack on commercial forestry species is being recorded in the Disease Diagnostic and Advisory Service database. [6] Low to moderate impact reflects the long time-scales before damage becomes apparent. [7] High impact anticipated because of root death interacting with drought. [8] Common species affected although the impact may be high but localised. [9] Losses likely to be increased by stems snapping in severe wind and rain storms. [10] High temperatures likely to favour not only the fungi but also the insect vectors.

Table 5.5
Analysis of the risk posed by bacterial diseases to British forests in the context of climate change.

Pathogen species	Host	Disease type	Known in UK	Risk of entry to Britain	Potential risk (losses) to British trees
Brenneria nigrifluens (syn. *Erwinia nigrifluens*)	*Juglans* spp. (walnut)	Shallow bark canker; affects stems and scaffold branches	Not in Britain but present in Europe	Moderate	Low
Brenneria quercina (syn. *Erwinia quercina*)	*Quercus* spp.	Drippy nut disease affecting acorns and twigs	Not in Britain or Europe	Moderate–high (dependent on host range)	Cross-pathogenicity is unknown but if cross-pathogenic then high risk
Brenneria rubifaciens (syn. *Erwinia rubifaciens*)	*Juglans regia*	Deep bark disease forming bark splits and bleeding lesions on stems	Not in Britain or Europe	Moderate	Low
Erwinia amylovora	Many *Rosaceae* including *Prunus*, *Crataegus* and *Sorbus* spp.	Fire blight, attacks blossoms, leaves, shoots and stems causing wilt, cankers and tree death in very susceptible species, e.g. *Sorbus*	Present and widespread in Britain and continental Europe	N/A	Disease incidence is expected to increase and losses would become moderate–high
Erwinia salicis	*Salix* spp.	Watermark disease. This disease is important on the cricket bat willow *Salix alba* var. *coerlea* but also affects other *Salix* spp.	Present and widespread in Britain and The Netherlands	N/A	Disease incidence is expected to increase but control measures should keep losses in check
Pseudomonas avellanae	*Corylus avellana*	Cankers and dieback		N/A	
Pseudomonas savastanoi	*Fraxinus excelsior*	Galls and cankers on stems	Present and widespread in Britain but at a low incidence	N/A	Although disease incidence may increase it is not expected to cause major losses
Pseudomonas syringae pv *aesculi*	*Aesculus* spp.	Bleeding stem canker	Present and widespread in Britain and continental Europe, especially central Europe	N/A	Disease incidence is expected to increase and losses, which are already moderately high, will increase
Pseudomonas syringae (including a range of pathovars such as pv. *morsprunorum* and pv. *syringae*)	Broad host range including: *Acer, Alnus, Cornus, Fraxinus, Pinus, Poplus, Prunus, Quercus, Salix* and *Tilia*	Bacterial blight (kills leaves and shoots), and bacterial canker (stems) and gummosis of fruit, twigs and stems	Present and widespread in Britain, especially damaging on wild cherry (*Prunus avium*)	N/A	Disease incidence is expected to increase but losses could be controlled with appropriate management practices
Xanthomonas populi	*Poplus* spp.	Bacterial canker	Present in Britain and continental Europe	N/A	Disease incidence is expected to increase but losses could be controlled by planting resistant cultivars

Information sourced from Phillips and Burdekin, 1982; Sinclair and Lyon, 2005.

Root pathogens

In contrast to foliar pathogens, the interaction between root rot pathogens and climate change may be much more cryptic with trees on drought-affected sites liable to show an increased predisposition to infection by root attacking pathogens such as *Heterobasidion*, *Armillaria* and *Collybia*. The most important root rot pathogen of conifer forestry, *Heterobasidion annosum*, has become increasingly common and damaging on drier sites and the risk of infection is considered to be greatest on well drained mineral soils (Redfern *et al.*, 2001; Pratt, 2003). A changing climate is likely to favour it even more. Another root rotter, *Armillaria*, affects both conifer and broadleaved tree species and is ubiquitous throughout the British Isles. Although some *Armillaria* species are only weakly pathogenic, they are opportunists and attack and kill trees already weakened by other biotic agents or abiotic factors such as drought (Gregory and Redfern, 1998). Consequently their potential to cause damage is expected to increase as trees suffer more frequent episodes of elevated temperatures and drought stress (Desprez-Loustau *et al.*, 2006). Long term (chronic) declines of mature trees such as beech, oak and ash (oak decline is perhaps the best known) are often another visible sign of root rot pathogen activity aggravated by climatic extremes (Auclair *et al.*, 1992). As some *Armillaria* species as well as *H. annosum* have an optimum temperature for growth of around 25°C, increased ambient temperatures could well enhance the process of infection and spread. Moreover, the ability of these root rot fungi to persist on infected sites for decades and increase over successive rotations increases their potential for damage in the future, both through group killing in young restock sites as well as in older plantations and woodlands.

Phytophthoras

Climate models suggest that the impact of *Phytophthora* species is likely to be significantly enhanced under future climate scenarios. Many *Phytophthora* pathogens already have the capacity to be fast acting aggressive pathogens, often with a wide host range, and this behaviour makes them a formidable threat under conditions of climate change. *Phytophthoras* are a mainly introduced group of pathogens and around 10 species are now widespread in Britain with the potential to be highly damaging to tree species (Brasier, 1999; Jung *et al.*, 2009). Some attack aerial plant parts with activity favoured by mild moist springs, but the majority infect and kill the roots of susceptible tree species. The latter require moist soil conditions, even periods of flooding, for infection and spread. The damage they cause tends to be most visible in the summer, especially if water availability is limited. A build up of soil-inhabiting *Phytophthoras* can result in the death of fine feeder roots, even root and stem girdling, so trees may die suddenly or show signs of marked decline. In addition, many species of *Phytophthora* can over-winter in soil providing winters are mild but also persist in dry soils for decades in the form of resistant spores, becoming active again under more favourable conditions

Bacterial pathogens

Until recently, the number of bacterial pathogens known to be damaging to woodland and forest trees in Britain was considered to be relatively modest and the most significant pathogens mainly affected fruit and ornamental trees. However, over the past 5–8 years, bacterial diseases on trees appear to be more common. For example, both horse chestnut and native oak species now suffer from bacteria-related disorders (Webber *et al.*, 2008); symptoms on both include stem bleeding, and mortality is by no means uncommon. The extent to which these disorders have been exacerbated by changing weather regimes is uncertain. However, if more extreme rainfall and wind storms occur, there may be increased opportunities for dispersal and infection (Boland *et al.*, 2004).

Implications for woodlands in the UK

For aggressive pathogens such as Dutch elm disease, climatic effects on tree physiology are likely to be small compared with the importance of the susceptibility of some tree genera and species which is under genetic control. For other pathogens, climate change is likely to play a major role in defining levels of damage. With foliar pathogens such as *Dothistroma septosporum* and some aerial *Phytophthoras*, the impact is likely to be direct and rapid, especially when environmental optima for sporulation and infection are reached. Moreover, once disease levels reach critical thresholds due to combinations of extreme events, they may not return to earlier levels when the weather returns to more normal conditions. This may be most critical for tree pathogens, particularly root rot and decay fungi which are usually long-lived, highly persistent once established on site, and operate over long time periods. These include pathogens such as *Heterobasidion*, *Armillaria* and *Collybia*, so the serious damage they can cause may take years to appear. In these instances the indirect influence of climate change on tree physiology may be more crucial, although trying to predict how specific climate change scenarios are likely to influence tree resistance to pathogens is still largely a matter of speculation.

5.6.3 Deer

Climate change would be expected to favour the expansion of deer populations, as the carrying capacity of many habitats will increase due to longer growing seasons, rising carbon dioxide levels and increased warmth promoting plant productivity. Direct effects of climate change are also likely to increase recruitment and reduce winter mortality leading to further increase in deer populations. However, it is difficult to predict the effects on deer populations in any particular area of the UK. For example, as climate change progresses, southern England will experience conditions that may impact negatively on roe deer through reduced forage availability during the key reproductive period (Irvine et al., 2007). This is supported by current roe deer populations in southern Europe being largely restricted to higher elevations. Furthermore, average body sizes are smaller than north European populations and juvenile growth is slower (Andersen and Linnell, 1998). There is also evidence (Raganella-Pelliccioni et al., 2006) that juvenile survival is positively correlated with spring rainfall suggesting that forage availability is limiting population size. Broadmeadow (2004) suggested that, based on current European distribution, the effect of climate change on red and roe deer populations is uncertain. However, climate envelope modelling suggests that climate space for Roe deer may be lost in drier areas of east England by the 2050s, and across much of England by the 2080s under the UKCIP02 medium–high climate change scenarios. The population ecology of non-native species of deer (fallow, muntjac and sika) has been less well studied. However, in the UK they have been introduced at more northerly latitudes than their native ranges and are thus likely to benefit from climate change.

Where deer populations do increase, their impact on the establishment and growth of young trees and shrubs will increase. Deer browsing also alters the composition and structure of vegetation resulting in a more open understorey and increasing dominance of grasses at the expense of forbs (Kirby, 2001). Even if deer populations do not increase significantly, climate change may affect their habitat selection with, for example, the potential for deer to concentrate on areas where soil moisture is not limiting (Irvine et al., 2007).

5.7 Impacts on timber quality

Potential impacts of climate change on timber quality can be separated into those that relate to the direct effects of climate change (temperature, carbon dioxide concentration, water availability, wind speed) and those that are the result of changes in growth rate. It should be noted that different species and provenances will be affected in different ways, particularly those impacts relating to water availability.

5.7.1 Direct effects of climate change on timber quality

Temperature

Milder winters could have a negative impact on tree growth form through late or incomplete winter hardening making trees more susceptible to frost damage. The same impact could result from early flushing, placing the frost tender new growth at risk of spring frost damage, as outlined in 5.1 above. Selecting more southerly, earlier flushing, provenances in anticipation of climate change could enhance this risk in the short to medium term and care should be taken in provenance selection for this reason. Earlier flushing may, however, have a positive impact on timber quality through countering the apparent relationship between late flushing and shake in oak (Savill and Mather, 1990). This would clearly be of economic benefit for producing high value logs over the longer term. A particular issue for conifer species is the tendency to produce lammas growth during mild, wet autumns, particularly if following summer drought. The shoots formed in this way are especially vulnerable to early autumn frost, which could lead to stem forking. Furthermore, the lammas growth sets a second whorl on the stem leading to increased knottiness in sawn timber.

Water availability

Apart from affecting timber quality through changes in growth rate, projected changes in water availability may lead to drought crack and ring shake in some species, as has been reported for Sitka spruce in eastern Scotland in recent years (Green and Ray, 2009). In contrast, Scots pine can produce poor form and lower timber quality as a result of forking and heavy branching on sites that are too wet. Since winter waterlogging is the principal problem, it is likely to become more widespread as a result of climate change.

Wind and exposure

The UK experiences a very windy climate that places limitations on forestry in some regions. Although projections of changes to the wind climate of the UK are far from certain (Chapter 2), potential impacts on timber quality are a concern alongside the more catastrophic effects of windthrow. Wind storms often result in leader loss which in turn leads to crooked stems as one of the side-branches takes over apical dominance from the lost leader. This poorer stem form in turn leads to a lower recovery of desirable straight 'green' logs used in construction grade timber production. Background wind loading can cause conifers to produce compression wood and broadleaved trees to produce tension wood. Both these types of wood lead to increased difficulties in processing and poorer performance in service. In particular compression wood can lead to increased distortion and failure under loading. Currently, compression wood is not a serious problem in British grown conifers but could be if windiness increases significantly.

5.7.2 Indirect, growth-rate mediated effects of climate change on timber quality

The most important climatic variable for plant growth is temperature, and when water and nutrients are not limiting, most species will produce increased growth (height or diameter or both) in warmer growing seasons. It has been suggested (Broadmeadow and Randle, 2002) that rising CO_2 levels and climate warming will result in an increase in productivity (defined here as Yield Class (YC): maximum mean annual wood volume increment over a rotation in cubic metres per hectare per year) of 1–2 units for most species (e.g. from YC 6 to YC 8 for oak; YC 14 to YC 16 for Sitka spruce). Trees growing faster will tend to have larger heights, diameters and inter-whorl distances. Changes in growth rate will affect wood properties and, in turn, timber performance. Properties affected include wood density, knot spacing and knot size. Typical values for Sitka spruce are given in Table 5.6 (Ray et al., 2008).

Table 5.6
Comparison of Sitka spruce timber quality properties with yield class

Property	Unit	Yield class		
		14	16	18
Wood density	kg m⁻³	425	422	420
Between whorl spacing	m	0.45	0.50	0.55
Knot size for largest whorl	cm	3.40	3.65	3.90

Although spruce and fir tend to have reduced wood density with increased growth rate, pine, larch and Douglas fir show little or no reduction in wood density with faster growth. Therefore, for these species, there is likely to be no change in the timber performance with increased growth rates resulting from climate change.

Hardwoods vary in response to increased rate of growth. The oaks, ash and elm are 'ring porous' producing harder and stronger timber when grown fast. In contrast, sycamore and birch are 'diffuse porous' and do not respond in this way. Chestnut and beech show an intermediate type of response.

Evidence on the direct effects of rising CO_2 levels on wood properties is conflicting, and it is difficult to distinguish between the direct effects of CO_2 and the indirect effects of CO_2 acting through enhanced growth. Evidence that CO_2 has little impact on wood anatomical properties is provided by Donaldson et al. (1987), Telewski et al. (1999) and Overdieck and Ziche (2005). However, this contrasts with the work of Conroy et al. (1990) and Ceulemans et al. (2002) who reported increases in tracheid wall thickness and tracheid width and resin canal density, respectively.

5.7.3 Interactions between timber quality and forest management

The negative impacts of climate change outlined above can, in part, be reduced by appropriate species or provenance and site selection. Equally, the opportunities presented by climate change can be taken through the same route of good silvicultural practice. If growth rates do increase as a result of climate change, this will clearly lead to increased sawlog production and higher value crops with more enduring uses. The future of commercial plantations will also be challenged in some regions – by either the direct effects of climate change or the impacts of pest and disease outbreaks. This may require the introduction of novel species for which there is a lack of processing and use information available in the UK. This should not be seen as a barrier for planting those species if there is clear evidence of the timber value of the species elsewhere in the world.

5.8 Preliminary commentary on the UK climate projections

In broad terms, the current UK climate projections (UKCP09) differ little in the magnitude of change from the UKCIP02 Climate Change scenarios (Chapter 2). The publication of new projections does not therefore invalidate previous analyses such as those outlined in 5.2 and 5.3 above. The main differences are that the changes in precipitation (both summer and winter) are reduced in UKCP09 compared with UKCIP02, while the projected changes in summer and winter temperature are slightly larger. There is also a small shift in the location of the more extreme changes in temperature, with the area of greatest change extending from the southeast towards the southwest peninsula. The implications of climate change for forestry in the UK are therefore unlikely to change significantly. However, the way in which the climate projections are presented together with the availability of the weather generator will enable uncertainty to be quantified and a risk-based approach to be applied to adaptation strategies. Although such an approach will be far from straightforward to apply to complex systems such as forests that are affected by interactions between a range of climate variables, it will provide justification for any modification of current biodiversity or forestry policy.

5.9 Research priorities

- Developing modelling capacity at the operational level. There is a need to integrate modelling capabilities for UK conditions. A modelling system should be developed that combines the practical applicability of knowledge-based decision support systems with the more theoretical stand-level process-based models that can represent the effects of changing atmospheric composition on tree physiology and be extended outside the evidence-base of empirical models.
- Identifying the possible effects of rising temperatures. Rising temperatures may mean that the chilling requirement for successful germination of some native tree species fails to be met. There is a need to (1) identify whether this is the case and (2) identify whether there is a relationship between the chilling requirement of specific provenances and ambient temperature. The impact of rising temperatures on other processes (e.g. winter hardening and leafing date/frost risk) should also be further considered.
- Learning from climate analogues. 'Climate matching' analysis can identify broad regions that currently experience a climate similar to that projected for the UK in the future. This provides an opportunity to explore tangible impacts of likely climate change to accompany predictions based on model simulations. The approach should be adopted to explore likely changes in

woodland ecosystems, the suitability of tree species for commercial forestry, and to inform changes to forest management that might be required in response to the changing climate. It must, however, be understood that such an approach can only provide broad guidance as complete analogues of future conditions, particularly when the distribution of extremes is considered, do not exist.

- Understanding risks to biosecurity. Pest risk analysis is required for a range of insect pests that potentially represent a risk to British forests if they are introduced. The *ex-situ* temperature response of growth should also be determined for a range of tree pests and pathogens to provide the basis for epidemiological modelling of future outbreaks under a changing climate.

References

AITKEN, S.N., YEAMAN, S., HOLLIDAY, J.A., WANG, T. and CURTIS-MCLANE, S. (2008). Adaptation, migration or extirpation: climate outcomes for tree populations. *Evolutionary Applications* 1, 95–111.

ANDERSEN, R. and LINNELL, J.D.C. (1998). Ecological correlates of mortality of roe deer fawns in a predator-free environment. *Canadian Journal of Zoology* 76, 1217–1225.

ARCHIBALD, S. and BROWN, A. (2007). *The relationship between climate and the incidence of Dothistroma needle blight in East Anglia.* Poster presented at IUFRO Conference on: Foliage, Shoot and Stem Diseases of Forest Trees; 21–26 May, Sopron, Hungary.

AUCLAIR, A.A., WORREST, R.C., LACHANCE, D. and MARTIN, H.C. (1992). Climatic perturbation as a general mechanism of forest dieback. In: Manion, P.D. and Lachance, D. (eds) *Forest decline concepts.* American Phytopathological Society Press, St. Paul, Minneapolis.

BECKER, M., NIEMINEN, T.M. and GÉRÉMIA, F. (1994). Short-term variations and long-term changes in oak productivity in Northeastern France. The role of climate and atmospheric CO_2. *Annals of Forest Science* 51, 477–492.

BERRY, P., DAWSON, T., HARRISON, P. and PEARSON, R. (2002a). Impacts on the distribution of plant species found in native beech woodland. In: Broadmeadow, M. (ed.) *Climate change: impacts on UK forests.* Forestry Commission Bulletin 125. Forestry Commission, Edinburgh. pp 169–180.

BERRY, P.M., DAWSON, T.P., HARRISON, P.A. and PEARSON, R.G. (2002b). Modelling potential impacts of climate change on the bioclimatic envelope of species in Britain and Ireland. *Global Ecology and Biogeography* 11, 453–462.

BERRY, P.M., JONES, A.P., NICHOLLS, R.J. and VOS, C.C. (eds) (2007a). Assessment of the vulnerability of terrestrial and coastal habitats and species in Europe to climate change. Annex 2 of *Planning for biodiversity in a changing climate* – BRANCH project Final Report, Natural England, Peterborough.

BERRY, P.M., O'HANLEY, J.R., THOMSON, C.L., HARRISON, P.A., MASTERS, G.J. and DAWSON, T.P. (eds) (2007b). *Modelling natural resource responses to climate change (MONARCH)*. MONARCH 3 Contract report. UKCIP Technical Report, Oxford.

BERRY, P.M. and PATERSON, J.S. (2009). *Impacts of climate change on Burnham beeches*. Report to the Corporation of London.

BEVAN, D. (1987). *Forest insects.* Forestry Commission Handbook 1. HMSO, London.

BIRKS, H.J.B. and WILLIS, K.J. (2008). Alpines, trees and refugia in Europe. *Plant Ecology and Diversity* 1, 147–160.

BOLAND, G.J., MELZER, M.S., HOPKIN, A., HIGGINS, V. and NASSUTH, A. (2004). Climate change and plant diseases in Ontario. *Canadian Journal of Plant Pathology* 26, 335–350.

BRASIER, C.M. (1999). Phytophthora *pathogens of trees: their rising profile in Europe*. Forestry Commission Information Note 30. Forestry Commission, Edinburgh.

BROADMEADOW, M. (2000). *Climate change – implications for UK forestry*. Forestry Commission Information Note 31. Forestry Commission. Edinburgh.

BROADMEADOW, M.S.J. (2004). *A review of the potential effects of climate change for trees and woodland in Wales*. Report prepared for Working Group 4 of the Wales Woodland Forum. Forest Research, Farnham.

BROADMEADOW, M.S.J. and JACKSON, S.B. (2000). Growth responses of *Quercus petraea, Fraxinus excelsior* and *Pinus sylvestris* to elevated carbon dioxide, ozone and water supply. *New Phytologist* 146, 437–451.

BROADMEADOW, M. and RANDLE, T. (2002). The impacts of increased CO_2 concentrations on tree growth and function. In: Broadmeadow, M. (ed.) *Climate change: impacts on UK forests*. Forestry Commission Bulletin 125. Forestry Commission, Edinburgh. pp. 119–140.

BROADMEADOW, M. and RAY, D. (2005). *Climate change and British woodland*. Forestry Commission Information Note 69. Forestry Commission, Edinburgh.

BROADMEADOW, M.S.J., RAY, D. and SAMUEL, C.J.A. (2005). Climate change and the future for broadleaved tree species in Britain. *Forestry* 78, 145–161.

BROWN, A. and MACASKILL, G. (2005). *Shoot diseases of pine*. Forestry Commission Information Note 68. Forestry Commission, Edinburgh.

BROWN, A. and WEBBER, J. (2008). *Red band needle blight of conifers in Britain*. Forestry Commission Research Note. Forestry Commission, Edinburgh.

BURGESS, T.I., WINGFIELD, M.J. and WINGFIELD, B.D. (2004). Global distribution of *Diplodia pinea* genotypes revealed using simple sequence repeat (SSR) markers. *Australasian Plant Pathology* 33, 513–519.

CEULEMANS, R., JACH, M.E., VAN DE VELDE, R., LIN, J.X. and STEVENS, M. (2002). Elevated atmospheric CO_2 alters wood production, wood quality and wood strength of Scots pine (*Pinus sylvestris* L.) after three years of enrichment. *Global Change Biology* 8, 153–162.

CONROY, J.P., MILHAM, P.J., MAZUR, M. and BARLOW, E.W.R. (1990). Growth, dry matter partitioning and wood properties of *Pinus radiata* D. Don. after 2 years of CO_2 enrichment. *Plant, Cell and Environment* 13, 329–337.

CURTIS, P.S. and WANG, X. (1998). A meta-analysis of elevated CO_2 effects on woody plant mass, form and physiology. *Oecologia* 113, 299–313.

DEPARTMENT OF THE ENVIRONMENT (1994). *Biodiversity: the UK Action Plan*. HMSO, London.

DESPREZ-LOUSTAU, M-L., MARÇAIS, B., NAGELEISEN, L-M., PIOU, D. and VANNINI, A. (2006). Interactive effects of drought and pathogens in forest trees. *Annals of Forest Science* 63, 597–612.

DICKENSON, S. and WHEELER, B.E.J. (1981). Effects of temperature, and water stress in sycamore, on growth of *Cryptostroma corticale*. *Transactions of the British Mycological Society* 76, 181–185.

DONALDSON, L.A., HOLLINGER, D., MIDDLETON, T.M. and SOUTER, E.D. (1987). Effect of CO_2 enrichment on wood structure in *Pinus radiata* (D. Don). *IAWA Bulletins* 8, 285–289.

ECOCRAFT (1999). Predicted impacts of rising carbon dioxide and temperature on forests in Europe at stand scale. Final project report (ENV4-CT95–ENV4-0077). IERM, University of Edinburgh, Edinburgh.

EVANS, H., STRAW, N. and WATT, A. (2002). Climate change: implications for insect pests. In: Broadmeadow, M. (ed.) *Climate change: impacts on UK forests*. Forestry Commission Bulletin 125. Forestry Commission, Edinburgh. pp. 99–118.

EVELYN, J. (1729). *Silva: or a discourse of forest-trees, and the propagation of timber in his majesty's dominions*, 5th edn. Walthoe, London.

FERNANDEZ, M.M.F. (2006). Colonisation of fire-damaged trees by *Ips sexdentatus* (Boerner) as related to the percentage of burnt crown. *Entomologica Fennica* 17, 381–386.

FORESTRY COMMISSION (2003). *National inventory of woodlands and trees*. Inventory Report for Great Britain. Forestry Commission, Edinburgh.

GIBBS, J.N. and INMAN, A. (1991). The pine shoot beetle *Tomicus piniperda* as a vector of bluestain fungi to windblown pine. *Forestry* **64**, 239–249.

GOSLING, P. and BROADMEADOW, M. (2006). Seed dormancy and climate change. *Forest Research annual report and accounts 2004–2005*. The Stationery Office, Norwich.

GRACE, J., BERNINGER, F. and NAGY, L. (2002). Impacts of climate change on the tree line. *Annals of Botany* **90**, 537–544.

GRADWELL, G.R. (1974). The effect of defoliators on tree growth. In: Morris, M.G. and Perring F.H. (eds) *The British oak*. Botanical Society of the British Isles Conference Reports 14. Classey, Faringdon. pp. 182–193.

GREEN, S. and RAY, D. (2009). *Potential impacts of drought and disease on forestry in Scotland*. Forestry Commission Research Note. Forestry Commission, Edinburgh.

GREGORY, S.C. and REDFERN, D.B. (1998). *Diseases and disorders of forest trees*. Forestry Commission Field Book 16. The Stationery Office, London.

HARRISON, P.A., BERRY, P.M. and DAWSON, T.P., eds (2001). *Climate change and nature conservation in Britain and Ireland: modelling natural resources response to climate change (the MONARCH project)*. UKCIP Technical Report, Oxford.

HENDRY, S.J., LONSDALE, D. and BODDY, L. (1998). Strip-cankering of beech (*Fagus sylvatica*): pathology and distribution of symptomatic trees. *New Phytologist* **140**, 549–565.

HERBST, M., ROBERTS, J.M., ROSIER, P.T.W., TAYLOR, M.E. and GOWING, D.J. (2007). Edge effects and forest water use: a field study in a mixed deciduous woodland. *Forest Ecology and Management* **250**, 176–186.

HOSSELL, J.E., BRIGGS, B. and HEPBURN, I.R. (2000). *Climate change and UK nature conservation: a review of the impact of climate change on UK species and habitat conservation policy*. DETR, London.

HULME, M., JENKINS, G. J., LU, X., TURNPENNY, J. R., MITCHELL, T. D., JONES, R. G., LOWE, J., MURPHY, J. M., HASSELL, D., BOORMAN, P., MACDONALD, R., and HILL, S. (2002). *Climate change scenarios for the United Kingdom*. The UKCIP02 Scientific Report. Tyndall Centre for Climate Change Research, School of Environmental Science, University of East Anglia, Norwich.

HUNTLEY, B. and BIRKS, H.J.B. (1983). *An atlas of past and present pollen maps for Europe: 0–13,000 years ago*. Cambridge University Press, Cambridge.

IRVINE, R.J., BROADMEADOW, M., GILL, R.M.A. and ALBON, S.D. (2007). Deer and global warming: how will climate change influence deer populations? *Deer* **14**, 34–39.

JUNG, T., VANNINI, A. and BRASIER, C.M. (2009). Progress in understanding Phytophthora diseases of trees in Europe 2004–2007. In: Goheen E.M. and Frankel, S.J. (eds) *Phytophthoras in forests and natural ecosystem. Proceedings of the fourth meeting of the International Union of Forest Research Organizations (IUFRO) Working Party S07.02.09*. 26–31 August, *Monterey, CA*. General Technical Report PSW-GTR-221. Albany, CA: US Department of Agriculture, Forest Service, Pacific Southwest Research Station.

KARNOSKY, D.F., TALLIS, M., DARBAH, J. and TAYLOR, G. (2007). Direct effects of elevated carbon dioxide on forest tree productivity. In: Freer-Smith, P.H., Broadmeadow, M.S.J. and Lynch, J.M. (eds) *Forestry and climate change*. CAB International, Wallingford. pp. 136–142.

KIRBY, K.J. (2001). The impact of deer on the ground flora of British broadleaved woodland. *Forestry* **74**, 219–229.

KIRBY, K. (2009). *Guidance on dealing with the changing distribution of tree species*. Technical Information Note 053. Natural England, Peterborough.

KNERER, G. (1993). Life history diversity in sawflies. In: Wagner, M.R. and Raffa, K.F. (eds) *Sawfly life history adaptations to woody plants*. Academic Press, San Diego. pp. 33–59.

KÖRNER, C., ASSHOFF, R., BIGNUCOLO, O., HÄTTENSCHWILER, S., KEEL, S.G., PELÁEZ-RIEDL, S., PEPIN, S., SIEGWOLF, R.T.W. and ZOTZ, G. (2005). Carbon flux and growth in mature deciduous forest trees exposed to elevated CO_2. *Science* **309**, 1360–1362.

KULLMAN, L. and OBERG, L. (2009). Post-little ice age tree line rise and climate warming in the Swedish Scandes: a landscape ecological perspective. *Journal of Ecology* **97**, 415–429.

LÅNGSTRÖM, B. (1984). Windthrown Scots pine as brood material for *Tomicus piniperda* and *T. minor*. *Silva Fennica* **18**, 187–198.

LONSDALE, D. and GIBBS, J. (2002). Effects of climate change on fungal diseases of trees. In: Broadmeadow, M. (ed.) *Climate change: impacts on UK forests*. Forestry Commission Bulletin 125. Forestry Commission, Edinburgh. pp.83–97.

MASTERS, G.J., BROWN, V.K., CLARKE, I.P., WHITTAKER, J.B. and HOLLIER, J.A. (1998). Direct and indirect effects of climate change on insect herbivores: *Auchenorrhyncha (Homoptera)*. *Ecological Entomology* **23**, 45–52.

MITCHELL, R.J., MORECROFT, M.D., ACREMAN, M., CRICK, H.Q.P., FROST, M., HARLEY, M.; MACLEAN, I.M.D., MOUNTFORD, O., PIPER, J., PONTIER, H., REHFISCH, M.M., ROSS, L.C., SMITHERS, R.J., STOTT, A., WALMSLEY, C.A., WATTS, O. and WILSON, E. (2007). *England biodiversity strategy – towards adaptation to* climate change. Final report to Defra for contract CRO327.

MURPHY, J.M., SEXTON, D.M.H., JENKINS, G.J., BOORMAN, P.M., BOOTH, B.B.B., BROWN, C.C., CLARK, R.T., COLLINS, M., HARRIS, G.R., KENDON, E.J., BETTS, R.A., BROWN, S.J., HOWARD, T. P., HUMPHREY, K. A., MCCARTHY, M. P., MCDONALD, R. E., STEPHENS, A., WALLACE, C., WARREN, R., WILBY, R. and WOOD, R. A. (2009). *UK climate projections science report: climate change projections*. Met Office Hadley Centre, Exeter.

NAKICENOVIC, N. and SWART, R. (eds) (2000). *Special report on emissions scenarios.* Intergovernmental Panel on Climate Change.

NEGTAP (2001). *Transboundary air pollution: acidification, eutrophication and ground level ozone in the UK.* Report of the National Expert Group on Transboundary Air Pollution. Defra, London.

NORBY, R.J., DELUCIA, E.H., GIELEN, B., CALFAPIETRA, C., GIARDINA, C.P., KING, J.S., LEDFORD, J., MCCARTHY, H.R., MOORE, D.J.P., CEULEMANS, R., DEANGELIS, P., FINZI, A.C., KARNOSKY, D.F., KUBISKE, M.E., LUKAC, M., PREGITZER, K.S., SCARASCIA- MUGNOZZA, G.E., SCHLESINGER, W.H. and OREN, R. (2005). Forest response to elevated CO_2 is conserved across a broad range of productivity. *Proceedings of the National Academies of Science* 102, 18052–18056.

NORBY, R.J. and LUO, Y. (2004). Evaluating ecosystem responses to rising atmospheric CO_2 and global warming in a multi-factor world. *New Phytologist* 162, 281–294.

OVERDIECK, D. and ZICHE, D. (2005). Tree tissue properties under ambient and elevated CO_2. In: Randle, T. (ed.) *MEFYQUE – Forest and timber quality in Europe: modelling and forecasting yield and quality in Europe*. Final contract report (QLK5-CT-2001–QLK5-CT00345) to EU. Forest Research, Farnham.

PEARSON, R.G., DAWSON, T.P., BERRY, P.M. and HARRISON, P.A. (2002). SPECIES: a spatial evaluation of climate impact on the envelope of species. *Ecological Modelling* 154, 289–300.

PENUELAS, J., ESTIARTE, M. and LLUSIA, J. (1997). Carbon-based secondary compounds at elevated CO_2. *Photosynthetica* 33, 313–316.

PETERKEN, G.F. and MOUNTFORD, E.P. (1996). Effects of drought on beech in Lady Park Wood, an unmanaged mixed deciduous woodland. *Forestry* 69, 125–136.

PETERSON, G. W. (1967). Dothistroma needle blight of Austrian and Ponderosa pines in Britain. *Forestry* 35, 57–65.

PHILLIPS, D.H. and BURDEKIN, D.A. (1982). *Diseases of forest and ornamental trees.* Macmillan Press, London.

PRATT, J.E. (2003). Stump treatment against Fomes. *Forest Research annual report and accounts 2001–2002*. The

Stationery Office, Norwich.

PYATT, D. G., RAY, D. and FLETCHER, J. (2001). *An ecological site classification for forestry in Great Britain*. Forestry Commission Bulletin 124. Forestry Commission, Edinburgh.

QUINE, C. and GARDINER, B. (2002). Climate change impacts: storms. In: Broadmeadow, M. (ed.) *Climate change: impacts on UK forests*. Forestry Commission Bulletin 125. Forestry Commission, Edinburgh.

RAGANELLA-PELLICCIONI, E., BOITANI, L. and TOSO, S. (2006). Ecological correlates of roe deer fawn survival in a sub-Mediterranean population. *Canadian Journal of Zoology* 84, 1505–1512.

RAY, D. (2008a). *Impacts of climate change on forestry in Wales*. Forestry Commission Research Note. Forestry Commission Wales, Aberystwyth.

RAY, D. (2008b). *Impacts of climate change on forestry in Scotland – a synopsis of spatial modelling research.* Forestry Commission Research Note. Forestry Commission Scotland, Edinburgh.

RAY, D. and BROOME, A. (2003). Ecological Site Classification: supporting decisions from the stand to the landscape scale. *Forest Research annual report and accounts 2001–2002*. The Stationery Office, Norwich.

RAY, D., PYATT, G. and BROADMEADOW, M. (2002). Modelling the future climatic suitability of plantation forest tree species. In: Broadmeadow, M. (ed.) *Climate change: impacts on UK forests*. Forestry Commission Bulletin 125. Forestry Commission, Edinburgh. pp. 151–167.

RAY, D., WAINHOUSE, D., WEBBER, J. and GARDINER, B. (2008). *Impacts of climate change on forests and forestry in Scotland*. Report compiled for Forestry Commission Scotland. Forest Research, Roslin.

REDFERN, D. and HENDRY, S. (2002). Climate change and damage to trees in the UK caused by extremes in temperature. In: Broadmeadow, M. (ed.) *Climate change: impacts on UK forests*. Forestry Commission Bulletin 125. Forestry Commission, Edinburgh. pp. 29–39.

REDFERN, D.B., PRATT, J.E., GREGORY, S.C. and MACASKILL, G.A. (2001). Natural infection of Sitka spruce thinning stumps in Britain by spores of *Heterobasidion annosum* and long-term survival of the fungus. *Forestry* 74, 53–71.

REDFERN, D.B., STOAKLEY, J.T., STEELE, H. and MINTER, D.W. (1987). Dieback and death of larch caused by *Ceratocystis laricicola* sp. nov. following attack by *Ips cembrae*. *Plant Pathology* 36, 467–480.

ROBERTS, J. and ROSIER, P. (2006). The effect of broadleaved woodland on chalk groundwater resources. *Quarterly Journal of Engineering Geology and Hydrogeology* 39, 197–207.

ROYAL SOCIETY (2008). *Ground-level ozone in the 21st century: future trends, impacts and policy implications*. Science Policy Report 15/08. Royal Society, London.

SAVILL, P.S. and MATHER, R.A. (1990). A possible indicator of shake in oak: relationship between flushing dates and vessel sizes. *Forestry* **63**, 355–362.

SCHAFER, K.V.R., OREN, R., ELLSWORTH, D.S., LAI, C.-T., HERRICK, J.D., FINZI, A.C., RICHTER, D.D. and KATUL, G.G. (2003). Exposure to an enriched CO_2 atmosphere alters carbon assimilation and allocation in a pine forest ecosystem. *Global Change Biology* **9**, 1378–1400.

SINCLAIR, W.A. and LYON, H.H. (2005). *Diseases of trees and shrubs*, 2nd edn. Cornell University Press, Ithaca.

SPARKS, T. and GILL, R. (2002). Climate change and the seasonality of woodland flora and fauna. In: Broadmeadow, M. (ed.) *Climate change: impacts on UK forests*. Forestry Commission Bulletin 125. Forestry Commission, Edinburgh. pp. 69–82.

STANOSZ, G.R., BLODGETT, J.T., SMITH, D.R. and KRUGER, E.L. (2001). Water stress and *Sphaeropsis sapinea* as a latent pathogen of red pine seedlings. *New Phytologist* **149**, 531–538.

STRAW, N.A. (1995). Climate change and the impact of the green spruce aphid, *Elatobium abietinum* (Walker), in the UK. *Scottish Forestry* **31**, 113–125.

STRIBLEY, G.H. and ASHMORE, M.R. (2002). Quantitative changes in twig growth pattern of young woodland beech (*Fagus sylvatica* L.) in relation to climate and ozone pollution over 10 years. *Forest Ecology and Management* **157**, 191–204.

TELEWSKI, F.W., SWANSON, R.T., STRAIN, B.R. and BURNS, J.M. (1999). Wood properties and ring width responses to long-term atmospheric carbon dioxide enrichment in field-grown loblolly pine (*Pinus taeda* L.). *Plant, Cell and Environment* **22**, 213–219.

UNITED NATIONS (1992). Convention on Biological Diversity. International Legal Materials 31 (1992): 818.

WALMSLEY, C.A., SMITHERS, R.J., BERRY, P.M., HARLEY, M. STEVENSON, M.J. and CATCHPOLE, R. (eds) (2007). *MONARCH – Modelling natural resource responses to climate change: a synthesis for biodiversity conservation*. UKCIP, Oxford.

WAINHOUSE, D. (2005). *Ecological methods in forest pest management*. Oxford University Press, Oxford.

WAINHOUSE, D. (2008). *Forest insect pests and global warming: a review*. Contract report to the Forestry Commission. Forest Research, Farnham.

WATT, A.D. and LEATHER, S.R. (1988). The pine beauty moth in Scottish lodgepole pine plantations. In: Berryman, A.A. (ed.) *Dynamics of forest insect populations: patterns, causes and implications*. Plenum Press, London. pp.

243–266.

WEBBER, J.F., PARKINSON, N.M., ROSE, J., STANFORD, H., COOK, R.T.A. and ELPHINSTONE, J.G. (2008). Isolation and identification of *Pseudomonas syringae pv. aesuculi* causing bleeding canker of horse chestnut in the UK. *Plant Pathology* **57**, 368.

WESCHE, S. (2003). *The implications of climate change for the conservation of beech woodlands and associated flora in the UK*. Research Report 528. English Nature, Peterborough.

WILSON, S.M., BROADMEADOW, M., SANDERS, T.G. and PITMAN, R. (2008). Effect of summer drought on the increment of beech trees in southern England. *Quarterly Journal of Forestry* **102**, 111–120.

WOODS, A., COATES, D. K. and HAMANN, A. (2005). Is an unprecedented Dothistroma needle blight epidemic related to climate change? *BioScience* **55**, 761–769.

YOUNG, C.W.T. (1978). *Sooty bark disease of sycamore*. Arboricultural Leaflet 3. HMSO, London.

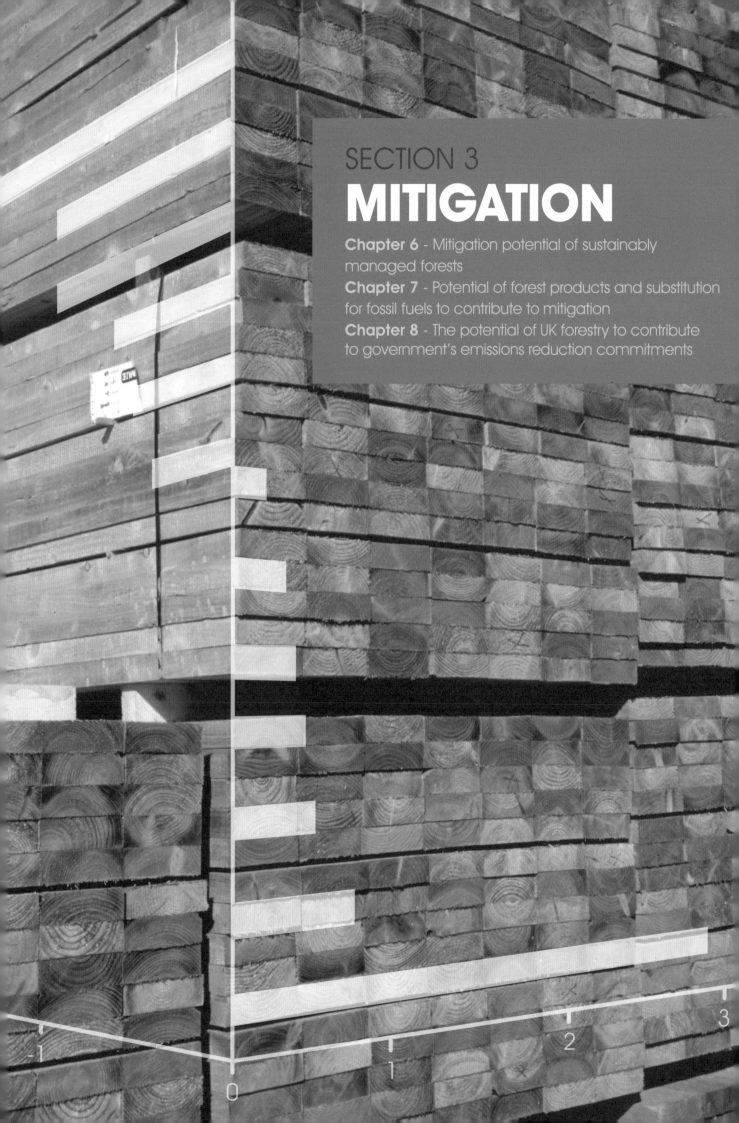

MITIGATION POTENTIAL OF SUSTAINABLY MANAGED FORESTS

W. L. Mason, B. C. Nicoll, M. Perks

Chapter

6

Key Findings

Many areas of British conifer forests will be due for felling in the next 10–20 years, which represents a major opportunity for adapting forests to future climate. Greater use of more southerly provenances is possible for all the major conifers so reducing their vulnerability to climate change. Managers should use a mix of forest management alternatives to obtain the best out-turn in terms of climate change mitigation and adaptation. Intimate species mixtures are only likely to be viable in broadleaved woods and mosaic mixtures should be preferred elsewhere.

Current policies seeking to diversify plantation forests through changes in species and structure may cause a decline in carbon sequestration unless this is offset by the use of more productive genotypes. When planning future planting programmes, greater emphasis should be given to species carbon content, and to their rates of carbon sequestration.

Forest planning and management must take uncertainty and risk into account. Current policies are resulting in more stands being retained for longer which will increase the risk of windthrow and disturbance to forest and soil carbon stocks. Stands which are currently marginal because of soil moisture requirement are likely to prove vulnerable to climate change adding to the threat from established and new pests and pathogens. Improved methods of forest planning are needed that take uncertainty into account, increase the resilience of British forests to climate change and enhance their role in carbon sequestration. These methods will need to be supported by appropriate training.

Mitigation involves all actions that help reduce net emissions of greenhouse gases or otherwise stabilise their concentration in the atmosphere. These actions include maintaining and enhancing long-term carbon stocks in trees, woodlands and forests and the use of woodfuel from sustainably managed forests as a substitute for fossil fuels. These represent additional management objectives for forestry and introduce the need to review those forest practices that may detract from the carbon storage by forests.

Mitigation will also involve an examination of energy use, particularly in timber transport. Adaptation of forests and forestry is examined in Section 4, but it is important to acknowledge that the two are closely linked: the mitigation contribution of forests and forestry in the future will depend on how well they are adapted to changing climate.

The potential role of sustainable forest management in combating change is clear (see Chapter 11). At a European level, a major means by which forestry can mitigate climate change is through alteration in management practices to increase the carbon density (i.e. the tonnes of carbon per ha) of forests (Nabuurs et al., 2007: Table 9.3). However, the potential for enhancement of carbon sinks within British forests is influenced by stand development phase, the approach to forest management, soil type, other site factors, species and provenance choice. Therefore, the subsections which follow describe the relative impact of each of these variables on forest carbon in the UK, and the interaction with afforestation, reforestation, machine operation and engineering. Future research needs are identified which will improve our understanding of sustainable forest management in relation to climate change mitigation.

6.1 Stand development phases

The benefits provided by forests vary with tree age and forest structure. Wood yields are usually maximised in stands of regularly-spaced stems of similar size, perhaps 15–20 m tall, which can be efficiently harvested by modern machinery. By contrast, public preference is for tall, large trees, and varied spacing with multilayered canopies (Ribe, 1989; Lee, 2001). Certain animal and plant species prefer the open habitat that occurs after the felling of one generation of trees and which lasts until the young regenerating trees have closed canopy. This succession from open conditions to a closed canopy and then the gradual break-up to a more open stand structure lasts at least a century in most British forests. In carbon budget terms, this amounts to a stand moving from being a carbon source (due to stand and soil disturbance during harvesting and ground preparation), to a sink during tree regrowth, and then progressively to being a growing carbon store with a reduced sink strength (see Chapter 3).

This succession can be described using the four phases of stand development that form the basis of a widely-used model of temperate forest stand dynamics (Oliver and Larson, 1996). The first phase ('stand initiation') covers the period when trees are establishing on a site and have not formed a canopy so that grasses, herbaceous plants, and other vegetation are still present. In the subsequent 'stem exclusion' phase, the trees have formed a continuous canopy and compete with each other for light, moisture and nutrients. Ground vegetation only persists if the canopy trees allow sufficient light to penetrate to the forest floor. In the third phase ('understorey reinitiation'), the canopy starts to open up as a result of competition ('self-thinning') and other processes. Light transmittance through the stand canopy increases, and tree saplings and woody shrubs can colonise and develop in the understorey. Stands in the last phase ('old growth') are characterised by the presence of big old trees, substantial canopy gaps, groups of saplings that regenerated in the previous phase

now reaching the canopy, and appreciable numbers of standing dead mature trees.

Species composition and management practice affect the duration of each phase and this also affects the forest carbon cycle. Thus, planting of fast-growing species combined with effective weed control results in rapid dominance of the site by the planted trees and a quick return to a situation where the stand (including the soil) is a net carbon sink. By contrast, poor establishment practice can result in an extended stand initiation phase and a lengthy period where the stand is at best 'carbon neutral'. Stand initiation is the phase when managers can best change species or introduce a range of provenances to increase forest resilience to climate change. Management intervention in the subsequent phases is essentially a means of manipulating the developing carbon store (Millar et al., 2007). Regular thinning of stands in the stem exclusion phase provides wood products to substitute for fossil fuels or to displace more energy intensive construction materials (see Chapter 7). Thinning also maintains the growth rate of the remaining trees for longer so that the period when the forest is sequestering carbon is extended.

Afforestation is best considered as a special case of the stand initiation phase which takes place on agricultural or other land far from suitable seed sources where planting is the most reliable way to develop forest conditions. Past management may also have depleted the soil nutrient reserves, particularly in the uplands, so that remedial fertiliser may be required to facilitate the start of the forest cycle. At the other end of the forest development cycle, old growth stands can revert to open ground if browsing pressure or vegetation change reduce regeneration success and result in the loss of the forest habitat.

Forests managed solely for wood production tend to have a high proportion of stands in the stand initiation and stem exclusion phases whereas those managed for multifunctional objectives (including timber) will tend to have representation across all phases. The estimated distribution of these phases in British forests in 2000 (Table 6.1) shows that they are currently dominated by fast-

Table 6.1
Area (thousand ha) of conifer and broadleaved high forest within Great Britain by four stand development phases and percentage distribution (brackets) in 2000 (adapted from Mason, 2007: Table 3).

Forest type	Stand initiation	Stem exclusion	Understorey reinitiation	Old growth
Conifers	219.5 (17)	1067.8 (77)	73.0 (5)	19.1 (1)
Broadleaves	57.9 (7)	269.6 (30)	376.9 (43)	176.3 (20)

growing conifer stands in the stem exclusion phase where carbon sequestration rates are highest. However, many of these stands will be due for felling in the next 10–20 years (Mason, 2007) which represents a major opportunity for directing forestry management towards addressing climate change issues. This will require assessment of the trade-offs between management of forests as carbon sinks vs management for other objectives such as recreation, biodiversity and timber (see Section 5).

6.2 Forest management alternatives and their implications for carbon budgets

Forest operations such as soil cultivation, weed control, thinning and timing and extent of felling, all result in spatial and temporal variations in forest carbon budgets. Forest management alternatives (FMAs) consist of a particular pattern of stand development supported by characteristic forest operational processes. FMAs can be defined by the general management objectives and a corresponding intensity of forest resource manipulation. Each FMA will have different carbon stocks and rates of sequestration (Table 6.3).

Duncker *et al*. (2008) identified five FMAs in Europe which, arranged in order of increasing intensity of wood biomass removal, are:

- Unmanaged forest nature reserve (FMA 1);
- Close-to-nature forestry (FMA 2);
- Combined objective forestry (FMA 3);
- Intensive even-aged forestry (FMA 4);
- Wood biomass production (FMA 5).

Features used to distinguish between these FMAs include: species composition, management of stand density and/or pattern, age pattern/phases of development, stand edges/boundaries, amount and intensity of timber and biomass removal, and site conditions. The amount of external energy used in operational processes also differs between management alternatives. The five FMAs encompass the three options for forest carbon management outlined by Broadmeadow and Matthews (2003). Thus, an unmanaged forest nature reserve is equivalent to 'carbon reserve management', where there is a gradual accumulation of carbon stocks primarily within deadwood and soils. Close-to-nature forestry is a 'selective intervention carbon management' approach

with the harvesting of high-quality timbers to replace more carbon intensive structural materials in housing (see Chapter 7). Combined objective forestry contains elements of both 'selective intervention carbon management' and 'carbon substitution management'. The latter involves an emphasis on managing forests for products which reduce net fossil fuel consumption in the wider economy such as construction timbers, boards and paper. Intensive even-aged management and wood biomass production are FMAs which focus on carbon substitution.

The salient characteristics of each FMA are outlined below.

6.2.1 Unmanaged forest nature reserve

The main objective of an unmanaged forest nature reserve is to allow natural processes and disturbances (e.g. windthrow) to create natural, ecologically valuable habitats. It will tend to be dominated by stands in the old growth and understorey reinitiation phases. The trees continue to accumulate carbon as a multilayered canopy ensuring continued woody biomass growth after local disturbance (*cf.* Luyssaert *et al*., 2008). FMA 1 represents a 'saturated' carbon stock with balanced above-ground fluxes but continued soil carbon accumulation. No operations are allowed in a forest reserve that might change the nature of the area. Examples in British forests include National Nature Reserves or long-term biological retentions. The soil disturbance after windthrow will result in some loss of carbon (see 6.3 below).

6.2.2 Close-to-nature forestry

The aim here is to manage a stand with the emulation of natural processes as a guiding principle. Financial return is important, but management interventions must enhance or conserve the ecological functions of the forest. Timber can be harvested and extracted, but some standing and fallen deadwood is left, which may reduce productivity. Only native or site adapted tree species are chosen. Natural regeneration is the preferred method of establishing new seedlings. The rotation length is generally much longer than the age of maximum mean annual volume increment (MMAI – see Glossary) and harvesting uses small-scale removals resulting in the development of an irregular and intimately mixed stand structure. The understorey reinitiation phase features prominently in FMA 2 and high long-term carbon stocks will result from this form of management. Stands in forests such as Glentress, Fernworthy and Clocaenog managed under a continuous cover forestry (CCF) or low impact silvicultural system (LISS), would fall within this FMA.

6.2.3 Combined objective forestry

In this FMA, management explicitly pursues a combination of economic (timber production) and non-market objectives. Mixtures of tree species are often promoted, comprising both native and introduced species suitable for the site. Natural regeneration is the preferred method of restocking, but planting is also widely used. Site cultivation and/or fertilisation may be carried out to speed up the development of a young stand. The rotation length is either similar to (in conifers) or longer than (broadleaves) the age of MMAI and the harvesting system is generally designed around small-scale clearfelling with groups of trees retained for longer periods to meet landscape and biodiversity objectives. Annual carbon sequestration benefits are lower than in other FMAs but this option provides a diverse species mix reducing risk. Characteristic stands are in the understorey re-initiation or late stem exclusion phase. Forest management aims to produce sawlogs as a primary timber product. Forests in areas of high landscape value such as the Trossachs, Snowdonia, and the Forest of Dean conform to FMA 3.

6.2.4 Intensive even-aged forestry

The main objective in intensive even-aged forestry is to produce timber, although landscape and biodiversity issues may feature as secondary objectives. Typical stands tend to be even-aged, in the stem exclusion phase, and composed of one or very few species. Any species can be suitable provided it is site adapted and non-invasive, and planting is the preferred method of regeneration. Intensive site management including cultivation and weed control is used to ensure rapid establishment. Genetically improved material is often planted where available. The rotation length is often less than or similar to the age of MMAI and clearfelling is normal practice. FMA 4 returns high annual sequestration for sites with medium-high productivity, while thinning strategy affects the total carbon budget. Stands in areas of high wind risk may not be thinned. In some cases, whole tree harvesting may occur but residues are normally left on site. This type of management is typical of many planted forests in Britain such as in Thetford, Kielder and Argyll.

6.2.5 Wood biomass production (or short rotation forestry or energy forestry)

The main objective is to produce the highest amount of small dimension wood biomass or fibre. Tree species selection depends mainly on the economic return, as long as the species is not invasive. Pure stands of single species are generally favoured and intensive site management may occur to ensure rapid canopy closure. Stands cycle between stand initiation and stem exclusion with a short rotation period, i.e. from 5–25 years depending on species characteristics and the economic return. The intensity of harvesting is at its maximum compared with the other alternatives. The final felling is a clear-cut with removal of all woody residues, if there is a suitable market. This represents the most intensive version of 'carbon substitution' management and maximum carbon sequestration rates for highly productive sites (see Chapter 8). If managed like traditional coppice crops with only stem wood harvested, this alternative can also combine reasonable rates of carbon sequestration with careful management of soil carbon stocks. FMA 5 is currently rare in British forestry but examples include poplar and willow short rotation crops grown to produce biomass, sweet chestnut coppice in parts of lowland Britain, dense stands of naturally regenerated conifers which are cleared for wood fuel, and experimental trials of species such as eucalypts, birch and aspen for energy production.

6.2.6 Current distribution of FMAs

In Table 6.2 we estimate the current distribution of these FMAs across the UK forest resource. They indicate predominance of intensive even-aged forestry and, to a lesser extent, combined objective forestry which reflects the expansion of plantation forests in the UK during the last century. The estimates may not allow for the recent under-management of smaller private woodlands in parts of the UK so the proportion of the 'unmanaged' FMA may be higher than is suggested. An important point highlighted by this analysis is the need to obtain better data on the types of forest management being practised in the UK and their distribution.

There will be variation between FMAs both in terms of the carbon stocks retained in the trees and in the rates of sequestration that can be expected (Table 6.3). In general terms, the higher the carbon stock, the lower will be the rate of sequestration and vice versa. The prevalence of intensive even-aged forestry in the UK means that data for other forest management alternatives are limited. For example Patenaude et al. (2003) quote values of around 400 tCO$_2$e ha^{-1} in the carbon stocks of the tree and shrub component of a broadleaved semi-natural unmanaged nature reserve which is half the value predicted for a Sitka spruce stand on an equivalent regime (Morison et al., 2009). The latter is extrapolated from models derived from intensive even-aged forestry and the estimated carbon

Table 6.2
Estimated percentage distribution of UK forests by FMA in 2005 and possible changes by 2025 (see Notes for further detail).

Year	Unmanaged forest nature reserve (FMA 1)	Close to nature forestry (FMA 2)	Combined objective forestry (FMA 3)	Intensive even-aged forestry (FMA 4)	Wood biomass production (FMA 5)
2005	2.5	7	35	55	0.5
2025	5	15	50	25	5

Notes: 1. The prime data source for these estimates is the report on the State of Europe's Forests (MCPFE, 2007) with interpretation by the authors. 2. It is assumed that all single species stands are predominantly intensive even-aged forestry, and that 2–3 species stands indicate combined objective forestry, although some of the latter may represent close-to-nature forestry. 3. Unmanaged forest nature reserves are calculated as those forests falling into MCPFE classes 1.1 and 1.2 (i.e. 'no active intervention' and 'minimum intervention') plus an allowance for the areas of commercial forests set aside as non-intervention areas under the UKWAS protocols. 4. Close-to-nature forestry is taken as being equivalent to MCPFE class 1.3 (i.e. conservation through active management) but with some increase to allow for the increasing commitment to this type of management in Forestry Commission forests. This MCPFE class may contain a small area of nature reserves which are managed under coppice systems, but there is no easy way of identifying these separately. 5. Wood biomass production is thought to have been little practised in 2005. 6. The estimates for 2025 represent the authors' estimates of the impact of current policy trends.

Table 6.3
Indicative estimates of whole tree carbon stocks (tCO$_2$eq ha^{-1}) and annual mid-rotation rates of carbon sequestration (tCO$_2$eq ha^{-1} year^{-1}) that may apply to each FMA (values in parentheses are extrapolated from other measures – see Notes for further detail).

Forest management alternative	Unmanaged forest nature reserve (FMA 1)	Close-to-nature forestry (FMA 2)	Combined objective forestry (FMA 3)	Intensive even-aged forestry (FMA 4)	Wood biomass production (FMA 5)
Carbon stocks	800	500	(450)	400	(200)
Annual rates	6	(11)	(16)	22	29

Notes: 1. Principal data sources used are Morison et al. (2009: Table 3.1 and Figures 3.2–3.6) for carbon stocks, Jarvis and Linder (2007) and Luyssaert et al. (2008) for rates. Values in brackets are extrapolations made by the authors. Sitka spruce is assumed as the species. 2. Extrapolations are based on the assumptions that: (a) carbon stocks in wood biomass production will be a function of the shorter rotation – half or less that of intensive even-aged forestry; (b) carbon stocks in combined-objective forestry are higher than intensive even-aged forestry because of a longer rotation, but the amount of increase is reduced because of likely admixture with less productive species; (c) similarly rates in close to nature and combined objective forestry are likely to be lower than for intensive even-aged forestry because of the greater age of the trees and the presence of less productive species mixtures.

stocks will vary with tree age and site productivity (e.g. Black et al., 2009). The validation and refinement of such estimates is an urgent research requirement.

One consequence of recent forest policies in the UK is a shift in the balance of FMAs away from the dominance of intensive even-aged forestry, based on single species stands towards a greater representation of combined objective and close-to-nature forestry regimes (Table 6.2). While the diversification of species and stand structures that will result from this change is likely to increase the resilience of the forests to climate change, there is potentially a decline in carbon sequestration rates, unless this is offset by the use of more productive genotypes (see below) and/or extended afforestation programmes. Greater use of fast growing species on short rotations as part of wood biomass production may also help to maintain the current rate of carbon sequestration in British forests.

There are also abiotic risks associated with this change in the balance of FMAs, since all the less 'intensive' regimes will result in trees being retained for longer before harvesting, thereby increasing the risk of wind damage. This risk would be compounded by any deterioration in the wind climate, e.g. an increased frequency of major storms (Ray, 2008; Schelhaas et al., 2003). This is of considerable concern as substantial areas of UK forests are sited on exposed sites and/or shallow rooting soils where the risks of wind damage are substantial (Quine et al., 1995). Obtaining better understanding of the potential changes in wind climate and adapting existing wind risk models (e.g. ForestGALES; Gardiner et al., 2000) to cope with more varied stand structures will help manage the risk across FMAs. Furthermore, increased winter rainfall could increase soil waterlogging and so reduce tree stability in stormy weather (Ray, 2008; see 5.1.5, Chapter 5).

It is unlikely that any single FMA can be considered as the optimum solution for adapting British forests to climate change and the FMAs should be considered as options to be used in combination depending upon site, management objectives, and species composition (Millar et al., 2007). Guidance should be developed for forest managers, which will outline the interactions between FMA, wood utilisation, and forest carbon management (Matthews et al., 2007) This guidance will require an improved understanding of the ways that management can affect the carbon contained in forest soils.

6.3 Carbon management and forest soils

6.3.1 Soil carbon stocks

Forest soils can contain more carbon than that retained within tree woody biomass, particularly in the case of peat-based soils common in the upland areas of the UK (Broadmeadow and Matthews, 2003; Janzen, 2004, see also Chapters 3 and 8). For instance, Greig (2008) estimated that the carbon stored in the soils of Kielder forest was 3.5–4.5 times that found in the above ground tree biomass. The stability of this store is of primary importance to climate change mitigation and therefore there is a need for an accurate inventory and monitoring programme.

Estimates of carbon content in forest soils vary between 90 and 2500 tCO_2e ha^{-1}, depending on soil depth, soil density, site type and management (Morison et al., 2009). Soils can essentially be split into non-organic (mineral) and organo-mineral/peaty soils (peaty-gleys, peaty podzols and deep peats). Organo-mineral soils have been reported to contain between 235 and 418 tCO_2e ha^{-1} in the horizons between 5 and 20 cm depth. However the carbon stock can reach between 620–1400 tCO_2e ha^{-1} depending on the age of the stand and the depth of the organic horizon. In peat soils (i.e. peat layer depth >40 cm) up to 1000 tCO_2 eq ha^{-1} can be held in the peat of 0–40 cm depth (Morison et al., 2009).

Measured soil carbon stocks for different soil and forest types from 167 forest plots across Great Britain in 2007 ranged between 400 and 1800 tCO_2e ha^{-1} (Figure 6.1; Vanguelova pers. comm.). Carbon content varied with soil depth, soil type, forest type and stand age. Carbon stocks across the different soil types decreased in the order: deep peats > peaty gleys > rendzinas and rankers > ground

water gleys > surface water gleys > podzols and ironpans > brown earths. The average carbon content across the non-organic soils was 539 tCO_2e ha^{-1} while on peaty soils and deep peats carbon stocks of 460–2000 tCO_2e ha^{-1} were found depending on peat layer depth. These stocks are in line with other measured carbon stocks for a range of forest soil types (Zerva and Mencuccini, 2005a; Zerva et al., 2005, Carey et al., 2008, Benham, 2008).

Figure 6.1
(a) Total soil carbon stocks (tCO_2e ha^{-1}, excluding litter) for each main soil group measured to a depth of 80 cm. Bars represent averages of total of 127 UK forest plots. Error bars represent the standard errors of the mean.
(b) Soil carbon stocks (tCO_2e ha^{-1}) for deep peat soils are related to peat layer depth.

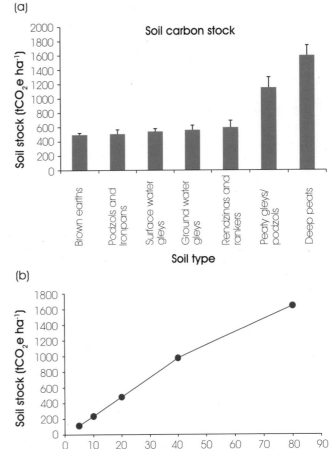

Scaling up these values for particular soil types across the forested areas of the UK (Figure 6.2) shows that forest soil carbon stocks were greater under conifers compared to broadleaves and highest in Scotland and lowest in Wales. Peaty gleys contributed most to the total carbon stock in Scotland, while brown earths and podzolic soils made the largest contribution in Wales and brown earths and surface water gley soils in England.

Figure 6.2
Total carbon stocks (in CO_2 equivalents) in forest soils in England, Wales and Scotland for the two main forest types, estimated by up-scaling UK BioSoil results.

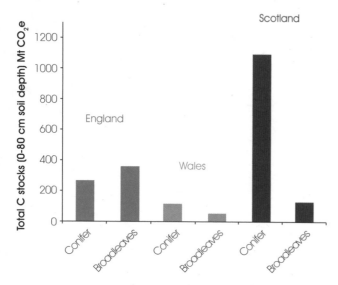

6.3.2 Impacts of disturbance on soil carbon

The effect of soil disturbance, whether due to anthropogenic (e.g. forest management activities) or climatic (e.g. catastrophic windthrow) causes, needs to be included in the calculations of forest carbon soil stocks. Site cultivation to provide a weed-free position for planting a young tree is a characteristic feature of establishment practice in British forestry, particularly in the uplands. However, the soil disturbance associated with this practice results in carbon losses, primarily through enhanced decomposition. The site will remain a net carbon source until uptake by growing biomass exceeds the soil losses which vary with the intensity of cultivation and soil type (Table 6.4 and Johnson, 1992).

There are few data on the potential contribution of soil disturbance on forest carbon balances. A study of root-soil plate volumes in an intensive even-aged (40-year-old) Sitka spruce forest in Scotland (Nicoll et al., 2005) indicated that a volume of soil of 1882 m^3 ha^{-1} would be disturbed if all trees were uprooted by windthrow. Stump harvesting operations will cause similar soil disturbance to windthrow, but current guidance on good practice for stump harvesting (Forest Research, 2009) recommends limiting disturbance to 60–70% of the site. Windthrow and stump harvesting would therefore be expected to result in around 750% and 450%, respectively of the soil disturbance from site preparation by excavator mounding reported by Worrell (1996). In Scandinavia, windthrow in the 2005 Gudrun storm resulted in a carbon sink reduction of around 3 million tonnes carbon, while the larger Lothar storm of 1999 may have resulted in losses of 16 million tonnes carbon (Lindroth et al., 2009).

6.4 Consequences of woodland creation for soil carbon content

The current rate of afforestation in the UK is around 7500 ha per annum (see Chapter 1). If the objective of afforestation is to sequester the maximum quantity of carbon in the short term, the choice of species and site is paramount. Land-use change can result in dramatic changes in soil carbon stocks, with, for example, conversion of agricultural crop land to forest plantation having a positive effect and pasture to forest plantation having a negative effect on soil carbon (Guo and Gifford, 2002). Other reviews (Polglase et al., 2000; Paul et al., 2002) have found that changes in soil carbon after afforestation were generally limited in magnitude.

The major determinants of the extent of soil carbon change

Table 6.4
The soil disturbance of site preparation treatments typical in upland UK forestry (after Worrell, 1996).

Treatment	Area affected (%)	Soil volume (m^3 ha^{-1})	
		Mean	Range
Ploughing	44–60	510	370–850
Mounding	26–35	250	170–340
Disc trench scarifying	20–32	170	110–280
Hand turfing	4–7	60	40–60
Hand screefing	Negligible	Negligible	

under afforestation are soil type and previous land use. British studies have concentrated on peat soils where afforestation could cause significant initial carbon loss from the soil due to drainage and ploughing (see Chapter 3). This loss can be about 20–25% of the total carbon in the peat (Harrison et al., 1997; Jones et al., 2000). However, there are difficulties in these comparisons, particularly the assumption that soil carbon is in equilibrium prior to the disturbance. Hargreaves et al. (2003) measured fluxes on a deep peat site in Scotland following afforestation, and found the soil became a net source of carbon peaking with a flux of 14.6 tCO_2e ha^{-1} $year^{-1}$, two years after planting. This net emission then fell to become a net sink of carbon with a maximum value of 7.3 tCO_2e ha^{-1} $year^{-1}$ occurring seven years after planting, before the size of the sink began to shrink. However, these data do not agree well with other studies of soils with lower carbon contents. An analysis of four upland UK afforestation sites by Reynolds (2007), coupled with modelling of biomass carbon accumulation showed that, despite a loss from the peat (soil) of 1.83 tCO_2e ha^{-1} $year^{-1}$, the forest stand net ecosystem productivity (NEP, see Chapter 3 for definitions) was around 165 tCO_2e ha^{-1} over a 26-year period (6.3 tCO_2e ha^{-1} $year^{-1}$), Zerva and Mencuccini (2005b), working on a peaty-gley site in northern England, found that the first 40-year rotation resulted in a decrease in soil carbon of 12.5 tCO_2e ha^{-1} $year^{-1}$. They attributed this decline to accelerated decomposition caused by drainage and cultivation. Subsequently, in the second rotation there was a recovery of soil carbon (see Chapter 3). However, the estimates from these studies have a large degree of variation associated with them (cf. Conen et al., 2005).

By contrast, no studies have been published on the afforestation of mineral soils in the UK. On a mineral soil site afforested with Norway spruce in Denmark, Vesterdal et al. (2002) found that although in the top 5 cm of soil, carbon content increased over the first few years, the lower layers of soil lost carbon, leading to an overall loss of 0.73 tCO_2e ha^{-1} $year^{-1}$. In Ireland, Black et al. (2009) report a mean annual increase in soil carbon content of 8.1–9.6 tCO_2e ha^{-1} $year^{-1}$ over the first 16 years of the rotation of high yield class (20–24 m^3 ha^{-1} $year^{-1}$) first rotation Sitka spruce stands established on surface water gley mineral soils. In Canada the attractiveness of afforestation for carbon sequestration was found to be highly sensitive to stand growth and yield (McKenney et al., 2004). However, the impacts of afforestation on site carbon balance, specifically the effects of cultivation on soil carbon, are poorly defined and understood.

6.5 Species and provenance choice in British forests as affected by anticipated climate change

The choice of tree species that are planted and the resulting stand composition may have a major impact on the carbon sequestration capacity of the forest ecosystem (Hyvönen et al., 2007). Broadmeadow et al. (2005) highlighted the need to select and use provenances and species that are more suited to the future climate, noting that sites which are currently marginal because of a species' soil moisture requirement are likely to prove problematic. Predictions of species response to climate change based on the Ecological Site Classification (ESC) decision support system (e.g. Broadmeadow and Ray, 2005; Ray, 2008) suggest that species suitability will change across Britain (see Section 2). For example, Corsican pine is anticipated to become more suitable across parts of southern and eastern England as a result of warmer temperatures, but this species is not currently recommended because of its susceptibility to red band needle blight (see Section 2). On drier sites in eastern Scotland, Sitka spruce will prove less well-adapted because of its sensitivity to summer drought, which raises questions concerning suitable replacement species. These predictions are based on average climate trends and do not allow for extreme events (e.g. the 2003 high temperatures in central Europe), which are more likely to influence species survival. The warming climate may permit the wider use of species that were previously not reliably cold-hardy within the British Isles. Examples include maritime pine and a range of other conifers, southern beeches, new poplar clones, various eucalypts and other broadleaved species. Information from existing trials suggests some of the species that might be suitable (Table 6.5) but systematic trials of potential new species are urgently needed to provide the knowledge base to underpin future planting programmes. However, it will not be possible to fully test some species because of the long lead times for forest development.

Choice of tree species should also reflect variation in carbon content between species as well the more traditional measures of volume increment found in current British yield tables (Edwards and Christie, 1981). Table 6.6 lists average carbon content of a number of major species used in Britain alongside the respective range of MMAI. The MMAIs quoted are for the valuable stemwood component and make no allowance for other components

Table 6.5
Potential species that might be considered in climate change adaptation strategies for production forestry in Britain.

Species for which there is existing UK-based knowledge of performance from operational trials/ forest gardens/arboreta	Species for which there is little or no UK trials data but expert knowledge suggests that they merit screening for UK potential
Conifers	
Abies alba	Abies bornmuelleriana
Abies amabilis	Abies cephalonica
Abies nordmanniana	Pinus armandii
Cedrus atlantica	Pinus ayacahuite
Cedrus libani	Pinus brutia
Cryptomeria japonica	Pinus elliottii
Picea omorika	Pinus koraiensis
Picea orientalis	Pinus monticola
Pinus peuce	Pinus strobus
Pinus pinaster	Pinus taeda
Sequoia sempervirens	Pinus wallichiana
Thuja plicata	Pinus yunnanensis
Broadleaves	
Acer macrophyllum	Betula papyrifera
Acer saccharinum	Carya ovata
Alnus rubra	Eucalyptus spp.
Alnus viridens	Fagus orientalis
Eucalyptus gunnii	Fraxinus americana
Eucalyptus nitens	Fraxinus angustifolia
Juglans regia	Fraxinus pennsylvanica
Nothofagus obliqua	Juglans nigra
Nothofagus alpina (syn. N. procera)	Liriodendron tulipfera
Nothofagus pumilio	Quercus alba
Platanus spp.	Quercus frainetto
Populus spp.	Quercus pubescens
	Quercus pyrenaica

such as branchwood and stumps. The carbon contents are derived from a limited number of samples and there may be variation due to growth rate or latitude as reported for Sitka spruce (Macdonald and Hubert, 2002). The ages of MMAI are more theoretical than practical, since many conifers are felled at younger and broadleaves at older ages than those cited below.

The figures in Table 6.6 suggest that greater emphasis should be given to species carbon content when planning future restocking programmes if carbon sequestration is the primary objective. For example, Sitka spruce stands growing at less than 12 m^3 ha^{-1} year^{-1} would be sequestering less carbon than Scots pine growing at 8 m^3 ha^{-1} year^{-1}, although the carbon stock in a given stand will

also depend upon the volume produced by each species. Similarly, the introduction of Douglas fir could increase the carbon density of upland spruce forests on higher yielding sites. In general the higher carbon content of most broadleaved species is offset by their much lower rate of growth, although in species with very long rotations (i.e. >100 years) such as oak, the carbon stocks averaged over time can be higher than in faster growing conifer stands (Vallet et al., 2009). Increasing the growth rate of any given forest type will also increase the rate of carbon storage (Cannell and Milne, 1995), while the wider benefits of substitution should also be considered (see Chapter 7).

The afforestation programmes carried out in the last century, particularly in northern and western Britain,

Table 6.6
Timber carbon content (tCO$_2$e m^{-3}), typical ranges of maximum mean annual volume increment (MMAI: m^3 ha^{-1} year^{-1}) and ages of MMAI for a range of conifers and broadleaves grown in Britain or which might be considered for planting under anticipated climate change (after Edwards and Christie, 1981; Lavers, 1983).

Conifers					Broadleaves				
Species	Scientific name	Carbon content	MMAI	Age	Species	Scientific name	Carbon content	MMAI	Age
Sitka spruce	*Picea sitchensis* (Bong.) Carr.	0.62	8–24	64–46	Oak	*Quercus robur* L., *Q. petraea*. (Matt.) Liebl.	1.12	4–8	90–68
Norway spruce	*Picea abies* L. Karst.	0.64	8–20	84–65	Birch	*Betula pendula* (Roth.), *B. pubescens* (Ehrh.)	1.10	4–12	49–40
Scots pine	*Pinus sylvestris* L.	0.84	6–12	82–69	Sweet chestnut	*Castanea sativa* Mill.	0.84	4–10	50–41
Corsican pine	*Pinus nigra* var. *maritima* (Ait.) Melville.	0.77	8–16	64–55	Ash	*Fraxinus excelsior* L.	1.10	4–12	49–40
Douglas fir	*Pseudotsuga menziesii* (Mirb.) Franco.	0.81	10–24	64–50	Beech	*Fagus sylvatica* L.	1.14	4–10	107–80
Japanese larch	*Larix kaempferii* (Lamb.) Carr.	0.81	6–14	56–41	Wild cherry	*Prunus avium* L.	1.03	4–12	50–40
European larch	*Larix decidua* Mill.	0.88	6–12	60–47	Hornbeam	*Carpinus betulus* L.	1.19	4–10	107–80
Hybrid larch	*Larix x eurolepis* Henry	0.74	6–14	56–41	Lime	*Tilia cordata* Mill., *T. platyphyllos* (Scop.)	0.92	4–10	50–41
Maritime pine	*Pinus pinaster* Ait.	0.79	6–14	71–54	Black poplar	*Populus nigra* L.	0.70	6–1	39–35
Grand fir	*Abies grandis* Lindl.	0.59	12–28	60–51	Rauli	*Nothofagus alpina* (Poep. and Endl.) Oerst.	0.77	8–18	45–35
European silver fir	*Abies alba* Mill.	0.73	12–22	73–64	Common alder	*Alnus glutinosa* L.	0.83	4–12	50–40

were largely based on a number of non-native conifers. Substantial knowledge has been accumulated on the site preferences of these species and the suitability of particular provenances for different regions of Britain (e.g. Samuel *et al.*, 2007 for Sitka spruce). For all the major conifers, greater use of more southerly provenances is possible (e.g. Oregon or Washington seed sources replacing Queen Charlotte Islands for Sitka spruce), and would be an effective means of adapting to predicted climate change. There may be a risk of unseasonal frosts affecting more southerly material, so careful matching of species and provenances to sites will be essential. The faster growth rates that will be obtained from more southerly provenances are likely to result in timber with lower density, at least in Sitka spruce (Macdonald and Hubert, 2002) and therefore in lower carbon content, with potential

implications for substitution. Tree improvement strategies can compensate for any decline in carbon content by selecting for higher wood density, as is possible in Sitka spruce (Moore *et al.*, 2009), but the benefits have yet to be fully explored.

By contrast, until recently, very limited work had been undertaken on provenance selection or other aspects of tree improvement for many native broadleaved species (Savill *et al.*, 2005). This situation is further complicated by the preference for using 'local' seed sources in many broadleaved woodlands (Hemery, 2008), since it is arguable that material from the near continent should be introduced (at least in southern Britain) to increase woodland resilience to climate change. Until recently, our ability to predict the likely impacts of climate change on species suitability, growth rates and consequently carbon sequestration rates has been greater for introduced rather than native species (see recent work presented in Section 2).

A strategy that is often proposed to enhance the resilience of a stand or forest to climate change is increasing the use of species or even provenance mixtures (Broadmeadow and Ray, 2005). However, this requires that all species in a mixture to have compatible growth rates and are capable of growing to maturity without intensive intervention. While this may often be the case in broadleaved woodlands, experience of conifer-broadleaved mixtures shows that the former tend to out-compete and suppress the latter on most sites throughout upland Britain (Mason, 2006). In conifer forests, it is better to aim for small clumps of single species ('mosaic' mixtures) rather than to try and create stem by stem or line by line mixtures of species or provenances ('intimate' mixtures). Intimate mixtures are most likely to be successful where the site conditions are suboptimal for all the species being considered for planting; in all other situations mosaic mixtures are likely to prove more reliable.

6.6 Machine operations and carbon impacts

The majority of forestry operations during forest management in the UK are now mechanised, including road building and maintenance, site cultivation, and thinning and harvesting. Each has a cost in terms of primary energy use, and an associated release of GHG. In addition, each operation involves some soil disturbance and will consequentially lead to a release of GHG. For a given energy input, emission figures can be derived. Carbon emissions arising from the use of diesel fuel in forestry operations can be calculated (Defra, 2007). These are equivalent to 0.071 tC MWh^{-1} and are used in calculating the fuel emission values in Table 6.8.

There has been very little investigation into fuel use and resultant GHG emissions of forestry operations in the UK. What information there is predominantly examines

carbon losses, and not production of other GHG. Here, we describe primary energy use and GHG release from each major machine operation, and then attempt to estimate soil carbon loss following each operation based on the volume of soil disturbed.

6.6.1 Road building and maintenance

Forest roads in the UK are essentially constructed as 'water bound macadam'. They are classed as 'Type A' for arterial roads that are in regular use, 'Type B' for infrequently used spur roads for access to stands, and 'Type C' for other purposes. Soil disturbance from road building may lead to accelerated loss of soil carbon through decomposition, especially from the higher carbon soils. The production of material for road building involves the release of substantial amounts of GHG from quarrying and preparation, not least of which is the release of N$_2$O from use of explosives in a quarry. Road building operations also require heavy machinery that consumes fossil fuel in extracting, preparing, transporting and laying road stone. Forest roads can require around 10 000 tonnes of rock per km (Whittaker 2008, Dickerson pers. comm. 2009). Class A roads require frequent maintenance. There is occasionally a need for reconstruction or upgrading work, but a well-built forest road with a good 'sacrificial' surface layer that is maintained and replenished, should have a good economic life (Dickerson, 1996). Life cycle analysis of UK forest roads has been conducted and the data have been used to estimate UK figures presented in Table 6.7.

Modelling the impacts of erecting turbines on deep peat soils (Nayak et al., 2008), along with forest civil engineering standards, allows calculation of the potential impact of floating road construction. For floating roads developed on a deep peat site the impact is estimated to be approximately 345 tCO$_2$e km^{-1} (assumed width 3.4 m, depth 0.5 m) for construction at afforestation.

6.6.2 Site cultivation

The use of machinery in site preparation will involve

Table 6.7
Estimated Type A and B forest road lengths in UK forests, total primary energy use in their construction and maintenance and associated GHG emissions (CO$_2$ equivalent), based on a life cycle analysis (Whittaker et al., 2008).

Estimated length of forest roads ('000 km)			Total annual primary energy use (000 GJ year^{-1})	Total GHG emissions (tCO$_2$e year^{-1})
Type A forest road	Type B forest road	Total forest road length		
15.2	28.1	43.3	556.5	42 493

Table 6.8

UK operational fuel use figures for standard establishment and harvesting procedures (includes direct and indirect emissions).

Operation	Machinery	Average fuel emissions	Units
Establishment	Excavator	0.6550	$tCO_2e\ ha^{-1}$
	Scarifier	0.2382	$tCO_2e\ ha^{-1}$
Agricultural conversion	Agricultural plough	0.073	$tCO_2e\ ha^{-1}$
Thinning	Harvester	0.0046	$tCO_2e\ m^{-3}$
Felling	Harvester	0.0036	$tCO_2e\ m^{-3}$
	Forwarder	0.0027	$tCO_2e\ m^{-3}$
Stump extraction (70% removal)	Modified excavator	0.0470	$tCO_2e\ odt^{-1}$
	Forwarder	0.0186	$tCO_2e\ odt^{-1}$
	Shred	0.1240	$tCO_2e\ odt^{-1}$

$tCO_2eq\ ha^{-1}$ = tonnes of carbon dioxide equivalent per hectare; odt = oven-dry tonnes.

carbon emissions. The figures in Table 6.8 are based on a preliminary survey of UK contractors working in forest establishment. The figures are higher than those from other European studies. For site preparation in Finland, Karjalainen and Asikainen (1996) reported an average carbon cost of 0.35 $tCO_2e\ ha^{-1}$ for fuel emissions in scarification, ditch clearing and remedial drainage. Similar figures were obtained in Sweden where standard site preparation practice had a carbon cost equating to ~0.6 $tCO_2e\ ha^{-1}$ (Berg and Lindholm, 2005).

6.6.3 Thinning and harvesting

UK derived 'on-site' fuel use is calculated as full life cycle analysis (LCA) carbon equivalent costs of forest harvesting operations (Table 6.8). Note that while it is possible to estimate fuel-derived emissions for some operations (e.g. harvesting) as C or CO_2e per timber volume, for other operations (e.g. woody biomass provision from stumps) the appropriate measure is CO_2e emitted per oven-dry tonne (odt) harvested.

Similar figures have been derived by Greig (2008) for Kielder with calculated harvesting emissions (harvesters and forwarders combined) of 0.00599 $tCO_2e\ m^{-3}$. The most directly comparable data on forest operational emissions comes from Scandinavia where Berg and Lindholm (2005) calculated that harvesting caused emissions of 0.0044 $tCO_2e\ m^{-3}$ and forwarding 0.0036 $tCO_2e\ m^{-3}$. Karjalainen and Asikainen (1996) performed a comprehensive assessment of the energy use and resultant GHG emissions in Finnish forestry. They calculated that harvesting caused emissions of 0.0039

$tCO_2e\ m^{-3}$, thinning 0.0082 $tCO_2e\ m^{-3}$ and forwarding 0.0041 $tCO_2e\ m^{-3}$. These higher values relative to British estimates may reflect differences in stand structure and operational efficiency.

6.6.4 Overall emission estimates

We combined figures from the sections above to provide preliminary estimates of GHG emissions for road construction/maintenance, cultivation, thinning operations, and harvesting in British forests (Figure 6.3). These calculations are based on theoretical yields and management regimes for a Sitka spruce stand of average productivity. These were adjusted for each FMA by appropriate rotation lengths, operations, out-turns and estimated forest area covered by each FMA

Figure 6.3

Estimated total annual GHG emissions ($tCO_2e\ year^{-1}$) from UK forest machine operations.

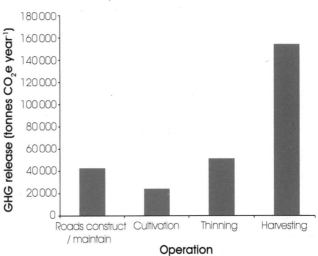

(Table 6.2). The figures do not include timber transport or GHG emissions following soil disturbance associated with these operations. Clearly, GHG emissions are considerably greater from harvesting than from any other forest operation and the total emission associated with forest machine operations in the UK is estimated to be 0.26 $MtCO_2e$ year^{-1} or an overall average of 0.09 tCO_2e ha^{-1} year^{-1}. Further work is needed to see how sensitive these results are to different policy scenarios, such as an increase in thinning to produce woodfuel.

The intensity of machine operations varies considerably between forest management alternatives and the yearly GHG emissions would be expected to increase with greater amounts of biomass removal as shown in Figure 6.4. This trend also reflects less use of site cultivation in FMAs which rely upon natural regeneration such as close-to-nature forestry.

A key finding from a recent analysis of the carbon budget for Kielder Forest (Greig, 2008) suggests that the annual carbon emission from all forest machine operations (e.g. harvesting, haulage, cultivation, roading) was around 0.21 tCO_2e ha^{-1}, or nearly 40 times less than the annual sequestration in the above ground tree biomass. The discrepancy between this and the UK average is in part explained by the exclusion of haulage from the UK figures.

Figure 6.4
Estimated GHG emissions per hectare per year from each forest management alternative (FMA).

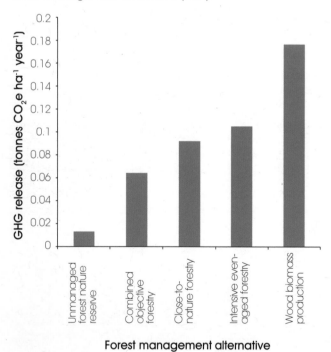

6.7 Incorporating mitigation strategies into forest management

The total UK forest carbon stock in trees is approximately 550 $MtCO_2e$ over 2.8 Mha with an average stock of approximately 200 tCO_2e ha^{-1}. This includes an allowance for open ground not in production and under-managed stands (Morison et al., 2009). Average soil carbon stocks for woodland soils in the UK vary greatly with soil type (see above), but a UK average (including litter) is approximately 830 tCO_2e ha^{-1} (Morison et al., 2009). Under current trends, the UK forest carbon stock will continue to increase (MCPFE 2007, Nabuurs et al., 2008), while annual growth increment exceeds losses and removals. However, forest carbon sequestration over coming decades will vary, due primarily to the fluctuation in afforestation rates during the last century, and the rates of carbon uptake to UK forests are now declining (see 8.1.1, Chapter 8 and Figure 8.1).

British forest managers are now being challenged to integrate mitigation strategies into forest planning to increase the potential for forestry to sequester atmospheric CO_2 and reduce overall GHG emissions. Assessments of forest management mitigation strategies should also include carbon storage in products and carbon substitution effects (Lindner et al., 2008, see Chapters 7 and 8). Sensitivity analyses of a model-based approach showed that parameters exhibiting the highest influence on carbon sequestration are carbon content, wood density and current annual increment of stems (Nabuurs et al., 2008).

Emissions reduction in all phases of the management cycle need to be identified and quantified in order to maximise the contribution of sustainable forest management to national climate change mitigation strategies. For example, extending rotation length and therefore moving into the understorey reinitiation phase, and perhaps creating old-growth characteristics under a close-to-nature FMA, can diversify habitat structure (Kerr, 1999), while also helping to adapt forests to climate change and favouring long-term carbon sequestration (Liski et al., 2001). The net carbon benefit of transition from even-aged Norway spruce to continuous cover management has been estimated at 1.65–2.75 tCO_2e ha^{-1} year^{-1} (Seidl et al., 2008). However, the application of this FMA in the UK is constrained by the risk of windthrow, which limits the number of sites where conditions permit the transformation of existing even-aged stands to more complex structures (Mason and Kerr, 2004). When carbon stocks are compared between unmanaged and managed forest stands, unmanaged stands typically show higher stocks. However, these results generally exclude the effects of disturbance (Lindner et al., 2008) and also do not take into account the full life cycle analysis including the GHG balance and the benefits of substituting forest products for fossil fuels and other materials. If the primary management objective is the maximisation of forest carbon sequestration rate, more intensive FMAs may be more favourable, as proposed for New Zealand (Turner et al., 2008).

Shorter rotation lengths may decrease the risk of abiotic and biotic damage, while regular thinnings can maintain stand vigour and increase resilience by developing sturdier trees. In stands facing drought stress and reduced growth rates, thinning practices may be adapted to optimise water use and increase vitality and vigour of the remaining trees in the stand (Kellomäki et al., 2005). In highly productive forest stands, altered management practices (e.g. different thinning intensities) may be needed to reflect the increased growth and yield of the forest ecosystem under future climate scenarios (Garcia-Gonzalo et al., 2007). However, as yet few studies have analysed the effects of silvicultural strategies on carbon sequestration, timber production and other forest services and functions at the operational level of the forest management unit (Seidl et al., 2007).

A modelling approach can be helpful to explore the impacts of different FMAs on forest ecosystem carbon balances. For example, the hybrid process-based tree growth model 3PGN[1] (Xenakis et al., 2008) has been calibrated and independently validated using eddy covariance and biometric data from the UK for the

assessment of the carbon sequestration potential of Sitka spruce plantations (Minnuno, 2009). The model is based on tree eco-physiology, but with important statistical components included, such as allometric equations, which increase model robustness and calculate woody biomass (carbon) outputs at the stand level for even-aged forests and coupled carbon and nitrogen balances in the soil. It thus enables a complete ecosystem level analysis of biome fluxes such as NEP (net ecosystem productivity, see Chapter 3). When the model was applied to Sitka spruce under two FMAs and calibrated across soils of different productivity, it showed that carbon sequestration varied with site characters and management (Table 6.9).

The values obtained are similar to those reported elsewhere (see Chapter 3). A Sitka spruce stand of moderate productivity (YC 14, i.e. 14 m^3 ha^{-1} $year^{-1}$) on a peaty-gley soil produced a net carbon accumulation during the active growth phase of ca 27 tCO_2e ha^{-1} $year^{-1}$. Total ecosystem carbon, which accounts for changes in both timber and soil carbon stocks suggests that wood biomass production (FMA 5) may achieve greater sequestration than intensive even-aged management (FMA 4) at the most productive (YC 20) site. The differences in total ecosystem carbon between FMA 4 and 5 at YC 20 on mineral soils are not great, but there are major and important differences within each FMA associated with yield class and soil type. However, timber product lifespan was not evaluated in this simulation.

The results illustrate how the series of FMAs can to be used to compare the carbon impacts of different silvicultural strategies as part of adaptive forest

Table 6.9
Modelled values of NPP, NEP and total ecosystem carbon uptake (tCO_2e ha^{-1} $year^{-1}$) for Sitka spruce under 'intensive even-aged management' (FMA 4) and 'wood biomass production' (FMA 5) across sites of different productivity (yield class).

Forest management and site details					NPP		NEP		Total ecosystem carbon
FMA Number	Yield Class	Soil type	Thinning	Rotn. length	Number of rotation		Number of rotation		
					1	2	1	2	
4	10	Peat	1	50	37.5	37.5	12.9	8.4	0.53
4	14	Peaty-gley	1	50	47.0	47.0	23.8	16.6	6.55
4	20	Mineral	1	50	60.9	60.9	34.4	25.0	9.65
5	10	Peat	2	40	34.8	34.8	11.9	7.4	−2.79
5	14	Peaty-gley	2	40	44.0	44.0	23.1	16.0	6.32
5	20	Mineral	2	40	57.4	57.4	33.7	24.0	10.47

[1] 3PGN is composed of 3PG (Physiological Principles for Predicting Growth) model (Landsberg and Waring, 1997) and the ICBM/2N (Introductory Carbon Balance Model) model (Andrén and Kätterer, 1997).

management to meet the requirements of sustainable forest management. Evaluation of best choice management alternatives are, however, hampered by considerable uncertainty and difficulty in analysing net carbon balances (Cathcart and Delaney, 2006). The modelling of the forest carbon balance in different FMAs, involving the selection and combination of various treatments and practices to fit specific circumstances, will be most useful (Millar *et al*., 2007; Pretzsch *et al*., 2008). Furthermore, this approach recognises that strategies may vary based on the spatial and temporal scales of decision-making: planning at regional scales will often involve acceptance of different levels of uncertainty and risk than is appropriate at local scales (Saxon *et al*., 2005).

An urgent need is to develop improved methods of forest planning that take climate change into account and which will help managers take actions to increase the resilience of British forests (Forestry Commission, 2009). For instance, the current guidance for Forestry Commission staff on forest design planning contains no reference to climate change. It is critical that this and other relevant operational guidance documents throughout the sector are revised to allow managers to consider how to adapt their silvicultural practices to a changing climate or alter management to maximise forest mitigation potential. This adjustment would comply with the aspirations of the revised UK Forestry Standard (UKFS, see 1.5.4, Chapter 1). The revised guidance needs to be coupled with site-based training that will help foresters identify areas that may be particularly at risk.

The knowledge base for such guidance would use the Ecological Site Classification (Pyatt *et al*., 2001) approach to integrate species suitability and site characteristics, particularly soil moisture, and would be combined with predictions of climate change to derive a vulnerability ranking for stands and forests in different regions of the country. Stands with higher vulnerability would be those where remedial actions would be concentrated, involving either a change of species or of forest management alternative. Such a methodology would probably need to be developed using a case study approach to see how current knowledge about climate change could be linked to the GIS-based planning systems which underpin contemporary forest management in the UK. Designing, testing and monitoring a process of this type is probably the key to ensuring that forest management practices help adapt British forests to future climate change and maintain the carbon stocks that will help mitigate its impacts.

6.8 Research priorities

- Develop methodologies to help forest managers identify sites and stands most vulnerable to climate change;
- Trialling of species that may be suitable for the current and projected British climate.
- Provide more accurate data on the distribution of FMAs in the UK, develop the capability to model carbon impacts of FMAs and validate estimates in representative stands.
- Better understanding of rates of carbon sequestration and stocks in older stands that are retained for landscape or biodiversity reasons.
- Improve knowledge of the role of fast-growing species used in wood biomass production as a means of maintaining carbon sequestration rates in British forests.
- Improve predictions of changes in wind climate and adapt existing wind risk models to predict vulnerability of more varied stand structures.
- Better prediction of the potential impact of extreme climatic effects (storms, drought) upon British forests.
- Development of an accurate inventory and monitoring programme for forest soil carbon stocks.
- Understand the impact of disturbance (such as harvesting and windthrow) on soil carbon stocks.
- Obtain better understanding of forest soil and ecosystem fluxes of nitrogen in addition to greenhouse gases.
- Validation of models developed for intensive even-aged forestry when applied to other FMAs and/or provision of more flexible models.
- Quantification of the impacts of afforestation on site carbon balance, specifically the impact of cultivation on soil carbon.
- More investigation into fuel use and GHG emissions of forestry operations including the role of more traditional methods of extraction (e.g. horse logging).
- Select for tree progenies with higher wood (hence carbon) densities.

References

ANDRÉN, O. and KÄTTERER, T. (1997). ICBM: The introductory carbon balance model for exploration of soil carbon balance. *Ecological Applications* 7, 1226–1236.

BENHAM, S. (2008). *The Environmental Change Network at Alice Holt Research Forest*. Forestry Commssion Research Note 001. Forestry Commission, Edinburgh.

BERG, S. and LINDHOLM, E-L. (2005). Energy use and environmental impacts of forest operations in Sweden. *Journal of Cleaner Production* 13, 33–42.

BLACK, K., BYRNE, K., MENCUCCINI, M., TOBIN, B., NIEUWENHUIS, M., REIDY, B., BOLGER, T., SAIZ, G., GREEN, C., FARRELL, E.D. and OSBORNE, B.A. (2009). Carbon stock and stock changes across a Sitka spruce chronosequence on surface water gley soils. *Forestry*. doi:10.1093/forestry/cpp005

BROADMEADOW, M. and MATTHEWS, R. (2003). *Forests, carbon and climate change: the UK contribution*. Forestry Commission Information Note 48. Forestry Commission, Edinburgh.

BROADMEADOW, M. and RAY, D. (2005). *Climate change and British woodland*. Forestry Commission Information Note 69. Forestry Commission, Edinburgh.

BROADMEADOW, M.S.J., RAY, D. and SAMUEL, C.J.A. (2005). The future for broadleaved tree species in Britain. *Forestry* **78**, 145–161.

CANNELL, M.G.R. and MILNE, R. (1995). Carbon pools and sequestration in forest ecosystems in Britain. *Forestry* **68**, 361–378.

CAREY, P.D., WALLIS, S., CHAMBERLAIN, P.M., COOPER, A., EMMETT, B.A., MASKELL, L.C., MCCANN, T., MURPHY, J., NORTON, L.R., REYNOLDS, B., SCOTT, W.A., SIMPSON, I.C., SMART, S.M. and ULLYETT, J.M. (2008). *Countryside Survey: UK Results from 2007*. NERC/Centre for Ecology and Hydrology (CEH Project Number: C03259).

CATHCART, J. and DELANEY, M. (2006). Carbon accounting: determining offsets from forest products. In: Cloughesy, M. (ed.) *Forests, carbon and climate change: a synthesis of science findings*. Oregon Forest Resources Institute, Portland, Oregon. pp.157–176.

CONEN, F., ZERVA, A., ARROUAYS, D., JOLIVET, C., JARVIS, P.G., GRACE, J. and MENCUCCINI, M. (2005). The carbon balance of forest soils: detectability of changes in soil carbon stocks in temperate and boreal forests. In: Griffiths, H. and Jarvis, P.G. (eds) *The carbon balance of forest biomes*. Taylor and Francis, Oxford. pp. 235–249.

DEFRA. (2007). Guidelines to DEFRA's GHG conversion factors for company reporting. Online at: www.defra.gov.uk/environment/business/envrp/conversion-factors.htm

DICKERSON, A. (1996). Resourcing civil engineering works in forestry. Unpublished thesis. Joint Board for Engineering Management.

DUNCKER, P., SPIECKER, H. and TOJIC, K. (2008). *Definition of forest management alternatives*. Eforwood project PD 2.1.3. Online at: http://87.192.2.62/eforwood/Home/tabid/36/Default.aspx

EDWARDS, P.N. and CHRISTIE, J.M. (1981). *Yield tables for forest management*. Forestry Commission Booklet 48. HMSO, London.

FOREST RESEARCH. (2009). *Stump harvesting: Interim guidance on site selection and good practice*. Forest Research, Farnham. Online at: www.forestresearch.gov.uk/stumpharvesting

FORESTRY COMMISSION. (2009). *Forests and climate change guidelines*. Consultation draft July 2009. Forestry Commission, Edinburgh.

GARCIA-GONZALO, J., PELTOLA, H., BRICEÑO-ELIZONDO, E. and KELLOMÄKI, S. (2007). Changed thinning regimes may increase carbon stock under climate change: a case study from a Finnish boreal forest. *Climatic Change* **81**, 431–454.

GARDINER, B., PELTOLA, H. and KELLOMÄKI, S. (2000). Comparison of two models for predicting the critical wind speeds required to damage coniferous trees. *Ecological Modelling* **129**, 1–23.

GREIG, S. (2008). A carbon account for Kielder forest. *Scottish Forestry* **62** (3), 2–8.

GUO, L.B. and GIFFORD, R.M. (2002). Soil carbon stocks and land use change: a meta analysis. *Global Change Biology* **8**, 345–360.

HARGREAVES, K.J., MILNE, R., and CANNELL, M.G.R. (2003). Carbon balance of afforested peatland in Scotland. *Forestry* **76**, 299–317.

HARRISON, A.F., JONES, H.E., HOWSON, G., GARNETT, J.S. and WOODS, C. (1997). *Long term changes in the carbon balance of afforested peatlands*. Final report to the Department of the Environment, Contract EPG1/1/3.

HEMERY, G.E. (2008). Forest management and silvicultural responses to projected climate change impacts on European broadleaved trees and forests. *International Forestry Review* **10**, 591–607.

HYVÖNEN, R., ÅGREN, G.I., LINDER, S., PERSSON, T., COTRUFO, M.F., EKBLAD, A., FREEMAN, M., GRELLE, A., JANSSENS, J., JARVIS, P.G., KELLOMÄKI, S., LINDROTH, A., LOUSTAU, D., LUNDMARK, T., NORBY, R.J., OREN, R., PILEGAARD, K., RYAN, M.G., SIGURDSSON, B.D., STRÖMGREN, M., VAN OIJEN, M. and WALLIN, G. (2007). The likely impact of elevated $[CO_2]$, nitrogen deposition, increased temperature and management on carbon sequestration in temperate and boreal forest ecosystems: a literature review. *New Phytologist* **173**, 463–480.

JANZEN, H.H. (2004). Carbon cycling in earth systems – a soil science perspective. *Agriculture Ecosystems and Environment* **104**, 399–417.

JARVIS, P.G. and LINDER, S. (2007). Forests remove carbon dioxide from the atmosphere: spruce forest tales! In: Freer-Smith, P.H., Broadmeadow, M.S.J. and Lynch, J.M. (eds) *Forestry and climate change*. CABI, Wallingford. pp. 60–72.

JOHNSON, D.W. (1992). Effects of forest management on soil carbon storage. *Water, Air and Soil Pollution* **64**, 83–120.

JONES, H.E., GARNETT, J.S., AINSWORTH, G. and

HARRISON, A.F. (2000). Long-term changes in the carbon balance of afforested peat lands. Final Report. April. *Carbon sequestration in vegetation and soils report.* Section 4, part 1. Report to the Department of the Environment, Transport and the Regions. Centre for Ecology and Hydrology, Edinburgh. pp. 3–14.

KARJALAINEN, T. and ASIKAINEN, A. (1996). Greenhouse gas emissions from the use of primary energy in forest operations and long-distance transportation of timber in Finland. *Forestry* **69**, 215–228.

KELLOMÄKI, S., PELTOLA, H., BAUWENS, B., DEKKER, M., MOHREN, F., BADECK, F-W., GRACIA, C., SANCHEZ, A., PLA, E., SABATE, S., LINDNER, M. and PUSSINEN, A. (2005). European mitigation and adaptation potentials: Conclusions and recommendations. In: Kellomäki, S. and Leinonen, S. (eds) *Management of European forests under changing climatic conditions.* Research Notes 163. University of Joensuu, Finland. pp. 401–427

KERR, G. (1999). The use of silvicultural systems to enhance the biological diversity of plantation forests in Britain. *Forestry* **72**, 191–205.

LANDSBERG, J.J., WARING, R.H. (1997). A generalised model of forest productivity using simplified concepts of radiation-use efficiency, carbon balance and partitioning. *Forest Ecology and Management* **95**, 209–228.

LAVERS, G.M. (1983). *The strength properties of timber.* 3rd edn. Building Research Establishment Report. HMSO, London.

LEE, T.R. (2001). *Perceptions, attitudes and preferences in forests and woodland.* Forestry Commission Technical Paper 18. Forestry Commission, Edinburgh.

LINDNER, M., GREEN, T. WOODALL, C.W.PERRY, C.H., NABUURS, G-J. and SANZ, M.J. (2008). Impacts of forest ecosystem management on greenhouse gas budgets. *Forest Ecology and Management* **256**, 191–193.

LINDROTH, A., LAGERGREN, F., GRELLE, A., KLEMEDTSSON, L., LANGVALL, O., WESLIEN, P. and TUULIK, J. (2009). Storms can cause Europe-wide reduction in forest carbon sink. *Global Change Biology* **15**, 346–355.

LISKI, J., PUSSINEN, A., PINGOUD, K., MÄKIPÄÄ, R. and KARJALAINEN, T. (2001). Which rotation length is favourable to carbon sequestration? *Canadian Journal of Forest Research* **31**, 2004–2013.

LUYSSAERT, S., SCHULZE, E-D., BORNER, A., KNOHL, A., HESSENMOLLER, D., LAW, B.E., CIAIS, P. and GRACE, J. (2008). Old growth forests as carbon sinks. *Nature* **455**, 213–215.

MACDONALD, E. and HUBERT, J. (2002). A review of the effects of silviculture on timber quality of Sitka spruce. *Forestry* **75**, 107–138.

MASON, W.L. (2006). *Managing mixed stands of conifers and broadleaves in upland forests in Britain.* Forestry Commission Information Note 83. Forestry Commission, Edinburgh.

MASON, W.L. (2007). Changes in the management of British forests between 1945 and 2000 and possible future trends. *Ibis* **149** (Suppl. 2), 41–52.

MASON, B. and KERR, G. (2004). *Transforming even-aged conifer stands to continuous cover management.* Forestry Commission Information Note 40. Forestry Commission, Edinburgh.

MATTHEWS, R.W., ROBERTSON, K., MARLAND, G. and MARLAND, E. (2007). Carbon in wood products and product substitution. In: Freer-Smith, P.H., Broadmeadow, M.S.J. and Lynch, J.M. (eds) *Forestry and climate change.* CABI, Wallingford. pp. 91–104.

MCKENNEY, D.W., YEMSHANOV, D., FOX, G. and RAMLAL, E. (2004). Cost estimates for carbon sequestration from fast growing poplar plantation in Canada. *Forest Policy and Economics* **6**, 345–358.

MCPFE (2007). *State of Europe's forests 2007: the MCPFE report on sustainable forest management in Europe.* MCPFE/UNECE/FAO, Warsaw.

MILLAR, C.I., STEPHENSON, N.L. and STEPHENS, N.L. (2007). Climate change and forests of the future: managing in the face of uncertainty. *Ecological Applications* **17**, 2145–2151.

MINNUNO, F. (2009). *Parameterisation, validation and application of 3-PGN for Sitka spruce across Scotland: spatial predictions of forest carbon balance using a simplified process based model.* MSc Research, University of Edinburgh.

MORISON, J., MATTHEWS, R., PERKS, M., RANDLE, T., VANGUELOVA, E., WHITE, M. and YAMULKI, S. (2009). *The carbon and GHG balance of UK forests: a review.* Forestry Commission, Edinburgh.

MOORE, J.R., MOCHAN, S., BRÜCHERT, F., HAPCA, A.I., RIDLEY-ELLIS, D.J., GARDINER, B.A. and LEE, S.J. (2009) Effects of genetics on the wood properties of Sitka spruce growing in the United Kingdom. Part 1: Bending strength and stiffness of structural timber. *Forestry.* doi:10.1093/forestry/cpp018

NABUURS, G.J., MASERA, O., ANDRASKO, K., BENITEZ-PONCE, P., BOER, R., DUTSCHKE, M., ELSIDDIG, E., FORD-ROBERTSON, J., FRUMHOFF, P., KARJALAINEN, T., KRANKINA, O., KURZ, W.A., MATSUMOTO, M., OYHANTCABAL, W., RAVINDRANATH, N.H., SANZ SANCHEZ, M.J. and ZHANG, X. (2007). Forestry. In: Metz, B., Davidson, O.R., Bosch, P.R., Dave, R. and Meyer L.A. (eds) *Climate change 2007: mitigation.* Contribution of Working Group III to the Fourth Assessment Report of the

Intergovernmental Panel on Climate Change. Cambridge University Press, Cambridge. pp. 541–584.

NABUURS G.J., THURIG E., HEIDEMA N., ARMOLAITIS K., BIBER P., CIENCALA E., KAUFMANN E., MAKIPAA R., NISEN P., PETRISCH R., PRISTOVA T., ROCK J., SCHELHAAS M.J., SIEVANEN, SOMOGYI Z and VALLET P. (2008). Hotspots of the European forests carbon cycle. *Forest Ecology and Management* **256**, 194–200.

NAYAK D.R., MILLER, D., NOLAN, A., SMITH, P. and SMITH, J. (2008). Calculating carbon savings from wind farms on Scottish peat lands – a new approach. *Final Report*. Rural and Environmental Research and Analysis Directorate of the Scottish Government. Online at: www.scotland.gov.uk/Publications/2008/06/25114657/0

NICOLL, B.C., ACHIM, A., MOCHAN, S. and GARDINER, B.A. (2005). Does steep terrain influence tree stability? A field investigation. *Canadian Journal of Forest Research* **35**, 2360–2367.

OLIVER, C.D. and LARSON, B.R. (1996). *Forest stand dynamics*. McGraw-Hill, New York.

PATENAUDE, G.L., BRIGGS, B.D.J., MILNE, R., ROWLAND, C.S., DAWSON, T.P. and PRYOR, S.N. (2003). The carbon pool in a British semi-natural woodland. *Forestry* **76**, 109–119.

PAUL, K.I., POLGLASE, P.J , NYAKUENGAMA, J.G. and KHANNA, P.K. (2002). Change in soil carbon following afforestation. *Forest Ecology and Management* **154**, 395–407.

POLGLASE, P.J., PAUL, K.I., KHANNA, P.K., NYAKUENGAMA, J.G., O'CONNELL, A.M., GROVE, T.S. and BATTAGLIA, M. (2000). *Change in soil carbon following afforestation or reforestation: review of experimental evidence and development of a conceptual framework*. National Carbon Accounting System Technical Report No. 20. Australian Greenhouse Office, Canberra, ACT, Australia.

PRETZSCH, H., GROTE, R., REINEKING, B., RÖTZER, T.H. and SEIFERT, S.T. (2008). Models for forest ecosystem management: a European perspective. *Annals of Botany* **101**, 1065–1087.

PYATT, D.G., RAY, D. and FLETCHER, J. (2001). *An ecological site classification for forestry in Great Britain*. Forestry Commission Bulletin 124. Forestry Commission, Edinburgh.

QUINE, C.P., COUTTS, M.P., GARDINER, B.A. and PYATT, D.G. (1995). *Forests and wind: management to minimise damage*. Forestry Commission Bulletin 114. HMSO, London.

RAY, D. (2008). *Impacts of climate change on forestry in Scotland – a synopsis of spatial modelling research*. Forestry Commission Research Note. Forestry Commission Scotland, Edinburgh.

REYNOLDS, B. (2007). Implications of changing from grazed or semi-natural vegetation to forestry on carbon stores and fluxes in upland organo-mineral soils in the UK. *Hydrology and Earth System Sciences* **10**, 61–76.

RIBE, R. G. (1989). The aesthetics of forestry – what has empirical preference research taught us. *Environmental Management* **13**, 55–74.

SAMUEL, C.J.A., FLETCHER, A.M. and LINES R. (2007). *Choice of Sitka spruce origins for use in British forests*. Forestry Commission Bulletin 127. Forestry Commission, Edinburgh.

SAVILL, P.S., FENNESSY, J. and SAMUEL, C.J.A. (2005). Approaches in Great Britain and Ireland to the genetic improvement of broadleaved trees. *Forestry* **78**, 163–174.

SAXON, E., BAKER, B., HARGROVE, W., HOFFMAN, F. and ZGANJAR, C. (2005). Mapping environments at risk under different global climate change scenarios. *Ecology Letters* **8**, 53–60.

SCHELHAAS, M.J., NABUURS, G.J. and SCHUCK, A. (2003). Natural disturbances in the European forests in the 19th and 20th centuries. *Global Change Biology* **9**, 1620–1633.

SEIDL, R., RAMMER, W., JÄGER, D., CURRIE, W.S. and LEXER, M.J. (2007). Assessing trade-offs between carbon sequestration and timber production within a framework of multi-purpose forestry in Austria. *Forest Ecology and Management* **248**, 64–79.

SEIDL, R., RAMMER, W., LASCH, P., BADECK, F-W and LEXER, M.J. (2008). Does conversion of even-aged, secondary coniferous forests affect carbon sequestration? A simulation under changing environmental conditions. *Silva Fennica* **42**, 369–386.

TURNER, J.A., WEST, G. DUNGEY, H., WAKELIN, S., MACLAREN, P., ADAMS, T. and SILCOCK, P. (2008). *Managing New Zealand planted forests for carbon – a review of selected management scenarios and identification of knowledge gaps*. Ministry of Agriculture and Forestry, New Zealand.

VALLET, P., MEREDIEU, C., SEYNAVE, I., BELOUARD, T. and DHOTE, J-F. (2009). Species substitution for carbon storage: sessile oak versus Corsican pine in France as a case study. *Forest Ecology and Management* **257**, 1314–1323.

VESTERDAL, L., RITTER, E. and GUNDERSEN, P. (2002). Changes in soil organic carbon following afforestation of former arable land. *Forest Ecology and Management* **169**, 137–147.

WHITTAKER, C., KILLER, D., ZYBERT, D. and RUSSEL, D. (2008). *Life cycle assessment of construction of forest roads*. (Spreadsheet tool). Imperial College, London.

WORRELL, R. (1996). *The environmental impacts and effectiveness of different forestry ground preparation practices.* Scottish National Heritage Research, Survey and Monitoring Report No 52. SNH, Battleby.

XENAKIS, G., RAY, D., MENCUCCINI, M. (2008). Sensitivity and uncertainity analysis from a coupled 3-PG and soil organic matter decomposition model. *Ecological Modelling* **219**, 1–16.

ZERVA, A., BALL, T., SMITH, K.A. and MENCUCCINI, M. (2005). Soil carbon dynamics in a Sitka spruce (Picea sitchensis (Bong.) Carr.) chronosequence on a peaty gley. *Forest Ecology and Management* **205**, 227–240.

ZERVA, A and MENCUCCINI, M. (2005a) Carbon stock changes in a peaty gley soil profile after afforestation with Sitka spruce (*Picea sitchensis*). *Annals of Forest Science* **62**, 873–880.

ZERVA, A. and MENCUCCINI, M. (2005b) Short-term effects of clearfelling on soil CO_2, CH_4 and N_2O fluxes in a Sitka spruce plantation. *Soil Biology and Biochemistry* **37**, 2025–2036.

POTENTIAL OF FOREST PRODUCTS AND SUBSTITUTION FOR FOSSIL FUELS TO CONTRIBUTE TO MITIGATION

Chapter

7

E. Suttie, G. Taylor, K. Livesey and F. Tickell

Key Findings

Climate change will fundamentally alter the world market for wood products and energy. Regulation, taxation and other mechanisms will alter the competitiveness of materials and products. Wood products and wood fuel have a significant role to play in substitution to reduce greenhouse gas (GHG) emissions in the UK.

If the wood construction products sector continues to grow as it has for the past 10 years there is potential to store an estimated additional 10 MtC in the UK's new and refurbished homes. Without legislation or incentive it may take 10 years to reach this additional stored amount as the construction sector is slow to change.

Wood fuel has the potential to save between 2 and 4 MtC per year by substituting for fossil fuel in the near future. A complex regulatory framework is currently in place to support the development of new bioenergy projects. More than 17 different incentive schemes were identified by the Biomass Task Force, and yet uptake is still limited. Action is required to ensure these incentives work together in an effective way.

The UK has a significant biomass resource, estimated at an annual 22 million oven-dried tonnes (Modt), although only a fraction of this is effectively captured for energy, currently contributing approximately 3–4% to heat and electricity production in the UK. In the short term, it could be useful for the UK to focus on developing a limited number of bioenergy chains. Biomass for heat provides one of the most cost-effective and environmentally sustainable ways to de-carbonise the UK economy. This should be linked to a joined-up policy and regulatory framework.

Public perception, understanding and acceptance of biotechnological routes to tree improvement may be a key challenge for the deployment of future energy forests. Technological advances may provide the step-change necessary for improved yields and to alter wood quality. Urgent engagement with the public is required to enable further development of this complex area.

Co-firing of power stations with biomass has, to date, largely relied on imports, but new regulations and modifications to the renewable obligation certificates (banded ROCs) will lead to demand for more dedicated energy crops and this may stimulate the UK energy forestry sector; both short rotation forestry and short rotation coppice. Research is required to enable selection of species and genotypes which are correctly adapted to future environments.

Substitution offers an attractive opportunity for tackling climate change by storing carbon in our buildings and reducing fossil fuel consumption. In contrast to alternative materials which release GHG in their production, wood products enable carbon to be stored in buildings. Failure to accept and adopt wood products arises in part from conservatism in the construction industry and outmoded attitudes that need to be robustly challenged.

Adaptation of the built environment to changes in climate will be critical to the future success of all building systems. By 2016, all new homes will need to be zero carbon rated and the UK has the target to cut emissions of CO_2 by 80% of 1990 emissions by 2050. This places huge technical demands on our buildings and the products used to make them. Legislation provides an opportunity to develop tailored wood-based products that will result in the necessary substitutions.

A harmonised approach for the measurement of GHG balances for construction systems, that provides genuine comparability and transparency for all materials would stimulate more use of wood products.

It is likely that the world and UK national markets for wood products and wood-derived energy as a component of bioenergy will be completely altered by the changing climate. Here we consider the opportunities for the UK forestry sector to contribute to tackling climate change by providing wood-derived fuels as alternatives to fossil fuels and wood products in place of other more GHG-intensive materials, most notably for construction and packaging.

Such changes of source materials are referred to as substitution. Below, we consider the potential for wood and wood-derived materials to contribute to UK renewable energy generation. The key advantage of using energy from crops is that the CO_2 released during combustion is recaptured by the growth of subsequent crops. Suitable woody crop systems are short-rotation coppice (SRC, 2–5 year rotations) and short-rotation forestry (SRF, 10–20 year rotations). The chapter then goes on to examine the potential of wood products to replace other materials in construction and packaging. Current construction practice is heavily dependent on use of materials which require large amounts of fossil fuel combustion in their manufacture. In contrast, the use of woody materials captures and stores carbon.

To date, a major barrier to the effective exploitation of wood products and fuels in substitution is a widespread ignorance of the qualities and opportunities offered by these sustainable systems. This problem is exacerbated by a consequent failure, particularly in the construction sector, to develop technologies which will enable more effective use of these materials. Clearly, in addition to the needs for research to plug gaps in our technological knowledge, there is also a significant role for education to bring both the economic and environmental advantages of wood as a substitute for fossil fuels to the attention of a wider public.

7.1 Wood for bioenergy – heat, power and liquid transport fuel

Biomass for energy can be defined as any biological mass derived recently from plant or animal matter. This includes material from forests (round wood, cutting residues and other wood brashings), dedicated crop-derived biomass (timber crops, woody short-rotation energy crops such as willow and poplar, grass crops, e.g. *Miscanthus*), dry agricultural residues (straw, poultry litter) and wet waste (slurry, silage), food wastes, industrial and municipal waste (e.g. woody waste from paper manufacture and consumption). Energy derived from these biomass streams, in general, has a lower carbon intensity (ratio of CO_2 released per unit of energy produced) and better energy balance than fossil fuels and biomass to liquid conversions. Although variations exist depending on feedstock, it is generally recognised that perennial woody crops used for heat have one of the best whole life cycle carbon balances of any route for biomass conversion (Royal Society, 2008; Rowe *et al.*, 2009; Environment Agency, 2009). In 2007, renewables contributed approximately 5 million tonnes oil equivalent (Mtoe) to the UK primary supply of energy and of this 81.8% was biomass derived (Figure 7.1). If we assume half of all energy crops and half of co-fired (i.e. mixing of biomass and fossil fuel feedstock) biomass is woody, and that waste combustion is likely to include a significant waste wood component, then we can estimate that woody

Figure 7.1
The contribution of woody biomass to the total biomass (bioenergy) derived from renewable resources used in the generation of UK electricity in 2007, measured in primary input terms. Of the c. 82% contributed by biomass, around 7.5 (6.9+1.6)% are provided by direct combustion of wood, approximately 6.2% (around half of 12.4% – see text) is derived from wood used in co-firing, and approximately half of the energy derived from plant biomass (including energy crops) around 2.25% is wood-derived. Waste combustion is also assumed to contain a significant component of woody material. In total, woody products are calculated to have contributed around 16% or the equivalent of 1–1.2 million tonnes of oil (Mtoe) to the 5.7 Mtoe of renewables used in 2007. (Source DTI Digest of UK Energy Statistics, 2009).

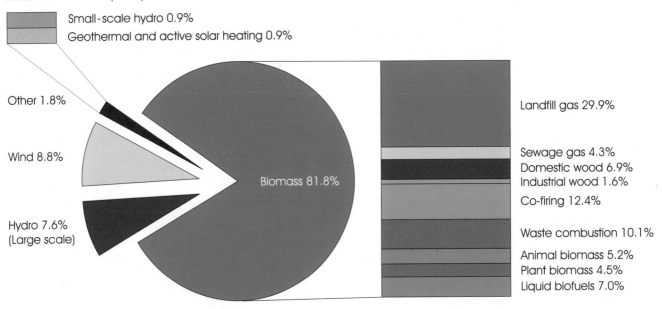

Small-scale hydro 0.9%
Geothermal and active solar heating 0.9%
Other 1.8%
Wind 8.8%
Biomass 81.8%
Hydro 7.6% (Large scale)

Landfill gas 29.9%
Sewage gas 4.3%
Domestic wood 6.9%
Industrial wood 1.6%
Co-firing 12.4%
Waste combustion 10.1%
Animal biomass 5.2%
Plant biomass 4.5%
Liquid biofuels 7.0%

Total renewables used = 5.17 million tonnes of oil equivalent

Source: DTI Digest of UK Energy Statistics, 2009.

biomass resources contributed approximately 1–1.2 Mtoe equivalent to the UK primary energy inputs which is approximately 0.5–1.0% of the primary energy supply to the UK of 236 Mtoe in 2007.

7.1.1 Contribution of biomass to the UK energy mix

Wood as a fuel can replace fossil fuels and offers an attractive route to reduction of net GHG emissions. The UK Biomass Strategy (Defra, 2007) called for expansion of wood use for fuel and the Renewable Energy Strategy (DECC, 2009a) also indicates a significant gain yet to be achieved from the use of biomass resources in the UK to produce energy. Woodfuel use for heat, electricity and in the future, transport fuel, are all highlighted in the strategy. However, emerging policy developments, in particular the effectiveness of the proposed renewable heat incentive and the feed-in tariff that allows excess energy to be sold back to the grid, will be key to the more effective exploitation of woodfuel. The implementation plan of the Forestry Commission Woodfuel Strategy (2009) is designed to improve the management of private woodland for energy.

The Biomass Task Force (Defra, 2005) reported that the UK biomass resource is approximately 22 million oven-dried tonnes (odt) annually and, of this, identified 5–6 million odt of wood waste generated per annum as a top priority for recovery and energy use. This is compared to 3 million odt annual production of cereal straw. The overall contribution of forestry to the UK biomass resource is unclear since discrepancies exist due to different assessment criteria and boundaries in the different reports (see 7.1.2 below and Table 7.1). However, the forestry component of the biomass resource remains significant and could contribute up to 7% of the biomass-fired heat market. Recent research depicting a number of UK future energy scenarios using MARKAL modelling (UKERC, 2009) all suggest that renewable heat from biomass will become increasingly important in the UK renewable energy landscape. Currently, only 1% of heat is derived from renewables (DECC, 2009a). The Renewable Energy Strategy has identified biomass to heat as a least-cost way to increase the share of renewable heat with a target of 12% by 2020. Deploying forest resources to achieve these renewable energy targets should be a priority over the coming decade.

Forest biomass can, potentially, be used for heat and electricity, biogas production and also as a liquid fuel, but some of these technologies for woody biomass remain at research scale only. As far as immediate commercial deployment is concerned, it has been suggested that heat, followed by small-scale combined heat and power (CHP), grid-fed electricity and then co-firing in large-scale power stations, represent the priorities for current use in the UK (Forestry Commission, 2007). Combustion technologies may be considered as mature, although their deployment for heat and power is still limited in the UK, despite the fact that they offer good GHG emissions savings, compared to liquid fuel routes. Liquid biofuels provide one of the few options for fossil fuel replacement for transport in the short to medium term. However, in recent months, with rising food prices and reportedly poor energy balances, the validity of their use has been questioned widely at both global (GBEP, 2009) and local levels (RFA, 2008). Current liquid biofuels include bioethanol, biodiesel and other biomass-based products such as biobutanol. Feedstocks are generally non-woody, oil-based crops such as oil seed rape for biodiesel, non-woody sugar and starch based crops such as sugar beet and sugar cane used for bioethanol. However, in future it is more than likely that lignocellulosic woody biomass will be converted to liquid fuels through biological processes (esterification, fermentation) or through thermochemical routes such as pyrolysis (Carroll and Somerville, 2009). Large projects on wood fuel for liquid fuels for transport are currently funded by the DOE in the USA, Genome Canada and by FP7 in Europe. At a national level in the UK there is a current research commitment to investigate SRC willow for bioethanol production as part of the BBSRC Sustainable Bioenergy Centre, created in 2009.

7.1.2 UK forestry biomass for bioenergy

Forest biomass resources for bioenergy in the UK can be defined as primary, secondary or tertiary. Primary forestry biomass resources include forest harvesting residues such as small roundwood logs and branches. Secondary residues are those from sawmills (chips and sawdust), while tertiary residues include paper, construction, recycling and material derived from urban tree and hedgerow maintenance (arboricultural waste). Several attempts to assess the UK forest biomass resource for bioenergy have been made in recent years. These include a joint assessment for the UK by McKay et al. (2003), the Defra UK Biomass Strategy (2007), and the Carbon Trust Biomass Sector Review (2004). The findings of these reports, as summarised by Whittaker and Murphy (2009), are given in Table 7.1. The contrasting estimates of forest biomass resource provided by these studies are consistent in revealing waste wood as the largest single source of woody resource which could become available for energy use (Defra, 2007). They include increasing amounts of waste wood that will arise as a consequence of the landfill directive. In future, the projected increase in harvest volumes can be expected to increase wood processing residues from primary and secondary sources over the next few years. The creation of new woodlands and restoration of management in neglected woodland can also be expected to make a significant contribution. Better management of existing woods could supply an

Table 7.1

Contrasting estimates of the biomass resources available from UK forests for bioenergy – as summarised in Whittaker and Murphy (2009). Data expressed as millions of oven dried tonnes.

Source of data	Total estimated forest biomass resource for bioenergy (Modt year⁻¹)	Details of the assessment	Reference
UK assessment	3.1	Primary and secondary sources as well as dedicated SRC	McKay, 2003
The Carbon Trust	11	Extensive overview including forestry (2.24 million odt year⁻¹), paper and card industry (2.52 million odt year⁻¹), wood packaging waste (970 000 odt year⁻¹), paper sludge (800 000 odt year⁻¹), and SRC (16 688 odt year⁻¹)	Carbon Trust, 2004
UK Biomass Strategy	1.3	Only includes sawmill co-products and arboricultural arisings	Defra, UK Biomass Strategy, 2007
UK Biomass report: Mirror to the US's billion ton report	13	Extensive analysis of all forest biomass sources and a forward prediction to suggest that a potential 23 million odt year⁻¹ could be available	Whittaker and Murphy, 2009

additional 1 Mt annually for energy purposes, bringing the annual wood fuel production in England to 2 Mt (Forestry Commission, 2007).

7.1.3 UK-sourced vs imported woody biomass

Statistics on imported woody biomass for energy are difficult to verify, since imports for co-firing may be subject to commercial confidentiality. This emphasises the complexity of the UK bioenergy system where there is a mix of home-grown feedstocks as well as those that are imported. In general, approximately half of the utilised feedstocks within the UK are derived from import, which includes approximately 1 Mt of biomass for co-firing. The co-firing market grew by over 100% between 2004 and 2006 and is likely to expand further. This will be driven in part by changes to the renewable obligation certificates (ROCs), that provide better incentives for home-grown biomass as compared to imported supplies.

7.1.4 Carbon capture by woody energy crops in the UK

The key advantage of using energy from crops compared to that derived from fossil fuels is that the CO_2 released during combustion of biomass can be recaptured by the growth of subsequent crops. Cannell and Dewar (1995) described a potential carbon saving from the use of wood-derived energy in the range of 5–19 MtC per year by substituting biomass for coal across the UK. These authors assumed that 1 t dry biomass used to generate electricity prevents 0.5 tC being emitted from coal, 0.44 tC from oil and 0.28 tC from natural gas.

Wood fuel use in Scotland in 2008 was recorded as 413 000 odt, of which 62% was virgin fibre (chip), 35% recycled fibre and the remaining 3% pellets. Woodfuel projects in Scotland were estimated to have saved 334 000 tonnes of CO_2 emissions in 2008. If all the projects in planning in Scotland were to go ahead in 2009, then woodfuel use would be around 1 400 000 odt, saving an estimated 1.1 $MtCO_2$ (0.3 MtC). With an increase in planning agreements, a realistic potential saving from biomass substituting for fossil fuels in the whole of the UK within the next five years could be up to 2 MtC per year. However there is potential to double this if the uptake of woody energy crops was as suggested in the UK Biomass Strategy (Defra, 2007).

A key question for the development of forest energy crops in both SRC and SRF systems is the impact that plantations may have on soil carbon stores and long-term

carbon sequestration (see Chapters 3, 6 and 8). The overall GHG balance of energy forests compared with other land uses such as arable, grassland and upland grazing is only now being quantified. The fertiliser applications required for bioenergy production in intensively managed annual crops are a major source of GHG emissions. Tree crops in contrast do not require annual fertilisation (St. Clair et al., 2008). Although data for N_2O and methane emissions from soil and crops are limited, when available data are coupled to models, the net positive advantage of woody energy crops is clear. St. Clair et al. (2008) showed that replacing arable and grassland with energy SRC had a net positive effect on GHG balance, while replacement of tall forest with SRC had a small negative impact on soil carbon and GHG emissions. These data and information of SRC yields for England and Wales have been used to develop maps assessing the potential for mitigation of GHG emissions (Hillier et al., 2009).

A recent model of the potential for carbon sequestration in SRC willow plantations suggests that within the UK, increases in soil organic carbon (SOC) under SRC alone could contribute around 5% of the emissions mitigation benefits of this crop. A US-based study of poplar plantations (Grigal and Bergson, 1998) similarly suggested that after an initial period of loss, carbon sequestration could be expected to result in gains equivalent to 1–1.6 tC ha^{-1} year^{-1} over a 10–15-year period. However, other studies provide different results. For example, an investigation on SOC sequestration at three sites in Germany (each with plots of SRC willow, poplar and aspen) reported an increase in SOC at one site of 20% compared to arable land, due mainly to increases in carbon in the top 10 cm of soil (Kahle et al., 2001). However, at the other two sites, no overall increase in SOC was seen, as increases in SOC in the top level of soil were balanced by a decrease in levels below 10 cm. A similar pattern was also seen in the study on SRC willow and poplar by Makerschin et al. (1999). This study also included a site on former grassland in which a loss of 15% of original SOC was reported, showing that former land use, and thus initial SOC levels, need to be considered when locating SRC plantations for maximisation of mitigation benefits (see 8.3, Chapter 8). This is certainly an area where further research is needed.

Despite these variations, there is a broad acceptance that while the conversion of arable land to SRC or Miscanthus will result in an increase in carbon sequestration in the soil, the conversion of grassland may not be as beneficial (Hillier et al., 2009). It is also important to note that in all cases, soil carbon concentrations will not increase indefinitely, as eventually a new higher equilibrium SOC will be achieved over some decades (Kahle *et al.*, 2001).

7.1.5 Woody bioenergy and climate change – adaptation and mitigation

If SRC and SRF are to be important elements in UK energy generation, then we need to ensure that they are suited for the likely future climate. Predicted changes in climate may have both direct and indirect effects on SRC and forests grown for energy, although again, empirical data in this area are limited. Direct effects include those of rising temperature, altered rainfall, increased CO_2 and tropospheric ozone on tree productivity, chemistry and morphology, while indirect effects include interactions with pests and pathogens and wider ecosystem impacts (see Chapter 3 and Section 2 for more detail). There is some empirical information on the climate impacts on SRC and SRF. In response to an experimentally increased CO_2 concentration to 550 ppm the productivity of stands of loblolly pine, poplar and aspen rose by an average of 23% but in some cases there were increases of up to 60% in total tree biomass (Karnosky *et al.*, 2007). Interactions with tree age, nutrition, climate and pests all influence this effect. In the UK climate, all evidence suggests that in future, yields of SRC are likely to increase as CO_2 concentrations continue to rise, this despite the fact that water may become limited (Oliver *et al.*, 2009). However, yield enhancement may eventually become limited by soil nutrient availability in these rotation systems.

7.1.6 Sustainability of woody-based bioenergy systems in the UK

The delivery of enhanced ecosystem services to the UK landscape (including carbon management, water and biodiversity preservation and amenity provision) will gain increasing importance in the UK, alongside the pressures to develop a low carbon society (DECC 2009b) and an 80% reduction in CO_2 emissions, as part of the legal requirements contained in the Climate Change Act. For dedicated woody energy crops such as fast-growing willow and poplar, there is now clear evidence from UK trials showing enhanced farm-scale biodiversity compared to arable land use, including increased small mammal breeding and bird populations (Rowe *et al.*, 2009). Some remaining questions exist regarding catchment-scale water resources but on-going research within the TSEC-BIOSYS and RELU research projects will answer the question of seasonal water use in SRC bioenergy cropping systems. All reported evidence suggests positive rather than negative impacts on water quality. The large unknown in many UK woody systems is the contribution of below-grown rhizosphere and soil processes to GHG balance.

Currently, liquid biofuel supply to the UK is regulated by the Renewable Fuel Agency. Minimum standards of reporting on GHG mitigation are required but there are no restrictions on land use or crop types, although this may change in the future. More than 70% of the current 2.5% by volume liquid transport fuel from biological sources is supplied from imported sources (RFA, 2008). The sustainability of this supply is largely unregulated, although bioethanol from Brazilian sugar cane is known to have one of the best energy balances of any bioenergy system. Both the UK and EU will address the issue of sustainability in the near future, with sustainability criteria emerging (EU, 2008). Future directives for liquid fuel are likely to include a minimum standard for GHG mitigation relative to fossil fuel, a consideration of prior land use and a ban on the use of pristine high carbon soils and ecosystems with high biodiversity. These changes are likely to encourage a move away from food crops for fuel and could favour the development of the woody biomass energy industry in the UK, since they will restrict feedstocks to those showing at least a 35% improvement in GHG balance relative to fossil fuels. Issues concerning the sustainability of biomass supply have been considered recently (Royal Society 2008).

To achieve sustainability standards for bioenergy supply, there is a requirement for comparison of contrasting and often complex bioenergy chains, including feedstock type, processing and end use. Life Cycle Analysis (LCA) assesses the complete GHG balance of the material or system under consideration and enables more valid comparisons (see Chapter 8). Several detailed studies now confirm that bioenergy chain efficiencies vary dramatically but that in general, woody biomass provides one of the least carbon intensive bioenergy chains, particularly when used for heat (Royal Society, 2008). Better tools are required for this type of LCA comparative analysis and considerable global research effort is on-going to develop these tools, particularly within the Global Bioenergy Partnership (www.globalbioenergy.org).

7.1.7 Drivers supporting the development of bioenergy in the UK

The Renewable Energy Strategy (DECC, 2009a) suggests that bioenergy can make an important contribution to the Government's energy and environment objectives, including energy security and the reduction of GHG emissions, relative to current practices. It particularly identified biomass heat as a cost-effective mechanism for decarbonisation of the energy sector. Scenarios presented for renewable energy (DECC, 2009a) confirm that bioenergy is likely to play an increasingly important role in contributing to renewable targets for heat, power and liquid fuel (Figure 7.2a). The Recent EU policy developments include the '20–20–20' policy that demands a 20% renewables deployment by 2020. Approximately 6% of all gross domestic energy requirements across Europe are provided by renewables, and bioenergy accounts for the largest share of this, providing about two-thirds of the supply. The European Environment Agency (EEA, 2006) identified the UK, Spain, Italy and France as having high potential for increased use of forests for bioenergy production. Currently, forestry contributes half of the European biomass supply. Despite this, attempts by the UK Government to stimulate the bioenergy sector in the UK have so far had limited success. Recently, however this market has seen increasing activity in micro CHP developments. Also the use of biomass to co-fire power stations such as Drax will be favoured by new 'banding' of the renewable obligations certificates leading to the use of UK-grown dedicated crops for co-firing.

Commercial interest in bioenergy is growing. Steven's

Croft in southern Scotland is one of the first large-scale (44 MW) dedicated biomass power stations in the UK, currently running exclusively on woody feedstocks with a requirement of over 400 000 tonnes each year. It provides energy for up to 70 000 homes. Several other dedicated biomass power stations are under consideration across the UK.

The report of the Royal Commission on Environmental Pollution (2004) on bioenergy and that of the Biomass Task Force (Defra 2005) report attribute the general lack of progress in the uptake of biomass energy crops to a focus on promoting specific technologies without full consideration of the wider market. There is also a lack of integration of biomass supply with its utilisation, and there are issues of public perception and planning, i.e. a whole-systems approach is needed requiring policy incentives and investment from several Government departments.

The European Commission has introduced the Biofuels directive to which the UK is committed. Thus the UK is moving towards development of transport biofuels with a target of 5.75% volume replacement of petroleum-based fuels by 2010 (EU, 2008) with a 2008 commitment of 2.5%, and a commitment to move towards 10% by 2020, regulated by the new Renewable Fuels Agency

Figure 7.2a
The possible relative contributions of different forms of renewable energy to the achievement of a proposed UK Government Target of 15% of total energy being derived from renewables by 2020. Wood fuel will be expected to make significant contributions to the major energy consumers in the electicity (blue labels), transport (green label) and heat generating (red labels) sectors. Data from DECC (2009a).

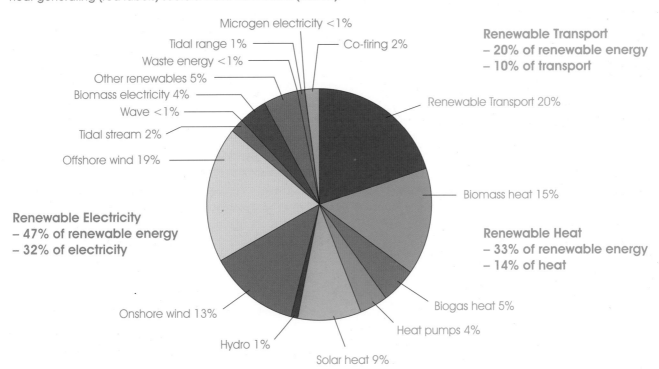

Figure 7.2b
Schematic flow diagram showing the development, over time, of European (dark blue) UK national (light blue) and UK regional (black) policy initiatives directed towards deployment of bioenergy at a range of scales. Solid connecting arrows indicate a direct link to emerging policy documents and broken arrows an emerging influence. Since 2003 these initiatives have contributed to the current Renewable Energy Strategy which places the target of 15% (see a. above) for the overall contribution of renewables to total energy demand by 2020. In the UK context, both the Energy White Paper (2003) and the Royal Commission on Environmental Pollution (2004) emphasised our failure to make use of bioenergy sources. In an attempt to rectify this situation the Biomass Task Force and later the UK Biomass Strategy, Woodfuel Strategy for England and Biomass Action Plans for Scotland were established. Modified from Slade *et al.* (2009).

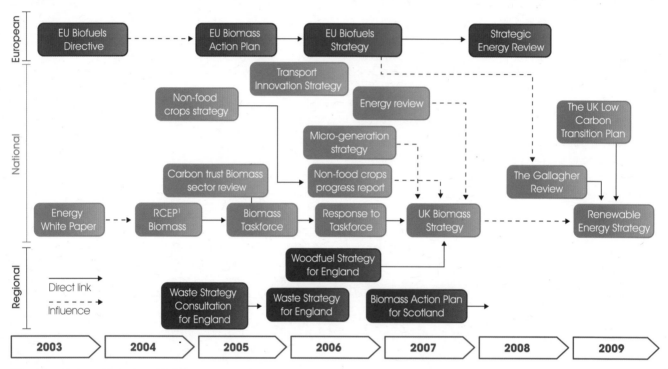

[1] Royal Commission on Environmental Pollution

(2008). It has been estimated that liquid fuel demand in the UK in 2010 will be 44.5 Mt, which will therefore require approximately 2.56 Mt of biofuel, providing a carbon saving of approximately 2 MtC.

Other drivers to support the deployment of bioenergy systems in the UK include capital grant support provided for infrastructure development, the 'banding' of the renewable obligations certificates and a new heat incentive, which is likely to be confirmed in late 2009. The Energy Crops Scheme, Woodland Grant Scheme and other regional support schemes are also available and are detailed in The Biomass Energy Centre website (www. biomassenergycentre.org.uk).

7.1.8 Barriers to the deployment of woody biomass for bioenergy

There is a plethora of incentives and schemes introduced to enable ambitious targets for bioenergy to be met in the UK and more widely across Europe (Figure 7.2b). More

than 17 different schemes were identified by the Biomass Task Force, and yet uptake is still limited. The effectiveness of these policies in delivering against bioenergy targets as defined by the Biomass Task Force and Biomass Strategy (summarised by Slade *et al.*, 2009), is questionable since the contribution of bioenergy to UK energy supply remains stubbornly low. Public perception of bioenergy schemes can often be negative with concerns over air pollution, the siting of major infrastructural changes and also changes to the landscape, all being noted as reasons for public rejection. Action is required to ensure these incentives work together in an effective way. Barriers to uptake are not only financial, although long lead-in times for perennial crops and contractual obligations between growers and energy producers remain a problem. In the long term, the European emissions trading scheme should help to give bioenergy a considerable boost.

Limited land area and the lack of planning to enable future management of UK land to deliver multiple benefits also inhibits widescale deployment of forest energy systems. A foresight activity on land use change and management is currently underway to address this issue and the 'Land Based Renewables' research initiative will be determining how ecosystems services might be valued in such a changing landscape. A second project will quantify the likely carbon benefits from increased bioenergy systems and a third will consider how the UK wind resource may be best utilised in a forested landscape alongside the deployment of other energy sources in areas of the UK, particularly Scotland. The UK land-based resource for farming and forestry deployment is modest at approximately 17–20 million ha. Since 2008, the steep rise in food prices has placed a new burden on the UK to deliver food crops and the possibility of any additional use of land to deliver liquid biofuels crops (e.g. from oil seed rape and sugar beet) as well as bioenergy heat and CHP (from forest and grass crops) remains open to speculation. Certainly the uptake of SRC and SRF by farmers and foresters as estimated from the Energy Crops Scheme, has remained sluggish, with only 5000–6000 ha of current plantations. In a series of farm-based surveys (Sherrington *et al.*, 2008), the reasons given include unwillingness to commit to contracts with power stations over several years and a general caution in growing a perennial crop that limits farm flexibility. There are major barriers to be overcome for any large scale forestry energy crop deployment. They require further Government incentives and farm-scale demonstration.

Varying amounts of land have been suggested for biomass feedstock production from the 20 million ha of UK agricultural and forest land available (Rowe *et al.*, 2009). For example, the Biomass Task Force (Defra 2005) suggested that approximately 1 million ha of bioenergy crops in the future could provide 8 million odt of energy crop annually, while the Biomass Strategy (Defra, 2007) proposed 350 000 ha of dedicated energy crops by 2020. There is no coherent current strategy for land deployment between food and energy crops. There will be large-scale changes in the landscape in the UK if specialist bioenergy crops are widely planted. If food prices remain high, there will be competition between these land uses. Perennial crops, such as trees, have a better energy ratio and are more effective at mitigation of GHG emissions than annual crops, and yet farming practice is such that this land use may be slow to develop (see Section 5). Co-firing of power stations is a market for biomass use that has developed since 2002 (growing by 150% between 2004 and 2006 and utilising 1.4 million odt of biomass) and could in future utilise a very large amount of dedicated biomass resource from energy crop supplies. At least half of the current

supply is sourced from outside the UK, with implications for sustainability.

Another barrier to be overcome is the current centralisation of power generation which leads to less favourable economics and a poorer GHG balance for biomass due to transport requirements. The development of microgeneration (small CHP units serving individual homes, businesses or communities) will alleviate the need to transport biomass from point of production to large regional power stations. Microgeneration is currently a small contributor to the UK energy economy but, with careful development, could become a very major one by 2030. In addition, no clear strategy currently exists in the UK to capture bioenergy from biomass 'waste' including municipal solid waste, and agricultural and forestry waste, and this should be an important future priority and is likely to be achieved through increase in the deployment of anaerobic digestion, given the maturity of this technology.

The expertise in woodfuel infrastructure is not well-developed in the UK compared with elsewhere in Europe. However, as wood fuel supply grows, spin-off growth in UK boiler manufacture, installation, maintenance and training is likely to occur.

7.1.9 Outstanding issues and research needs

A clear strategy for UK land management is required, since there are many competing land uses. This finite resource must be managed effectively.

There is considerable enthusiasm over the possibility in the future of new bioscience technologies (DoE, 2006) harnessed to improve photosynthetic gains for bioenergy, including the use of synthetic biology. Purpose-designed energy forestry could be an important part of a 'biorefinery' (a refinery using biomass for the production of liquid fuels), contributing energy streams linked to high quality chemical and other biofuel outputs. In future, biotechnology could deliver important improvements to current forest traits for energy, including:

- higher yielding forest energy crops that require minimal inputs (optimised not maximised), thus improving efficiency further;
- forest energy crops with different qualities – increased lignin for calorific combustion, or improved oils, starches and sugars for liquid biofuels;
- forest energy crops with improved resistance to biotic and abiotic stresses that are likely to occur in future.

The model bioenergy tree is poplar (for which the DNA sequence and genomic resources are already available). This is an important resource for future accelerated science advances but the UK has limited scientific investment in this model, in contrast to the USA, China, Canada and the rest of Europe. A strategic view on utilisation of the model tree for energy forestry should be established.

Second generation biotechnologies (molecular breeding in the absence of genetic modification), the use of genetically-modified trees with enhanced traits for carbon sequestration and energy production should be on the agenda for future research, in line with current efforts in Canada, USA, India and China. Trials of exotic species and new silvicultural practices designed to select systems best suited to emerging climate scenarios should be undertaken (see Section 2 and Chapter 8).

Development of new gasification technologies and other advances linked to the biorefinery concept and new technologies for conversion of biomass to fuel are likely to develop to commercial scale by 2020.

By 2020 and beyond, gasification and other technologies may be deployed to improve the efficiency with which wood-based energy supplies are processed and delivered. Advanced technologies for heat and power generation from green and woody plants may be possible at commercial scale using biological rather than thermochemical conversion pathways. Wood waste should be developed as an energy source.

7.2 Wood products

Climate change will almost certainly change world markets for the systems and products used in construction. It is likely that in some regions incentives will be developed to promote and support wood product integration in buildings in place of more energy intensive materials. The increasing emphasis on reducing the environmental impact of building has led to the development of so-called 'green' building specifications. These reflect the positive contribution of wood products. Increases in the volume of wood products used in construction by more timber-rich buildings combined with extended service life of wood products will contribute to reducing GHG emissions. However, such increased use of wood is not considered under Article 3.4 of the Kyoto protocol to the UNFCCC, and thus does not apparently contribute to the achievement of Kyoto Protocol targets, nor will improvements be reported in GHG

inventories. The scale of the contribution that substitution of wood products can make to tackling climate change is not clearly understood in the research and technical community because the measurement parameters vary from study to study as do the findings. A result is that the benefits of wood products have not been communicated in a meaningful way to the public.

The pathway of carbon flow from atmosphere into forests and through the different components to wood products that is the focus of this section, is shown in Figure 7.3. The GHG balance of different types of constructions is a highly complex issue. There are relatively few studies and estimates based on buildings, and the boundaries and assumptions in estimates are not always clearly stated and uniform. The variation in wood products industries across the world, and the differing building regulations and practices is a further layer of complexity.

Gustavsson (2008) presented a Scandinavian Carbon balance case study comparing a wood frame apartment building with a concrete framed apartment building that had similar costs to build. The study included primary energy used to produce the buildings, the electricity from fossil fuels and the CO_2 balance in cement reactions. It was concluded that production of materials for the wood frame building used less primary energy, reduced the net CO_2 emission and recovered more biomass residue to replace fossil fuels. The wood frame building gave a net CO_2 emission of −40 tC compared with the concrete framed building net CO_2 emission of approximately +25 tC over a 100-year life cycle. With such estimates it is essential to understand that there are considerable uncertainties regarding the amounts of materials used which will vary from country to country and design to design. In addition, the primary energy used to produce materials will vary with technology, time and country.

A significant complexity in assessing the contributions of wood products arises from the method of accounting for imported and exported material. For example, the accounting of carbon sequestered in the harvested tree could be credited in the country of growth or passed on to the country where the wood product is used. Similarly the decomposition or energy recovery of wood in one country may be credited to that country where the activity happens as an emission or passed back to the country where the tree was grown.

Further complexity is added by differing approaches to the key issue of how to dispose of the products, including:

Figure 7.3
The full life cycle pathways of carbon contained in trees and wood products showing their connections to the atmosphere. The wood products components of the pathway considered here are enclosed within the dashed red line. The consideration of long lasting wood-based construction products in carbon life cycle analyses is of fundamental importance for estimation of national carbon budgets.

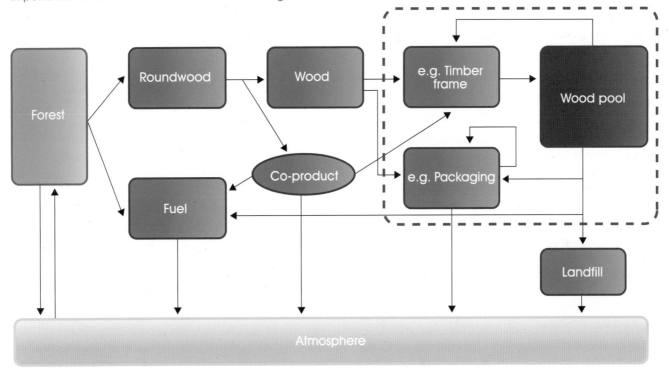

- recycling to extend the stored life of the carbon in the wood products;
- energy recovery for immediate release of carbon but displacement of fossil fuel;
- disposal in landfill where the decay rates are variable;
- accounting that considers short-term storage in, e.g. packaging and also long-term storage in the frame of a timber building or a panel product.

7.2.1 Wood and other materials in construction and packaging

The energy consumed in producing a construction material, from the extraction of raw material to its transport and manufacture, leads to associated CO_2 emissions which are known as the 'embodied CO_2' of the materials. For example, an estimated 5% of the total annual anthropogenic CO_2 emissions are associated with world concrete production. In addition, the way we extract, produce and transport our construction materials has huge impacts on the amount of embodied CO_2. By identifying and selecting low embodied CO_2 materials, we can therefore substantially reduce overall CO_2 emissions (Lazarus, 2005).

Of the 678 Mt of materials consumed in the UK annually, 420 Mt is in construction. A relatively small fraction, estimated to be 4% by weight, is wood and wood-based products (e.g. panel products). This fraction is growing. The construction of UK homes accounts for some 3% of our annual CO_2 emissions, because of the embodied CO_2 in the material used (Table 7.2).

Table 7.2
Showing the quantities of CO_2 contained (embodied) in a typical semi-detached house (as used in volume house building) normalised for a typical occupancy and over a 60 year working life. The values are calculated for house area (top row) and per person per year (second row). The total UK CO_2 equivalent emissions per person per year are shown in row three, so that the percentage of individual CO_2 emissions which are embodied by house building can be calculated (bottom row).

Embodied CO_2 for volume house builders	600–800 kg m^{-2}
Embodied CO_2 per person per year	286–381 kg
UK Total CO_2 equivalent emissions per person per year	12 300 kg
Embodied CO_2 of volume domestic dwellings as % of total CO_2 emissions	2.3–3.1%

In addition, approximately 70 Mt of waste is produced from construction and demolition every year in the UK. A large proportion (75%) of this is recycled, with only 25% going to landfill. But the recycling is generally as low-grade products such as crushed aggregate hardcore. For all products, the focus should be on a cascade down the hierarchy of reuse, recycle and energy recovery.

Construction in a low-carbon economy will need to be highly efficient, it should shift to use of materials that have low embodied CO_2 and use designs that reduce whole building life operational energy. At present, the operational energy used to heat space during the building's lifetime dominates over the buildings embodied energy. As our buildings are designed to more passive heating standards, with high air-tightness and low to zero energy input, then the embodied energy of the construction materials may begin to dominate.

The most promising opportunity for wood products to contribute to carbon storage in the UK is in buildings. FAO statistics from 2003 show that 20% of all wood consumption is used in construction, 30% as panel products and the remainder in packaging and communication. Of the 25 million m^3 of timber consumed in the UK per year about half is used in construction with 85% of the total being imported. In the construction of buildings, wood is a versatile material that can be used as:

- structure (frame, roof);
- engineered panels (sub-floors, joists, wall panels, SIP (structural insulated panel));
- thermal and acoustic insulation (wood wool or recycled paper insulation);
- high aesthetic items (floors, joinery, furniture, cladding);
- biomass boiler fuel.

Wood products can replace more energy intensive construction materials such as concrete and steel, which can result in carbon savings in embodied energy and also increase the carbon storage in buildings. However, a sound evidence base for the construction industry is far from established. Data is presented in fragments and the sector would benefit from robust data gathering to provide a summary of trends in wood product markets for construction and packaging, including the balance of imported and exported raw material and products. The growth in timber frame is well reported by the UKTFA from 7% (1997) of new build to 22% (2008), representing a growth of 300 000 new timber frame houses in 10 years. This amounts to an estimated 1.6 MtC stored.

For the purposes of this chapter, wood products include solid wood, wood-based panels, paper and board. Construction products include permanent and temporary works, public and private buildings and infrastructure. Broadly, this category has products of longer design life needs up to 100 years, which opens up an opportunity for long-term carbon storage and low life cycle impact in the majority of applications.

Packaging includes paper and board and typically is used in short life span applications of up to one year. It has inherent value as a recycled resource. Wood-based products such as cardboard and paper have been compared with glass, PVC, PET, steel and aluminium for packaging (Reid et al., 2004). Cardboard and glass represent the lowest GHG contribution per kilogram of packaging. Savings from using virgin card in gCO$_2$e kg^{-1} material are for glass (1100), PET (2950), PVC (2850), steel (2910) and aluminium (4040). For cardboard the net GHG emissions were negative (–0.4 kgCO$_2$e kg^{-1} material) due to energy recovery in one life cycle path.

Wood-based products represent sinks of carbon, whereas the uses of all other common construction materials are net sources of CO_2 (Figure 7.4). This accounting of emissions and storage is important. It is also important to note that these relative performance figures are changing as sectors seek to reduce the carbon footprint of their products and materials.

It is important to account for carbon in the whole life cycle from the raw material extraction to the end of life of the product. A 1 m^3 of steel I-beam does a very different job to 1 m^3 of sawn timber in all respects including structural span, service life and end-of-life use. The major complexity here is the need to capture the whole picture by considering material as part of functional units. A negative CO_2 emission for wood products is an excellent platform for increasing the use of wood-based products in construction providing the regulatory framework demands low carbon technologies. In addition, products must perform as predicted up until the end of the design life to avoid wasteful resource use and subsequent imbalance through premature failure and early replacement.

Wood products used in construction are stored and there is a degree of turnover as they are replaced. The wood products pool in the UK is estimated to be 80 MtC stored and growing at 0.44 MtC per year (Broadmeadow and Matthews, 2003). Wood products have favourable net negative CO_2 emissions as more carbon is stored

Figure 7.4
The net CO_2 emissions of the major products used in the construction industry. All non-woody components are seen to constitute net sources of CO_2 while all wood-based components are net sinks. LVL = laminated veneer lumber.

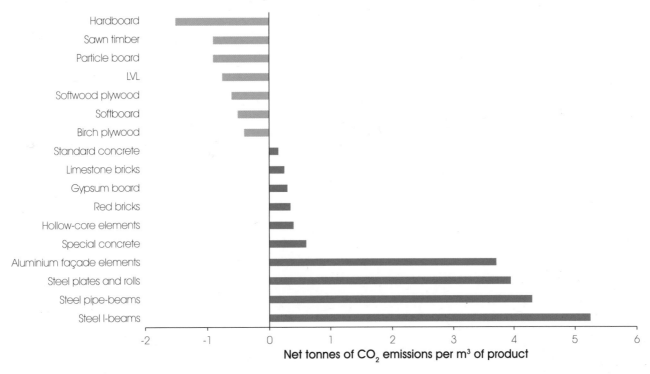

(Reproduced from WRI Trees in the Greenhouse Report using RTS data of 1998–2001 Aulisi A, Saucer A and Wellington F. World Resources Institute 'Trees in the Greenhouse')

than produced in manufacture. Whether the product is in packaging or construction, other materials are improving their position as they reduce impacts, strive for low energy processing, become resource efficient and apply new technologies through innovation and the application of life cycle analytical research. To keep pace, wood products also need to conduct this important work and continue to lead in low impact solutions for packaging and construction.

7.2.2 Challenges facing the increased use of wood-based products

Awareness of carbon storage potential in wood products is limited but improving. There remain, however, considerable challenges in realising a major shift towards increasing wood products in construction in the UK.

Lack of know-how and inexperience of use in the hands of engineers and construction professionals in the UK presents a primary barrier for widespread use of wood products. Traditional UK building practices have not extensively used wood as they do in Germany, Scandinavia and Austria, for example. Timber presents challenges to

use in UK markets as designers, architects, engineers and specifiers are not broadly as familiar with the material and its use as with steel and concrete. This knowledge gap is evident in architectural and engineering higher education. Limited know-how is overcome in part by publications and training but these are fragmented and new initiatives risk limited or low impact. Technical performance and capabilities need to be in appropriate formats for ease of use – design packages, software, systems not individual products – and to provide an easy route to meet building regulations and approvals bodies' requirements. Wood products need to work hard in a conservative construction industry. The mortgage lending and insurance sector trusts traditional construction because of its track record in delivery of service life of brick and block technology. New technologies and timber frame systems initially were hindered by lack of understanding of the robustness of the systems to deliver long-life buildings. These perceptions still persist, despite the substantial independent research evidence and the wide range of existing successful buildings.

Uncertainties about wood product service life exist as there are not systems widely available for its prediction in construction as there are for concrete and steel. Testing of new and existing wood products needs to be robust to enable prediction of service life against defined performance criteria. Maintenance systems need to be fully adopted, integrated and accompanied by training packages for all wood systems to extend service life in use.

The myths surrounding the use of wood, wood products and timber construction need to be challenged by effective communication of research findings and by the use of projects to pass on learning. Reid *et al.* (2004) record some of the misconceptions about wood. They note that the perception that wood use is non-sustainable arises because it is associated with felling and not with the replanting and regeneration of forests. Concerns around overheating, low thermal mass and durability of wood, together with fire safety and complex specification of fire safety issues, are sometimes overemphasised in the competition of different materials for market share in domestic housing. There is a need for example construction projects using wood and for the research findings about construction with wood to be centrally promoted. There are only a few mechanisms to enable such promotion. Among these are the 'Wood for Good', 'Wood for Gold' and some Forestry Commission activities. Independent research should be used to present the case for wood products.

There are skills shortages and gaps in training concerning the use of wood in the construction industry. Since there are so few data on wood product use in construction, so data gathering and analysis should be a priority to improve this and to monitor change.

The construction industry is relatively conservative and product substitution occurs slowly. Early adoption of new products is determined by perceived practicality, fashion and modernity (Reid *et al.*, 2004). The markets are often mature and rooted in traditional practice and products. Substitution of existing products by wood is more likely to be driven by product quality (e.g. straightness, freedom from defect), availability, price and ease of use or maintenance, than by innovation and considerations of environmental sustainability *per se*.

Furthermore, the forest industries are characterised by many small-to-medium enterprises (less than 250 employees), a large number of very small businesses (20 employees) and a large number of trade associations and representative bodies. Gaining a focus to national

research has been a challenge which has been somewhat alleviated by the UK's National Research Agenda (FTP, 2009). Overall, the approach to innovation and R&D leads to a focus on small step changes and immediate problem-solving. Focus on medium- to longer-term benefits is more difficult to achieve. The fragmented nature of the industry results in difficulties in reaching consensus as well as causing weak communication of opportunities.

7.2.3 Drivers supporting the specification of wood-based products

The key strengths of wood-based products need to be emphasised to get maximum penetration in major markets such as timber frame, and massive timber construction using cross-laminated timber. Offsite manufacture offers opportunity for wood-based construction systems to capture the efficiency and sustainability benefits that include fast build programmes, less disruption, reduced site waste and more cost-efficient processes. It is also important to communicate the additional functionality benefits of wood such as its thermal insulation and thus its savings in emissions through reduced space heating.

There are significant market drivers that support the promotion and use of wood and wood products in construction. Some of these are based in direct Government directives and codes and others are less formally driven.

The Green Guide for Specification (GGS) (BRE, 2009) is a tool which compares different building elements at a functional unit level for their environmental impacts. The GGS is based on full life cycle analysis (LCA) and yields an overall rating on a scale from A+ to E for any functional unit to be employed in a given construction. For timber, the extent of carbon sequestration over the growth period of the raw material is included as part of the rating. Without exception, wood and wood-based products contained in other functional units achieve either A or A+ ratings in the GGS system. In England, the Code for Sustainable Homes (CLG, 2006) is pushing for a step change in the delivery of sustainable new-build housing. This is underpinned by the Green Guide for Specification and is setting a framework in which all new-build housing will be zero carbon rated by 2016. At this time it remains unclear as to the impact of the economic downturn on our ability to innovate and meet these targets. House building has slowed to a level of less than a quarter of that predicted before the credit crunch unfolded.

Building regulations and codes can support the development and integration of long service life and durable wood and timber products into construction, but these must be facilitated by standards development. The smooth integration and transition to the latest European standards, among which Eurocode 5 covers the use of timber in buildings and civil engineering structures, will be critical for the future growth of the structural use of timber.

7.2.4 Potential for substitution and the scale of opportunity

The science of carbon accounting across the whole life cycle of wood products is still developing. However, life cycle assessments (LCA) such as those that underpin the Green Guide to Specification and which consider all stages of a product's life, consistently place wood products in low environmental impact categories. As LCAs develop, further competitive advantages can be expected to be for wood products in a low carbon economy.

The UK is starting from a varied baseline of wood product types which vary in their ability to penetrate, establish and sustain a presence in the various markets. Some are established and mature such as fencing and others where growth is anticipated, such as cladding, are very small.

Table 7.3 lists many wood products currently employed in the UK and provides estimates of their current carbon storage, service life and the extent of their use. The current level of demand, market prospects for the next 10 years and the issues faced in expanding their market share are also indicated. There are likely to be good prospects for growth of wood products in timber frame, walls, exterior cladding, floors and joinery. These will have a major role in storing carbon.

The markets for wood products that can be most readily captured should be identified with the impacts they will have on carbon storage and substitution. Options should be analysed and placed in the context of impact of carbon storage and ease of market capture and a scheme for achieving this is shown in Figure 7.5.

A traditional brick and block built, three bedroom, semi-detached house is estimated to contain wood products that store 4.4 tCO_2 equivalent in the roof, tiling battens, floors and studs (Davies, 2009). This figure can be as high as 15.0 tCO_2 stored in a timber frame dwelling. For traditional brick-built housing, the value scales up to approximately 92 $MtCO_2$ stored (25 MtC) in existing UK

Figure 7.5
A scheme to allow the identification of markets which would maximise the carbon store in buildings and thus which should be targeted for market development.

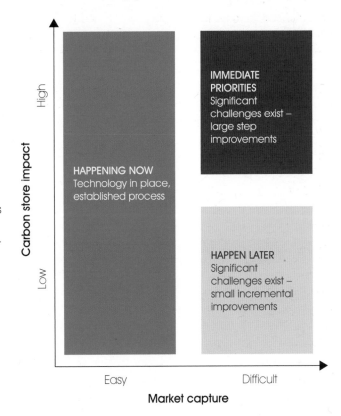

homes. This value can be compared with the total UK CO_2 emitted 150 MtC per year by burning fossil fuels alone (Cannell, 2003). The magnitude of this carbon store in buildings is made more impressive when combined with the carbon displaced by replacing more fossil fuel intensive products which is estimated to be between 15 t and 40 tCO_2 for a single dwelling.

Prior to the economic downturn, the UK had an ambitious target of building 250 000 new homes per year in which an estimated 1.6 $MtCO_2$ (or 0.44 MtC) would be stored. Recent predictions are that about 80 000 new homes were to be built in 2009 (Construction News, 2009) which could represent storage of 0.5 $MtCO_2$ (0.14 MtC), if an assumed 20% market share (UKTFA, 2009) is selected for timber frame construction.

There has been much debate about the carbon savings to be gained by substituting timber for brick and cement. A much quoted statistic is that 1 t of CO_2 is saved if 1 t of brick or concrete is replaced with timber. However, the assumptions in deriving this comparison can be challenged because the materials have considerably

Table 7.3

Representation of different wood products, their market demands, minimum service life, market prospects and estimated carbon stored in 2009.

Use for wood product	Market demands for product	Minimum service life (years)	Market prospect for next 10 years	Estimation of CO_2 stored (number of homes or wood products x convert wood to CO_2 x volume in typical home or wood product)
WOOD PRODUCTS IN THE CONSTRUCTION OF HOMES				
Timber frame	Improving acoustic and thermal insulation	60	↑	$2\,000\,000 \times 0.9 \times 6$ $= 10.8$ $MtCO_2$
Walls	Improving acoustic and thermal insulation	60	↑	$5\,000\,000 \times 0.9 \times 2$ $= 9.0$ $MtCO_2$
Structural floors and floor cassettes	Improving acoustic and thermal insulation	60	↑	$5\,000\,000 \times 0.9 \times 2$ $= 9.0$ $MtCO_2$
Floor covering	High aesthetic, affordability, long lasting and wear resistant	15	↑	$5\,000\,000 \times 0.9 \times 0.36$ $= 1.6$ $MtCO_2$
Trussed rafters	New designs utilising UK timber	60	→	$12\,600\,000 \times 0.9 \times 2.58$ $= 29.3$ $MtCO_2$
Exterior cladding	High aesthetic, extending maintenance intervals	30	↑	$2\,000\,000 \times 0.9 \times 0.33$ $= 0.6$ $MtCO_2$
Tiling battens	Improving treatment quality	60	→	$12\,600\,000 \times 0.9 \times 0.457$ $= 5.2$ $MtCO_2$
Exterior joinery (doors and windows)	Improving thermal performance and extending maintenance intervals	30	→	$8\,400\,000 \times 0.9 \times 0.26$ $= 2.0$ $MtCO_2$
Interior joinery	High aesthetic, affordability, long lasting and wear resistant	30	↑	$15\,000\,000 \times 0.9 \times 0.13$ $= 1.8$ $MtCO_2$
Wood wool insulation	Long lasting thermal performance	60	↑	$5000 \times 0.9 \times 1.5$ $= <0.01$ $MtCO_2$
			TOTAL	**69.2 $MtCO_2$ = 18.87 MtC**
ADDITIONAL WOOD PRODUCTS				
Fencing	Reliability, improving treatment quality, end of life options	15	→	$10\,000\,000 \times 0.9 \times 0.33$ $= 3.0$ $MtCO_2$
Furniture (indoor)	High aesthetic, affordability, long lasting and wear resistant, end of life options	15-30	→	$21\,000\,000 \times 0.9 \times 0.11$ $= 2.1$ $MtCO_2$
Furniture (outdoor)	High aesthetic, affordability, low maintenance	10	→	$10\,000\,000 \times 0.9 \times 0.04$ $= 0.4$ $MtCO_2$
Landscaping timber	Reliability, improving treatment quality, end of life options	15	→	$10\,000\,000 \times 0.9 \times 0.11$ $= 1.0$ $MtCO_2$
Foundations	Long service life	60-100	→	Not quantified
Scaffold boards	Maintaining preferred choice status	10	→	Not quantified
Transmission poles	Maintaining preferred choice status, serviceability, working at height, end of life	60-100	→	$14\,000\,000 \times 0.9 \times 0.7$ $= 8.8$ $MtCO_2$
			TOTAL	**15.2 $MtCO_2$ = 4.15 MtC**
PACKAGING				
Pallets, boxes, crates	Maintaining preferred choice status, improving hygiene	3	→	$20\,000\,000 \times 0.9 \times 0.02$ $= 0.4$ $MtCO_2$
Product packaging	Innovative form and function	0.1	↑	Estimated 20.0 $MtCO_2$
			TOTAL	**20.4 $MtCO_2$ = 5.55MtC**

Note: the conversion factor used is 900 kg CO_2 stored per m^3 of wood.

different properties, varying service lives (10–100+ years) and different roles in buildings. The amount of carbon associated with a defined functional unit is an important concept to ensure an appropriate comparison is made between products (Table 7.3). Work is needed here to gather better data of this nature.

Estimates of the annual carbon sequestration in the wood products pool (sawnwood, roundwood, panels, paper, board) for the UK over the period 1990–99 range from 2.4 to 4.9 MtC (Hashimoto, 2002). The estimates published for other Kyoto Protocol Annex I countries are approximately 4MtC per country. By adding the carbon stored in all wood products used in the construction process (based upon carbon storage values for products derived from Table 7.3), it is estimated that the total carbon stored in UK homes in 2009 is 19 MtC (Figure 7.6, left hand blue column). This is comparable with the figure 25 MtC estimated earlier, based on Davies (2009). Applying a scenario where wood products gain wider acceptance and there are increases in the percentage of the housing stock that will be timber rich (i.e. timber framed, clad, floors, wood wool insulation) shows potential significant impacts on carbon stored and carbon saved through substitution. Using the displacement of between 15 t and 40 t of CO_2 for a single dwelling for all UK homes this yields an estimated 43 MtC (for an estimated 50% of UK housing stock) displaced in 2009 and a theoretical 229 MtC maximum displaced in all housing. Figure 7.6 shows the theoretical maximum if all UK homes were timber rich. Somewhere between these two bars is realistically where we are targeting. If the wood construction products sector continues to grow as it has in the past 10 years there is potential to store as estimated additional 10 MtC in the UK's new and refurbished homes. This would save a further estimated 20 MtC as the substitution effect of displacing more carbon intensive materials. Without legislation or incentive it may take 10 years to reach this additional stored amount as the construction sector is slow to change.

One of the principle challenges is to upgrade the existing housing stock for a low carbon economy. This could be achieved through refurbishment to enrich timber contents. However, it must be recognised that the extent to which such upgrading can be achieved will be restricted by planning and conservation issues, plus the longevity of existing stock. On the positive side, the imposition of more exacting energy efficiency standards will increase incentives to employ wood products in the ongoing process of regeneration.

Figure 7.6
Estimates of the total carbon stored in wooden construction products in all UK homes standing in 2009 (left-hand column) and the theoretical maximum achievable carbon storage if the industry converted to construction of 100% timber-rich buildings (right-hand column). The values in each column are subdivided to show construction products themselves (dark blue) additional wood products (light blue) and the carbon saved by substitution effects of displacing more carbon-intensive materials using wood (grey). The carbon content of each product is derived from values shown in Table 7.3.

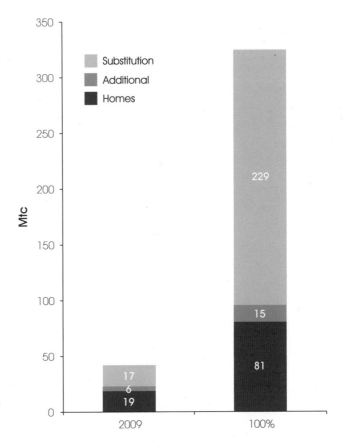

As new timber-rich buildings are constructed and existing buildings are refurbished in the UK, the total wood product pool in the housing stock can be expected to increase progressively. In addition, an extended service life will hold the wood products in the pool for longer before they pass to landfill and subsequently decompose or are recovered and recycled into new products. The end-of-life issues for construction products are of key importance. Wood needs to deliver products that can maximise reuse and recycling and not provide future problems. The recycling infrastructure needs to be developed and recovery opportunities maximised.

7.2.5 Climate change and the built environment

Climate change will affect the type and quality of products which are delivered from UK forests. As the regulatory frameworks which drive changes to buildings respond to climate change, the markets for construction materials will also change – increasingly buildings will be required to deliver flood resilience and provide in-built passive cooling. The service life of construction products may be altered by changed environmental weathering factors. Wood products will need to be well placed to deliver into new construction systems providing tailored solutions in composites and will also have the advantage of carbon storage which will becoming increasingly important.

7.2.6 Product life times, sustainability and future scope

Materials with long service life store carbon for longer, and wood products often have the additional advantage of reuse (second life) opportunities either as new wood products (chipped and integrated into panel products) or for displacement of fossil fuel (energy recovery). New technologies are emerging to extend the service life of wood products and, along with improved maintenance and the potential for reuse, this will further improve the carbon benefits of wood products. Examples are paints containing low or no volatile organic compounds for external use, which reduce the environmental impact of paint and require less energy in their manufacture.

Sustainable and ethical production are keys to the continued success of wood products in construction. The forest industries have led the way with sustainably sourced raw materials, and well recognised certified schemes such as the Forest Stewardship Council (FSC) and Programme for the Endorsement of Forest Certification (PEFC) marks have become established. Confirmation of sourcing from sustainably-managed forests (chain of custody) is a requisite for public procurement and will filter through to mainstream private construction in the next five years.

Economic instruments to reduce CO_2 emissions are likely to be in place soon which will influence the choice of materials for construction. A carbon tax would change the market for building materials and reduce competitiveness of materials produced by processes which result in high GHG emissions and are energy inefficient. As a result of the energy efficiency and low GHG emissions associated with their production, forest products may have

considerable opportunity in provision of feedstock for other sectors.

7.2.7 Research challenges and gaps

The UK has pockets of research and innovation talent focussing on forest products which provides small incremental improvements in our knowledge and understanding of the likely environmental benefits of wood products. A compelling case to use wood in construction is hampered by the evidence being incomplete and fragmented. Initiatives have yet to provide the step change needed to enable UK grown forest products to take a greater contribution to construction. The research proposed in the UK National Research Agenda (FTP, 2009) should be undertaken as a platform for taking forward commercially relevant studies designed to bring down the barriers to the use of home grown timber.

Wood products in construction can certainly contribute to the delivery of zero carbon homes in the UK, but more research is needed to determine the extent achievable. A clear timetable and process is required to change the building regulations so that standards are in place to achieve the zero carbon 2016 target for all new-build homes.

7.3 Conclusions

Substitution of wood and wood products for other construction materials and for non-renewable energy sources offers a major opportunity for tackling climate change by storing carbon in our buildings and reducing fossil fuel consumption. Greenhouse gas emissions from the production and use of wood products are lower than those from other materials commonly in use in construction. Provided the challenges can be met, substitution of wood for fossil fuels and for those materials used in construction which require high GHG emissions in their production presents an attractive solution for industry.

7.4 Research priorities

- Development of scenarios describing projected consumption of biomass energy and sustainable wood products are needed to determine how much more forest is required.
- In order to compare wood and other construction materials, GHG balances and energy efficiencies

for different construction systems using consistent assessment methods are required.

- Life cycle analyses of wood products and an understanding of the turn-over of carbon in different wood product pools are required.
- The built environment will need to adapt to climate change and the impacts of such adaptation on the use of wood products should be determined.
- Research on the optimal adaptation of our woodlands and forests should take into account the need for increased supplies of sustainable wood products and woodfuel.

References

AULISI, A., SAUER, A. and WELLINGTON, F. (2008). *Trees in the greenhouse: why climate change is transforming the forest products business.* World Resources Institute, Washington.

BIOTECHNOLOGY AND BIOLOGICAL SCIENCES RESEARCH COUNCIL (2006.) *Review on bioenergy research.* Online at: www.bbsrc.ac.uk/organisation/policies/reviews/scientific_areas/0603_bioenergy.html (accessed April 2008).

BRE (2009). *The green guide for specification.* 4th edn. Wiley-Blackwell, Oxford. Online at: www.thegreenguide.org.uk

BROADMEADOW, M. and MATTHEWS, R. (2003). *Forests, carbon and climate change; the UK contribution.* Forestry Commission Information Note 48. Forestry Commission, Edinburgh.

CANNELL, M.G.R. (2003). Carbon sequestration and biomass energy. *Biomass and Bioenergy* 24, 97–116

CANNELL, M.G.R. and DEWAR, R.C. (1995). The carbon sink provided by plantation forests and their products in Britain. *Forestry* 68, 35–48.

CARROLL, A. and SOMERVILLE, C.R. (2009). Cellulosic biofuels. *Annual Review of Plant Biology* 160, 165–182.

CLG (2006). The code for sustainable homes. Online at: www.planningportal.gov.uk/uploads/code_for_sust_homes.pdf

CONSTRUCTION NEWS (2009). Online at: www.cnplus.co.uk (accessed February 2009).

DAVIES, I. (2009). *Sustainable construction timber.* Forestry Commission Scotland.

DEFRA (2005). *Biomass Task Force Report.* Department for Environment, Food and Rural Affairs, London. Online at: www.defra.gov.uk/farm/crops/industrial/energy/biomass-taskforce/index.htm (accessed 3 August 2009).

DEFRA (2007). *UK biomass strategy.* Department for Environment, Food and Rural Affairs, London. Online at: www.defra.gov.uk/Environment/climatechange/uk/energy/renewablefuel/pdf/ukbiomassstrategy-0507.pdf (accessed 3 August 2009).

DEFRA (2008). *Strategy for waste to energy.* Department for Environment, Food and Rural Affairs, London. Online at: www.defra.gov.uk/environment/waste/strategy/factsheets/energy.htm

DEPARTMENT OF ENERGY (2006). *Breaking the biological barriers to cellulosic ethanol: a research roadmap from biomass to biofuels workshop.* Department of Energy, London.

DEPARTMENT OF ENERGY AND CLIMATE CHANGE (2009a). *The UK renewable energy strategy.* Cm. 7686. The Stationery Office, London. Online at: www.decc.gov.uk/en/content/cms/what_we_do/uk_supply/energy_mix/renewable/res/res.asp (accessed 3 August 2009).

DEPARTMENT OF ENERGY AND CLIMATE CHANGE (2009b). *The UK low carbon transition plan.* The Stationery Office, London. Online at: www.decc.gov.uk/en/content/cms/publications/lc_trans_plan/lc_trans_plan.aspx (accessed 3 August 2009).

DEPARTMENT OF TRADE AND INDUSTRY (2007). *Digest of UK energy statistics.* Online at: http://stats.berr.gov.uk/energystats/dukes07_c7.pdf (accessed April 2008).

ENVIRONMENT AGENCY (2009). Biomass - Carbon sink or carbon sinner? Online at: www.environment-agency.gov.uk/business/sectors/32595.aspx (accessed August 3 2009).

EUROPEAN ENVIRONMENT AGENCY (2006). *How much biomass can Europe use without harming the environment.* EEA Report No. 7. Online at: www.eea.europa.eu/publications/eea_report_2006_7 (accessed 3 August 2009).

EUROPEAN UNION (2008). *EU sustainability criteria for biofuels.* Online at: www.foeeurope.org/agrofuels/documents/Proposal_sustainability_criteria_biofuels_27Mar08.pdf (accessed 8 June 2008).

FOREST-BASED SECTOR TECHNOLOGY PLATFORM (2009). *A national research agenda for the UK forest-based industries.* Online at: www.forestplatform.org

FORESTRY COMMISSION (2007). *A woodfuel strategy for England.* Online at: www.forestry.gov.uk

FORESTRY COMMISSION SCOTLAND (2009). *Woodfuel: demand and usage in Scotland.* Update report, 2008. Online at: www.forestry.gov.uk

FTP (2009). The Forest Technology Platform National Research Agenda. Online at: www.forestplatform.org

GLOBAL BIOENERGY PARTNERSHIP (2009). Online at: www.globalbioenergy.org/ (accessed 3 August 2009).

GRIGAL, D.F. and BERGSON, W.D. (1998). Soil carbon changes associated with short rotation systems. *Biomass and Bioenergy* 14, 371–377.

GUSTAVSSON, L. (2008). Substitution effects of wood-based construction materials. Harvested wood products in the context of climate change, Palais des Nations, Geneva. IEA Bioenergy: T38: 2005: 05. Online at: www.ieabioenergy-task38.org/projects/task38casestudies/finswe-brochure.pdf

HASHIMOTO, S., NOSE, M., OBARA, T. and MORIGUCHI, Y. (2002). Wood products: potential carbon sequestration and impact on net carbon emissions of industrialized countries. *Environmental Science and Policy* 5, 183–193.

HILLIER, J., DAILEY, G., AYLOTT, M.J., WHITTAKER, C.J., RICHTER, G., RICHE, A., MURPHY, R., TAYLOR, G. and SMITH, P. (2009). Greenhouse gas emissions from four bioenergy crops in England and Wales: Integrating spatial estimates of yield and soil carbon balance in life cycle analyses. *Global Change Biology Bioenergy* 1, 267–281.

KAHLE, P., BEUCH, S., BOELCKE, B., LEINWEBER, P. and SCHULTEN, H.R. (2001). Cropping of *Miscanthus* in Central Europe: biomass production and influence on nutrients and soil organic matter. *European Journal of Agronomy* 15, 171–184.

KARNOSKY, D.F., TALLIS, M.J., DARBAH, J. and TAYLOR, G. (2007). Direct effects of elevated carbon dioxide on forest tree productivity. In: Freer-Smith, P.H., Broadmeadow, M.S.J. and Lynch, J.M. (eds) *Forestry and climate change.* CABI, Wallingford. pp. 136–144.

LAZARUS, N. (2005) BedZED Construction materials report. Online at: www.massbalance.org/downloads/projectfiles/1640–00195.pdf

MCKAY, H., BIJLSMA, A., BULL, G., COPPOCK, R., DUCKWORTH, R., HALSALL, L., HUDSON, J.B., HUDSON, R.J., JOHNSON, D., JONES, B., LIGHTFOOT, M., MACKIE, E., MASON, A., MATTHEWS, R., PURDY, N., SENDROS, M., SMITH, S. and WARD, S. (2003). *Woodfuel resources in Britain.* Final Report. Forestry Commission.

MAKERSCHIN, F, (1999). Short rotation forestry in central and northern Europe. *Forest Ecology and Management* 121, 1–7.

OLIVER, R., FINCH, J. and TAYLOR, G. (2009). Second generation bioenergy crops and climate change: Effects of elevated CO_2 on water use and yield. *Global Change Biology Bioenergy 1,* 97–114.

RENEWABLE FUELS AGENCY (2008). *The Gallagher Review of the indirect effects of biofuels. Renewable Fuels Agency UK.* Online at: www.dft.gov.uk/rfa/reportsandpublications/reviewoftheindirecteffectsofbiofuels.cfm (accessed 3 August 2009).

REID, H., HUQ, S., INKINEN, A., MACGREGOR, J., MACQUEEN, D., MAYERS, J., MURRAY, L. and TIPPER, R. (2004). *Using wood products to mitigate climate change: a review of evidence and key issues for sustainable development.* Online at: www.iied.org/pubs/display.php?o=10001IIED

ROYAL COMMISSION ON ENVIRONMENTAL POLLUTION (2004). *Biomass as a renewable energy source.* Online at: http://www.rcep.org.uk/bioreport.htm (accessed October 2006).

ROYAL SOCIETY (2008). *Sustainable biofuels: prospects and challenges.* Royal Society, London.

ROWE, R.L., STREET, N.R. and TAYLOR, G. (2009). Identifying potential environmental impacts of large-scale deployment of dedicated bioenergy crops in the UK. *Renewable and Sustainable Energy Reviews* 13, 271–290.

SLADE, R., PANOUTSOU, C. and BAUEN, A. (2009). Reconciling bio-energy policy and delivery in the UK: will UK policy initiatives lead to increased deployment? *Biomass and Bioenergy* 33, 679–688.

SHERRINGTON, C., BARTLEY, J. and MORAN, D. (2008). Farm-level constraints on the domestic supply of perennial energy crops in the UK. *Energy Policy* 36, 2504–2512.

ST. CLAIR, S., HILLIER, J. and SMITH, P. (2008). Estimating the pre-harvest greenhouse gas costs of energy crop production. *Biomass and Bioenergy* 32, 442–452.

THE CARBON TRUST (2004). *Biomass sector review.* Online at: www.carbontrust.co.uk/Publications/publicationdetail.htm?productid=CTC512 (accessed 3 August 2009).

UKERC (2009). *Making the transition to a secure and low-carbon energy system.* Synthesis report, The UKERC Energy 2050 Project.

UK TIMBER FRAME ASSOCIATION (2009). *Statistics.* Online at: www.timber-frame.org.uk

WHITTAKER, C. and MURPHY, R. (2009). *Assessment of UK Biomass Resource.* A mirror report to the US 'Billion Ton Report'. AtlannTic Alliance (personal communication).

THE POTENTIAL OF UK FORESTRY TO CONTRIBUTE TO GOVERNMENT'S EMISSIONS REDUCTION COMMITMENTS

Chapter

8

R. W. Matthews and M. S. J. Broadmeadow

Key Findings

Significant opportunities exist for the forestry sector in the UK to deliver GHG emissions abatement from woodlands planted since 1990, potentially amounting to 15 MtCO$_2$ per year by the 2050s and equivalent to 10% of the UK's total GHG emissions if current emissions reduction targets are achieved. The abatement that could be delivered is highly sensitive to the level and timing of woodland creation.

Within forestry, woodland creation is the most effective approach to GHG abatement in the medium to long term, but can deliver relatively little in the UK Government's first three carbon budgets (to 2022). However, by 2050, a 25 000 ha per year programme of woodland creation between now and 2025 could deliver 130 MtCO$_2$ abatement through sequestration in growing biomass, or total abatement (including fossil fuel and product substitution) of 200 MtCO$_2$.

There is limited scope for changes in forest management alone to deliver significant levels of emission abatement, implying that woodland creation should be the initial focus of activity. Optimising timber production in appropriate stands offers the largest opportunities for abatement through new approaches to forest management. Measures that focus solely on increasing forest carbon stocks are likely to limit the abatement potential because of lost opportunities for fossil fuel and product substitution.

If the forestry sector is to deliver the abatement that it can potentially provide its full contribution must be recognised including both carbon storage in forest biomass and abatement through wood and timber products substituting for fossil fuels directly and indirectly.

Currently, the UK's land use, land use change and forestry (LULUCF) GHG inventory does not adequately reflect emissions from, and uptake by, existing woodlands. Furthermore, the lack of attribution to the forestry sector of emissions abatement contributed by forestry products used by other sectors, e.g. energy and construction may limit the development of abatement strategies in the forestry sector.

Woodland creation (and subsequent management) in the UK can be a cost-effective approach to combating climate change; for a number of woodland creation scenarios, net social costs of abatement are negative, but rise to £70 per tonne CO$_2$ for the least cost-effective options. Production conifer plantations represent particularly cost-effective abatement, although if ancillary benefits were also included in the analysis, the cost-effectiveness of broadleaf and native woodland options, in particular, would increase.

Woodland creation provides a range of co-benefits (social, economic and environmental) that many other approaches to emissions abatement do not provide; if these co-benefits were included in the net cost calculations, abatement through woodland creation would appear even more cost-effective, particularly for those woodland creation options with lower revenue from sales of timber or woodfuel.

A significant woodland creation programme would require the existing regulatory requirements and standards to be maintained; it would also require a spatial planning framework to be established to identify where woodland creation could contribute most to other objectives.

Forest growth results in removal of CO_2 from the atmosphere into the carbon stock of the forest, and the provision of woodfuel and wood products that can be used to substitute for fossil-fuel derived energy sources and materials. At a global level, the potential of the forestry sector to reduce net GHG emissions (i.e. provide 'abatement' of GHG emissions) has been evaluated by the IPCC as a total of 6.7 $GtCO_2$ per annum in 2030 (Nabuurs *et al.*, 2007).

Approximately half of this amount would be achieved by substituting timber and wood products both directly and indirectly for fossil fuel use. The remaining 3.2 $GtCO_2$ of abatement was assessed to be evenly distributed between reduced deforestation and afforestation. The abatement potential for Europe has been estimated as 295 $MtCO_2$ per year in 2030 (Nabuurs *et al.*, 2007).

The concept of accounting for the fossil fuel substitution benefits associated with forest management and woodland creation was reviewed by Matthews (1996). However, in such analyses an over-emphasis on the carbon stored in forest biomass often hides the cross-sectoral contribution that can be made through forest management. Given the current policy focus on carbon budgets, both nationally and globally, it is therefore timely to reconsider the wider potential of forestry within the UK to provide carbon abatement. This chapter therefore provides an holistic assessment of the total GHG emissions abatement potential of the forestry sector in the UK at two scales: national and stand-scale. The contribution of those forest management practices that result in the largest changes in the carbon stocks of forest biomass (as discussed in Chapter 6) are estimated. The roles of harvested wood products, and the fossil fuel emissions that are avoided through utilisation of wood and timber products (see Chapter 7) are taken into account. Importantly, the long-term GHG emissions abatement potentials through to 2100 of a range of woodland creation options are presented. The cost-effectiveness of woodland creation is then evaluated in comparison with other abatement options available across other sectors of the economy.

8.1 Carbon abatement through forestry: UK and international perspectives

The UK Government has set an extremely challenging and legally binding emissions target of an 80% reduction on 1990 emissions by 2050 – meaning that current UK annual emissions (as for 2007) of 611 $MtCO_2e$ need to fall to 155 $MtCO_2e$. This target was set on the advice of the Committee on Climate Change. It represents the necessary contribution of the UK to restricting the global increase in temperature to 2°C, taking 'burden-sharing' between developed and developing countries into account. Interim targets have been set for the first three five-year budget periods that are required to be in place under the terms of the Climate Change Act (GB Parliament, 2008). The target for 2020 (2018–2022) is for a 34% reduction on 1990 GHG emissions, a figure that may rise to 42% if a comprehensive global agreement on emissions reduction is reached through UNFCCC negotiations (CCC, 2008; HMT, 2009).

An assessment has been made of the abatement that each sector (e.g. transport, energy, agriculture and forestry) could deliver (CCC, 2008) and, in the UK Low Carbon Transition Plan (DECC, 2009b), the UK Government announced how the emissions reductions committed to in the first three budget periods would be met. Although abatement resulting from woodland creation was not identified as contributing significantly in the first three budget periods, its long-term role in helping to achieve the ultimate target of an 80% reduction in emissions was outlined, and a programme of new woodland creation was recognised as having the potential to contribute to this target. In this chapter we explore the aspiration expressed in the Low Carbon Transition Plan by providing quantification of the extent to which forests, under a range of different scenarios, could indeed provide a significant contribution to the UK mitigation strategy.

The potential for decarbonisation of energy supply was explored in the UK Renewable Energy Strategy (DECC, 2009c). This included an assessment indicating that renewables could contribute 15% of energy requirements by 2020. Biomass was assessed as delivering 33% of the renewables target, with woodfuel making a significant contribution. Annual production of 2 million tonnes woodfuel from English woodlands is included in the assessment of the Woodfuel Strategy for England (FC,

2007), while the Scottish Forestry Strategy (Scottish Executive, 2006) has committed to delivering 1 $MtCO_2$ emissions abatement through renewable energy production by 2020 (see 1.5.1, Chapter 1; 7.1.3 and 7.1.4, Chapter 7).

8.1.1 The forestry contribution to the UK's GHG inventory: projections to 2020

In the UK, the greenhouse gas (GHG) balance of the forestry sector is reported as a component of the Land Use, Land Use Change and Forestry (LULUCF) GHG inventory using methodologies compatible with the Good Practice Guidelines published by the Intergovernmental Panel on Climate Change (IPCC, 2006), as described in Chapter 1. The GHG balance of UK woodland (using the UK definition of woodland: see Chapter 1) is reported under the terms of the United Nations Framework Convention on Climate Change (UNFCCC). For pragmatic reasons, the forestry GHG inventory is currently restricted to woodlands planted after 1919 because good data records are available for these woodlands. It is assumed that woodlands planted before this date are at equilibrium in terms of GHG balance. The inventory projections of net carbon uptake rates by forests are prepared using the C-FLOW carbon accounting model (Dewar, 1991; Dewar and Cannell, 1992; see Thomson, 2009 for the current formulation and parameterisation of the model). Annual carbon uptake/removals in tree biomass, soil carbon and harvested wood products are modelled on the basis of year of planting and assuming that the woodlands are managed conventionally, according to published forest growth and yield models (Edwards and Christie, 1981). Broadleaved woodland is modelled as Yield Class (YC) 6 beech (see Glossary for definition of Yield Class). Modelling of conifer woodland assumes YC 12 Sitka spruce (but YC 14 in Northern Ireland). The projections (Figure 8.1a,b) assume that annual rates of woodland creation (8360 ha year^{-1}) and removal (1128 ha year^{-1}) continue as reported for 2006 (although the rate of woodland removal is assumed to decline in the future). This is described as the 'business-as-usual' (BAU) scenario.

The dynamics of the current and projected UK forest carbon sink are thus largely determined by historic planting patterns involving large scale afforestation schemes operating through the 1950s to the 1980s. As a result of these, the estimated strength of the UK forest carbon sink increased from 12 $MtCO_2$ year^{-1} in 1990 to a peak of 16 $MtCO_2$ year^{-1} in 2004 (Figure 8.1c). However, because of the marked decline in new planting since the 30 000 ha per year in the late-1980s (see Figure 1 in 1.4, Chapter 1),

Figure 8.1
Values of net GHG (CO_2 equivalents) uptake for the UK GHG inventory projected to 2020. Values for: (a) forest land, harvested wood products (HWP), and the net change from the whole LULUCF sector (LULUCFnet); (b) net CO_2 uptake by forests in the four countries of the UK, including changes in stocks of harvested wood products; (c) forest net uptake assuming 'business-as-usual' (BAU) (as in forestland (a) above), high emissions scenario (no new afforestation) and low emissions scenario (increased afforestation). (See text for description of scenarios. All data as reported to UNFCCC; Thomson, 2009.)

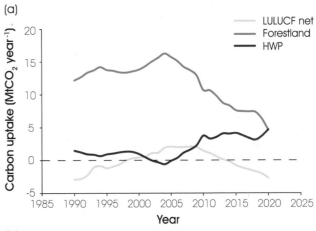

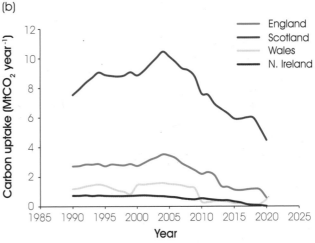

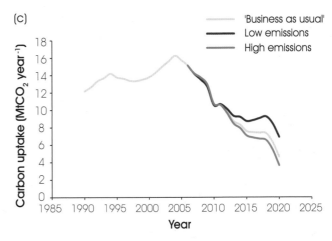

the estimated forest carbon sink has subsequently fallen to little more than 12 $MtCO_2$ year[-1] in 2009. A further dramatic decline is projected to 4.6 $MtCO_2$ year[-1] in 2020 (Figure 8.1c). This decline in the strength of the forest carbon sink has serious implications for the UK's GHG inventory, particularly in the light of the challenging targets for emissions reductions outlined first in the Climate Change Act (2008) and, subsequently, in the UK Low Carbon Transition Plan (DECC, 2009b). It is, however, important to recognise that despite the falling CO_2 uptake rates, UK woodlands remain a carbon sink, albeit at a much reduced level, through to 2020.

Largely because of the pattern of previous large scale planting, followed by reductions in the extent of woodland creation experienced in the UK over the last decade, there will be a rapid decline in the extent of abatement provided by forest land up to 2020 (Figure 8.1a). The decline in new planting contributes significantly to the LUCLUF GHG inventory becoming negative by 2020, i.e. it becomes a net source of carbon (Figure 8.1a). The pattern is broadly the same for all of the devolved administrations but the decline appears to be particularly marked in Scotland as a result of the relatively large proportion of new planting during the 1950s to 1980s in that country (Figure 8.1b). The UK GHG inventory has explored the impacts of two alternatives to the BAU scenario on the projected forest carbon sink (Figure 8.1c). A low emissions scenario considers the impact of increasing woodland creation between 2007 and 2020 to 25000 ha year[-1] (compared with the BAU assumption of 8360 ha year[-1]), with the assumption that broadleaf and conifer species are planted in the same ratio as at present. A high emissions scenario considers the impact of no new woodland creation between 2007 and 2020. As illustrated in the three projections in Figure 8.1c, for the for the low emissions scenario involving enhanced woodland creation, the decline in abatement is reduced relative to the BAU and high emissions (no woodland creation) scenarios.

The decline in the forest carbon sink evident in the projections to 2020 can only be modified a little by increases or decreases in new woodland planting. This is clearly demonstrated by the relatively small difference between the low and high emissions scenarios (Figure 8.1c) compared to the decline in the strength of the overall sink. However, a subsequent significant level of abatement would be provided by a sustained increase in woodland creation starting at the present as demonstrated in 8.2 and 8.3 below.

8.1.2 Forestry contribution to UK Kyoto protocol-reporting

The reporting of the GHG balance of forests under the terms of the UNFCCC's Kyoto Protocol is restricted to CO_2 emissions and uptake (i.e. removal from the atmosphere) associated with afforestation, deforestation and reforestation (ARD) that has taken place since 1990. This provides a much better indication of the potential contribution of the forestry sector through changes to rates of woodland creation and removal because the effects of the high planting rates in the 1950s to 1980s that dominate the inventory projections shown in Figure 8.1 are not included. Business-as-usual (BAU) projections for UK forests (assuming woodland creation and removal continue at 2006 levels) indicate that CO_2 uptake associated with the 'Kyoto forest' (i.e. new woodlands planted since 1990 and accounting for woodland removal) will rise to 2.5 $MtCO_2$ per year in 2012. Although meeting commitments made under the Kyoto Protocol is clearly a high policy priority at present, it is uncertain how LULUCF reporting, particularly for forestry, will be taken forward. It is important to recognise that unlike the GHG inventory reported to the UNFCCC, emissions and uptakes associated with ARD that are reported under the terms of the Kyoto Protocol do not include carbon stocks associated with harvested wood products.

8.1.3 The role of forestry in meeting the UK's GHG reduction commitments

Further development of the UK's LULUCF GHG inventory that is currently in progress (Thomson, 2009), coupled with the results being delivered by the National Forest Inventory (see Chapter 1), will improve its precision and enable the impacts of changes in forest management to be better reflected in the values reported. However, accounting methodologies mean that although the carbon stocks in harvested wood products may be allocated to the LULUCF/forestry sector, emissions reduction that result from wood substituting for fossil fuels are not. This includes both direct substitution in the form of woodfuel and indirect substitution (product displacement) through timber products replacing high energy materials such as concrete and steel. This approach to emissions accounting can result in the conclusion (see Chapter 6) that to maximise the forestry sector's contribution to emissions reduction, carbon stocks in forest biomass should be maximised and harvesting minimised (Forster and Levy, 2008). However, as demonstrated by Nabuurs (1996), Tipper et al. (2004) and Nabuurs et al. (2007) such a conclusion can be over-simplistic, with total abatement in the longer term maximised by maintaining high growth rates through managing woodlands and utilising the resulting timber products effectively. The need to supply a future low carbon society with sustainably produced wood products should also be recognised. Furthermore, the sustainable management of woodlands in the UK also has a potential role in reducing unsustainable harvesting of old growth forests and subsequent land use change elsewhere in the world.

A further complication in GHG accounting relates to GHG emissions abated through direct fossil fuel substitution (see above) when woodfuel is used to generate electricity. Emissions from electricity production are capped under the EU Emissions Trading Scheme (EU-ETS) – and are part of the so-called 'traded sector'. As such, emissions reductions in this sector would not be considered as abatement because a reduction in traded-sector emissions in the UK would (theoretically) result in higher emissions elsewhere in the EU. It would only be considered as abatement if such actions directly brought about a reduction in the cap (DECC, 2009a). However, we believe that it is important to acknowledge the contribution of the forestry sector to reducing emissions within the UK as a part of the global response to climate change. Emissions reduction through substituting directly for fossil fuels in electricity generation are included in the abatement potential reported in 8.2 and 8.3 below. However, for consistency with wider cross-sector studies they are not included in the subsequent evaluation of cost-effectiveness (see 8.4 below).

8.2 National level forest management scenarios

The way in which existing woodlands are managed has a significant impact on their carbon stocks (see 3.4 and 3.5, Chapter 3 and Chapter 6) and, consequently, on their ability to deliver emissions abatement. Although the current UK GHG inventory would not reflect any changes made in forest management, planned improvements to the forestry sector inventory may allow the impacts of any actions to be reflected in the future. The objective of this chapter is to evaluate the effect of forest management options at national scale on the GHG inventory, incorporating a limited number of country specific measures. The scenarios presented here are purely illustrative and do not reflect what is planned or might be achievable; they are intended to demonstrate the level to which the GHG inventory could

be affected by changes in forest management and to identify those measures that warrant further exploration.

8.2.1 The CARBINE carbon accounting tool

The following discussion describes simulations of GHG balances in the forest sector (as well as interactions with the energy and construction sectors) with the main objective of capturing indicative changes in emissions abatement under a number of different management scenarios and woodland creation options. The simulations have been undertaken using the Forest Research CARBINE model. CARBINE was the world's first forest carbon accounting model to be developed (Thompson and Matthews, 1989a,b) and has common features of structure and functionality with other forest carbon models such as C-FLOW (Dewar, 1990, 1991; Dewar and Cannell, 1992), CO_2fix (Mohren and Klein Goldewijk, 1990; Nabuurs, 1996; Mohren et al., 1999) and CBM-CFS3 (Kurz et al., 2009). From the outset, links between the forest sector and harvested wood products (HWP) were recognised to be important and are represented within the CARBINE model structure. The model was also one of the first to be applied to understanding impacts across the forest, energy and construction sectors (Matthews, 1994). Subsequently, CARBINE has been further developed into a national-scale scenario analysis tool and has been used to assess the impacts of current and alternative forestry practices on GHG balances in Great Britain (Matthews, 1996). The modelling system is based on conventional yield models (e.g. Edwards and Christie, 1981), coupled to models of carbon content, decomposition, soil carbon exchange, product utilisation and empirical data on the GHG balance of forestry operations, timber transport and timber processing.

The forest biomass and management components incorporated in CARBINE are shown in Figure 8.2. The modelling involves a sequential approach in which an appropriate yield model is selected and used to estimate biomass of various tree components employing the BSORT model (Matthews and Duckworth, 2005). Wood density (Lavers, 1983) is used within BSORT in order to convert wood volumes to dry weight, 50% of which is assumed to be carbon (Matthews, 1993). The biomass components are roots, stump, roundwood, sawlog, tips, branches and foliage. The biomass of the different compartments are considered either as standing (living) biomass, in-forest debris, or extracted material to be processed. Finally, the modelling system provides estimates of direct fossil fuel substitution (i.e. using woody

Figure 8.2
Schematic representation of the structure and components of the CARBINE forestry carbon accounting model.

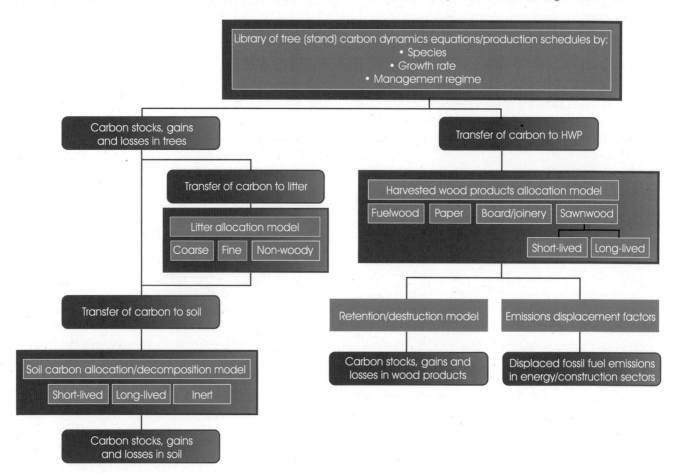

biomass for energy generation), and product substitution where, for example in construction, a wood product may replace other building materials with higher associated fossil fuel emissions (see Chapter 7).

Alongside growth and product estimations, the impacts on soil carbon and operational fossil fuel use involved in establishing, maintaining and harvesting trees, and processing extracted material of the forest are all considered. Model estimates have been produced for a number of species and site conditions broadly representative of existing and potential scenarios across the UK. While site conditions and species are key determinants of C stocks and sequestration potential of woodland, stand management (notably thinning, application of silvicultural systems and rotation length) also has profound impacts (Chapter 6). In this series of simulations, the GHG balance of the non-forest under-storey is not considered.

The comprehensive analysis associated with the CARBINE

modelling system represents a marked departure from many previous attempts to evaluate abatement potential, which, like C-FLOW, have concentrated on forest carbon stocks and/or carbon stocks in harvested products (e.g. Forster and Levy, 2008). The inadequacies of many previous studies were highlighted by Matthews (1996), who reviewed the studies of the impacts of forest management (including woodland creation) on carbon balance available at that time. Of the 43 studies evaluated, only three considered all forest carbon pools and the contribution from both direct and indirect fossil fuel substitution. Matthews (1996) concluded that the inconsistencies and differing viewpoints evident in the studies were the result of the differing carbon-budgeting methodologies used in the studies – and that 'a clear and correct picture of the extent to which forest management can be modified to enhance C sequestration by forests' can only emerge when conventions for the modelling and reporting of carbon budgets for forests and their management are agreed. The analysis presented here therefore provides clarity over the contribution of each

Chapter 8: The potential of UK forestry to contribute to Government's emissions reduction commitments

component to total emissions abatement, enabling an holistic assessment of the contribution of forest management and woodland creation to carbon abatement to be made.

8.2.2 Forest management scenarios

The forest management scenarios detailed below evaluate a range of measures that affect forest carbon stocks or the ability of forest products to deliver emissions abatement by substituting for fossil fuels, either directly or indirectly. A comparison of CARBINE simulations with those produced by the C-FLOW model used to produce the GHG inventory for the forestry sector is provided by Robertson *et al.* (2003) and Matthews *et al.* (2007).

Enhanced afforestation scenario (EAS)

This scenario explores the abatement potential that could be achieved in the four countries of the UK by means of enhanced levels of woodland creation over the period 2010 to 2050. It is based on published targets, aspirations and case studies produced by the respective countries (e.g. FC, 2009; DECC, 2009b), coupled in some cases with expert judgement of desirable and achievable rates of woodland creation. The nature of the woodland created differs between countries (Table 8.1), its composition in the simulations having been determined following discussions with policy representatives in each country. A total of four distinct woodland types are included for England, designated as: high yielding short rotation forestry, managed broadleaf woodland (Sycamore-Ash-Birch), unmanaged native broadleaf woodland (Native), and conventionally managed Douglas fir. For Scotland, Wales and Northern Ireland, managed broadleaf (Sycamore-Ash-Birch) and unmanaged native woodland categories are assessed, while Sitka spruce replaces Douglas fir as the modelled conifer crop species (but mixed Sitka spruce and Douglas fir in Wales). The combined total of new planting for the four countries is assumed to be 23 200 hectares year[-1] for the period 2010–2050, representing an enhancement of 14 840 hectares year[-1] over the baseline projection of new planting assumed in the UK inventory of 8360 hectares year[-1] (see 8.1.1 above). The recent rate of deforestation (1128 ha year[-1] for the UK) is assumed to continue, maintaining consistency in terms of woodland area with the UK's 'low emissions scenario' GHG inventory projections described in 8.1.1 above. For consistency with reporting conventions under the Kyoto Protocol, results for the EAS scenario include the contributions due to existing new planting between 1990 and 2009 as well

as the proposed new planting from 2010. The abatement potentials of a broader range of woodland creation options and species are considered later.

Carbon stock enhancement scenario (FMS-A)

FMS-A recognises that forest carbon stocks can be enhanced by: (1) ceasing the management (i.e. harvesting and/or thinning) of existing woodland, effectively creating 'carbon reserves' or (2) deferring harvesting by 20 years or 25% of the rotation length, whichever is the greater. Here, both approaches are included in the analysis and applied to half the managed forest area in each country (calculated from timber production statistics). The relative contributions of each of the enhancement components assumed in the analysis differs between countries (Table 8.1).

Enhanced management scenario (FMS-B)

FMS-B assumes increasing the level of management (e.g. thinning and harvesting) will reduce the carbon stocks in the standing biomass. If the timber products are used in substitution (either as woodfuel to substitute for fossil fuels or as wood products to substitute indirectly), the abatement that may be achieved could more than compensate for the lower forest carbon stocks, particularly in the longer term. This scenario assumes that in all four countries, 50% of the existing managed woodland (as calculated from timber production statistics) is managed closer to optimum rotation length for timber production (Table 8.1).

Improved productivity scenario (FMS-C)

FMS-C considers the impact of increasing productivity at restocking, based upon the apparently logical assumption that increases of productivity (i.e. yield class) will enhance both rates of sequestration in biomass and, to a larger extent, the abatement potential through fossil fuel substitution. Increased productivity could be achieved through species/provenance selection (including suitability for the projected impacts of climate change, see Sections 2 and 4) or through the continued development of improved planting stock. It is assumed that YC is increased by 2 m^3 ha[-1] year[-1] on 50% of sites at restocking by using Douglas fir to replace Corsican pine; Japanese larch to replace Scots pine; and Western red cedar to replace Sitka spruce (Table 8.1). Improved productivity is thus assumed to be implemented across 11% of woodlands in England, 30% in Scotland and 18% in Wales, based on species breakdown recorded in the National Inventory of Woodland and Trees

(Forestry Commission, 2001a,b, 2002). FMS-C supplements either FMS-A or FMS-B to give two scenarios: FMS-C(A) and FMS-C(B).

Reinstating management scenario (FMS-D)

FMS-D considers the impact on abatement of bringing a proportion of woodlands that are currently unmanaged, or under-managed, back into productive management through re-instatement of regular thinning and harvesting (and re-planting as necessary). As in scenario FMS-B, the reduction in carbon stocks needs to be set against the increased abatement that could be delivered through direct and indirect fossil fuel substitution. This scenario assumes that 50% of unmanaged woodland (calculated from timber production statistics) in each of the countries is brought back into management (Table 8.1).

8.2.3 Illustration of the contribution of different components to total abatement potential from forestry

In order to understand the differences in projected emissions abatement potential of the various forest management scenarios, it is helpful first, to illustrate the way the individual components contribute to total emissions abatement. A number of components constitute the total carbon balance for the forestry sector: trees, litter and soils, harvested wood products, direct fossil

fuel substitution ('fuel'), indirect fossil fuel substitution ('materials') and emissions arising from woodland removal ('deforestation') (Figures 8.3a and b). The dynamics of each of these components can affect the overall abatement potential as outlined for one of the forest management scenarios, C-stock enhancement (FM-A), in Figure 8.3b. Up to 2020, the analysis presented in Figure 8.3a for the business-as-usual (BAU) scenario is similar to the LULUCF GHG inventory projections presented in 8.1.1 above, but additionally considers abatement through fossil fuel substitution and within forest uptake and emissions in greater detail. The BAU assumes that UK afforestation and woodland removal rates continue into the future at 2006 rates (8360 ha year[-1] and 1128 ha year[-1], respectively). Figure 8.3b estimates the forest sector GHG inventory, additionally assuming the implementation of forest management scenario FMS-A. The results demonstrate:

- the relative contributions of each component of forest carbon to overall abatement;
- the changes over time in the relative contribution from sequestration and substitution to abatement;
- the relative impact of implementing FMS-A compared with the carbon balance of the forest sector assuming current (BAU) approaches to woodland management and levels of afforestation and deforestation;
- the details of the five forest management scenarios are given in Table 8.1.

Table 8.1
Assumed changes to forest management in England, Scotland, Wales and Northern Ireland for the five scenarios used to project changes in CO_2 uptake values reported in Figures 8.3 and 8.4. Enhanced afforestation – areas are annual rates of new planting. Combined total new planting for UK is 23 200 ha year[-1] (14 840 ha year[-1] additional to 8360 ha year[-1] assumed under BAU).

Country	EAS: Enhanced afforestation	FMS-A: C-stock enhancement	FMS-B: Enhanced management	FMS-C: Improved productivity	FMS-D: Reinstating management
England	SRF: 1500 ha SAB: 2000 ha Native: 5000 ha DF: 1500 ha	30% of managed forest to carbon reserves 20% deferred fell	50% of managed forest to optimum rotation length	FMS-A or FMS-B plus: 50% CP to DF 50% SP to JL 50% SS to WRC	50% of unmanaged woodland brought into production
Scotland	Native: 2000 ha SAB: 2500 ha SS: 5500 ha	20% of managed forest to carbon reserves 30% deferred fell	50% of managed forest to optimum rotation length	FMS-A or FMS-B plus: 50% SS to WRC	50% of unmanaged woodland brought into production
Wales	Native: 500 ha SAB: 1000 ha DF/SS: 1000 ha	10% of managed forest to carbon reserves 40% deferred fell	50% of managed forest to optimum rotation length	FMS-A or FMS-B plus: 50% SS to DF 100% SP to JL	50% of unmanaged woodland brought into production
Northern Ireland	Native: 100 ha SAB: 300 ha SS: 300 ha	10% of managed forest to carbon reserves 40% deferred fell	50% of managed forest to optimum rotation length	FMS-A or FMS-B plus: 50% SS to WRC	50% of unmanaged woodland brought into production

SRF, short rotation forestry; SAB, sycamore-ash-birch mixture; Native, native woodland species (as appropriate for country) managed with amenity as main priority; SS, Sitka spruce, DF, Douglas fir, SS/DF, Douglas fir and Sitka spruce in mixture; CP, Corsican pine; SP, Scots pine; JL, Japanese larch; WRC, Western red cedar.

Figure 8.3
Comparison of the net carbon uptake by the components (see key) UK forests (sink = positive values; source = negative values) for (a) 'Business-as-usual' GHG inventory projections for the forestry sector with (b) projections that assume that the C-stock enhancement scenario (FMS-A) is implemented.

(a)

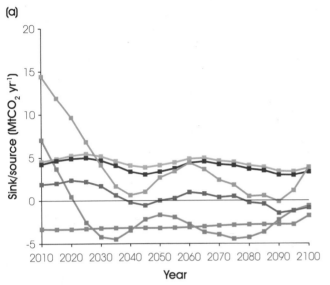

(b)

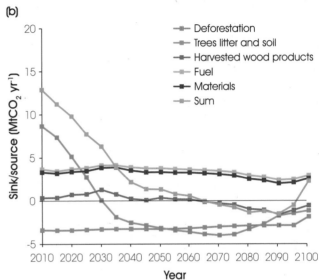

A comparison of Figure 8.3a with Figure 8.3b shows that the FMS-A scenario gives a change in the projected year that the trees, litter and soil component becomes a net source of CO_2 (i.e. becomes negative) from 2020 to 2030. The strength of the trees, litter and soil sink is enhanced by FMS-A relative to the BAU in the near term because of reduced levels of harvesting. However, between 2030 and 2100, forest biomass and soils become a larger source if FMS-A is implemented, due to the decline in growth rate in mature over-stocked stands. Furthermore, when the contribution to emissions reductions in other sectors is also considered (HWP and fuel), it is evident that as a result of reduced abatement through fossil fuel substitution and product displacement, the abatement potential of forestry in its entirety ('sum') is reduced by implementation of FMS-A. This is particularly the case for the period between 2040 and 2080, during which there is a rise in total abatement ('sum') in the BAU scenario, but not FMS-A. Importantly, in 2050, the point at which emissions reduction commitments are most challenging, implementing FMS-A is projected to result in a reduction in total abatement of 1.3 $MtCO_2$. Drawing conclusions on the optimum approach to achieving maximum emissions abatement without considering the contribution outside the forest therefore risks defining management prescriptions that reduce rather than optimise abatement. It should be noted that the decline in the projected strength of the tree, litter and soil sink is, as discussed in 8.1.1 above, primarily a result of the age structure of British woodlands,

particularly the high levels of woodland creation in the 1950s to 1980s. A further point to recognise is that this FMS-A scenario applies to only 50% of managed woodland in the UK, of which only 10–30% are managed as carbon reserves. The impacts on emissions abatement could be significantly larger under different assumptions for FMS-A.

8.2.4 Evaluation of the abatement potentials of different forest management scenarios

For comparative purposes, the contribution of trees, litter and soil as net CO_2 sinks or sources for each of the forest management scenarios expressed as differences in net CO_2 uptake from the business as usual scenario (Figure 8.3) are shown in Figure 8.4a alongside the total abatement including by fossil-fuel and forest product substitution in Figure 8.4b. The quantitative significance of trees, litter and soil as contributions to the overall abatement is evident. Longer projections for each of the scenarios over different time periods are quantified as average abatement potentials and are presented in Table 8.2.

EAS: enhanced afforestation

The clearest conclusion from the comparison of afforestation and forest management scenarios presented in Figure 8.4 is that the enhanced afforestation scenario provides, by far, the greatest potential for additional

Figure 8.4
Impacts of each of the forest management scenarios (see key) on projected CO_2 uptake compared to the BAU scenario in Figure 8.3 (sink = positive values; source = negative values). (a) Net CO_2 emissions associated with trees, litter and soils; (b) total abatement also including carbon stored in harvested wood products, fossil fuel substitution and product substitution.

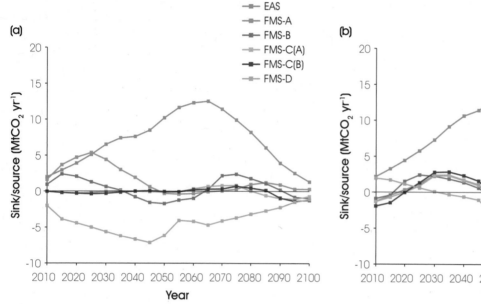
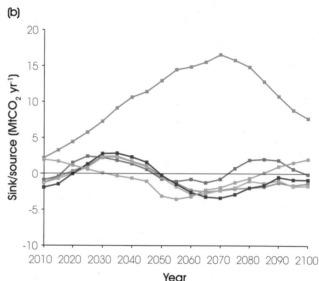

abatement, particularly in the longer term and when abatement through substitution is also considered. The enhanced afforestation scenario results in the total emissions abatement by all woodland created since 1990, rising to more than 17 $MtCO_2$ in 2070 (Figure 8.4b). During the 2050s (by which time the UK aims to achieve an 80% reduction in GHG emissions) the total abatement reaches 15 $MtCO_2$ year[-1], equivalent to 10% of the UK's total emissions if current reduction targets are achieved. This assumes a total of 23 200 ha year[-1] is planted between 2010 and 2050, representing 14 840 ha year[-1] over and above the BAU assumption that afforestation continues at 8360 ha year[-1] (the area of woodland created in 2006). This large level of abatement is primarily the result of the increase in the forest biomass and soil sink strength, but carbon stored in harvested wood products, fossil fuel substitution and product displacement also make a contribution, particularly towards the end of the simulation period.

FMS-A: carbon stock enhancement

Of the forest management scenarios, FMS-A (carbon stock enhancement) appears in Figure 8.4a to provide a consistent and significant additional abatement potential by the uptake of CO_2 into trees, litter and soil over the period 2025 to 2050. This is particularly the case when the FMS-A scenario is combined with FMS-C (improved

productivity). However, as discussed in 8.2.3 above, when total abatement is considered (Figure 8.4b) the benefits of FMS-A are smaller and, even when combined with FMS-C lead to significant emissions relative to the BAU scenario in the longer term (see below for explanation).

FMS-B: enhanced management

In contrast to the FMS-A scenario, FMS-B (enhanced management) appears to contribute little to additional abatement when only forest biomass and soil carbon stocks are considered. However, the contribution of enhanced management to abatement through substitution results in additional abatement of up to 2.5 $MtCO_2$ year[-1] between 2020 and 2040, and 2075 and 2095. The additional abatement is further enhanced between 2020 and 2050 by combining FMS-B with FMS-C, but the opposite result is apparent towards the end of the simulation. Furthermore, FMS-B is the only forest management scenario that delivers net abatement over all time periods presented in Table 8.2, even when the assessment of abatement is restricted to forest biomass and soils components. Optimising rotation length therefore appears to offer real opportunities for abatement, although the level of abatement is limited, largely because forests in the UK that are managed for production are often already close to optimum rotation length.

FMS-C: improved productivity

In this analysis, FMS-C is assumed to be implemented in combination with either FMS-A or FMS-B. When combined with FMS-A, FMS-C appears to lead to a significant reduction in the uptake into forest biomass and soils (Figure 8.4a), although the impact on total abatement is reduced when substitution and storage in harvested wood products is also considered. When combined with FMS-B, FMS-C appears to reduce fluctuations in the uptake into forest biomass and soils with little impact when averaged over the long term (Table 8.2). However, when abatement through substitution is also considered, there is a marked reduction after 2050. In both cases, the impacts of FMS-C can be largely attributed to complex changes in patterns of production, including rotation lengths and consequent changes to the dynamics of forest carbon stocks that are assumed for more productive stands.

FMS-D: reinstating management

Bringing unmanaged woodlands back into management (FMS-D) leads to significant net emissions (up to 5.5 $MtCO_2$ year^{-1}) from forest biomass and soils relative to the BAU scenario. However, this impact is reduced when total abatement is considered, with the result that over the full course of the simulation to 2150 (Table 8.2), FMS-D provides a small amount of additional abatement (0.3 $MtCO_2$ year^{-1}). An important point to recognise is that the majority of unmanaged woodland is slow-growing, broadleaved woodland for which both levels of production (and therefore substitution) and rates of recovery in carbon stocks following harvesting are smaller than for faster growing conifer species. The age and current growth rate of a stand brought back into management will also have a profound effect on the balance between substitution benefits and recovery of carbon stocks, requiring more detailed knowledge than available as input to this national scale evaluation.

Table 8.2
Comparison of the impact of enhanced afforestation and forest management scenarios on emissions abatement (compared to the business-as-usual scenario) over different time periods and for the different countries of the UK (sink = positive values; source = negative values). (See 8.2.2 above for definition of forest management scenarios.)

Average impact of scenario on BAU emissions abatement potential over different periods (MtCO₂ year⁻¹)										
	England		Scotland		Wales		N. Ireland		UK	
	Forest	Total	Forest	Total	Forest	Total	Forest	Total	Forest	Total
2010 to 2050										
EAS	2.3	2.6	3.0	4.2	0.4	0.5	0.2	0.3	6.0	7.7
FMS-A	0.7	0.2	2.0	0.5	0.3	0.0	0.1	0.0	3.1	0.7
FMS-B	0.0	0.1	0.4	0.7	0.0	0.0	0.0	0.0	0.5	0.9
FMS-C(A)	0.0	0.2	−0.1	0.4	0.0	0.0	0.0	0.0	−0.1	0.6
FMS-C(B)	0.0	0.2	−0.1	0.6	0.0	0.0	0.0	0.0	−0.1	0.8
FMS-D	−3.3	0.2	−1.5	−0.1	−0.7	0.0	−0.1	0.0	−5.5	0.1
2010 to 2100										
EAS	3.7	4.7	2.9	5.3	0.6	0.9	0.2	0.4	7.3	11.2
FMS-A	0.4	0.0	0.9	−0.6	0.1	−0.2	0.1	−0.1	1.6	−0.9
FMS-B	0.1	0.1	0.3	0.4	0.0	0.0	0.0	0.0	0.4	0.4
FMS-C(A)	0.0	0.0	0.0	−0.7	0.0	−0.2	0.0	−0.1	−0.1	−0.9
FMS-C(B)	0.0	0.0	−0.1	−0.6	0.0	−0.2	0.0	−0.1	−0.1	−0.9
FMS-D	−2.6	−0.4	−1.1	−0.3	−0.5	0.0	−0.1	0.0	−4.3	−0.7
2010 to 2150										
EAS	2.5	3.5	2.0	4.5	0.4	0.8	0.1	0.3	5.0	9.1
FMS-A	0.3	−0.1	0.6	−1.0	0.1	−0.2	0.0	−0.1	1.1	−1.4
FMS-B	0.1	0.0	0.2	0.2	0.0	0.0	0.0	0.0	0.2	0.3
FMS-C(A)	0.0	−0.1	0.0	−0.9	0.0	−0.2	0.0	−0.1	0.0	−1.4
FMS-C(B)	0.0	−0.2	0.0	−0.9	0.0	−0.2	0.0	−0.1	0.0	−1.4
FMS-D	−1.8	0.2	−0.8	0.0	−0.3	0.1	0.0	0.0	−2.9	0.3

Optimising forest management

These calculations provide a suitable framework for determining which forest management options deliver abatement in both the longer and shorter term, and thus which of them should be prioritised. When only the tree, litter and soil component is considered, only forest management scenarios enhanced afforestation (EAS) and FMS-A lead to substantial emissions abatement relative to BAU (Table 8.2). However, of these two scenarios, only EAS continues to provide significant abatement over the longer term.

Even more importantly, when total abatement in the forestry sector including fossil fuel substitution is calculated, the potential abatement increases significantly only in the enhanced afforestation case, with implementation of FMS-A resulting in reduced abatement relative to BAU in the longer term. FMS-B (enhanced management) is the only forest management scenario that results in increased emissions abatement (relative to BAU) over all three periods considered for both the forest component alone and total abatement, although at a much lower level than for EAS. Enhanced afforestation (EAS) is thus the only scenario that can deliver high levels of emissions reduction, whether only the forest component or total abatement are considered. These same conclusions can be drawn at individual country level, particularly that the largest level of abatement can be achieved through woodland creation.

Earlier analysis of emissions abatement by UK forests over a 50-year period by Matthews (1996) are presented in Table 8.3. The conclusions were broadly consistent with those from the forest management scenarios analysed above, however, Matthews (1996) also considered a longer, 500-year period which provided clarity over the long-term implications of changing approaches to forest management. Over this longer period, the impact of stopping all harvesting activity was particularly stark (a reduction in abatement of 23 $MtCO_2$ year[-1]), while an increase in UK forest area of 20% (500 000 ha) was projected to increase average abatement by 8 $MtCO_2$ year[-1] over this same 500-year timeframe. In both cases, these estimates of changes to potential abatement included substitution benefits ('whole forestry sector'); if sequestration in the forest only, was considered ('forest only'), ceasing harvesting activity resulted in annual abatement falling by 5.5 $MtCO_2$ year[-1] and the additional abatement potential of the increased forest area falling to 4 $MtCO_2$ year[-1].

8.3 Emissions abatement potential of different woodland creation options

The above text established that, across the various forest management scenarios examined, enhanced afforestation is the only option that could greatly increase the emissions abatement potential of the forestry sector. This discussion therefore explores the abatement potential for a range of different woodland creation options and provides an evaluation of the cost-effectiveness of each, based on the

Table 8.3
Evaluation of forest policy or management options for enhancing emissions abatement potential of British forests, through comparison of two alternative calculation methods (sink = positive; source = negative). Modified after Matthews (1996).

Example management option	Change in annual abatement by British forests compared to the business-as-usual scenario ($MtCO_2$ year[-1])			
	Forest only (excluding substitution)		Whole forestry sector (including substitution)	
	Over 50 years	Over 500 years	Over 50 years	Over 500 years
Shorter rotations (–20 years)	–4.4	–0.7	–4.8	–2.9
Longer rotations (+20 years)	2.2	0.4	–2.2	–2.9
Utilise unmanaged forests	0.0	0.0	0.4	1.1
Improve timber quality	–1.1	0.4	–0.4	0.4
Stop all harvesting and felling	19.8	2.6	–5.5	–23.5
Increase forest area by 20%:				
with conifers	2.6	0.4	4.0	8.1
with broadleaves	1.8	0.4	4.0	8.1
Business as usual absolute rates	–0.7	–1.1	28.2	27.9

analysis of Crabtree *et al.* (2009). An illustrative example is then provided of how different options incorporated into a woodland creation programme could contribute to the UK's GHG emissions reduction commitments.

While the growth rate and nature of the woodland created is clearly important, the emissions abatement associated with harvested wood products is crucial to the total abatement potential and its delivery. It should be noted that, at this stage, differences in abatement that would result from the different woodland creation options and abatement delivered through fossil fuel substitution would not be registered in the UK's GHG inventory for the LULUCF sector. This is likely to remain the case for fossil fuel substitution (although the abatement would be registered indirectly in other sectors as described in the introduction to this chapter).

8.3.1 Woodland creation options

New woodlands can be planted to deliver three principal objectives:

1. **Energy forestry:** production of biomass primarily for use in energy generation.
2. **Productive conifer/mixed forestry:** primarily for timber and other harvested wood products.
3. **Low impact/multi-purpose forestry, including native woodland:** for conservation, amenity and landscape.

Each of these objectives could be achieved by a range of planting and management options, and would only be appropriate on particular sites. We have chosen a number of possible options in each of the three above categories.

Table 8.4
Details of the modelled woodland creation options, for which emissions abatement potentials are shown in Table 8.5.

Option	Soil	Trees				
		Species	Spacing (m)	Yield class (m^3 ha^{-1} $year^{-1}$)	Management regime	Rotation (years)
B1	Sand	Eucalyptus	2.0	36	No thinning (short rotation forestry)	7
B2	Gley	Eucalyptus	2.0	20	No thinning (short rotation forestry)	7
C1	Loam	SAB	1.5	6	Standard thinning	80
C2	Loam	SAB	1.5	8	Standard thinning	80
D1	Gley	SAB	1.5	4	No thinning	No felling
D2	Gley	SP	3.0	4	No thinning	No felling
E1	Loam	SS/DF mix	2.0	16	Standard thinning (synchronised for 2 species)	50
E2	Loam	DF	2.0	20	Standard thinning	50
F	Loam	OK/SAB/DF/ JL mix	1.5	4/4/14/10	ACF (selection)	No final clearfell
G	Loam	SS/DF mix	1.7	12	Standard thinning (synchronised for 2 species)	60
H	Loam	Sitka spruce	2.0	12	ACF (shelterwood)	Final removal of over-storey at 60 years
I	Loam	SS/DF mix	2.0	12	ACF (selection, synchronised for 2 species)	No final clearfell
J	Peaty-gley	Willow	1.0	20	No thinning (short rotation coppice)	6 (harvesting) 24 (re-planting)
K	Peaty-gley	SAB	1.5	12	No thinning (short rotation forestry)	15
L	Peaty-gley	Eucalyptus	2.0	16	No thinning (short rotation forestry)	12

SAB, combined sycamore, ash, birch yield model; SS, Sitka spruce; DF, Douglas fir; OK, oak; JL, Japanese larch; SP, Scots pine; ACF, alternative to clearfell. 'Selection' is defined as the harvesting of individual trees or groups of trees within a regime of continuous cover; shelterwood is defined as the felling and re-planting of small blocks or stories of trees.

These are (Table 8.4):

- Energy forestry: options B, J, K, L
- Productive conifer/mixed forestry: options E, G, H
- Low impact/multi-purpose forestry: options C, D, F, I.

The options evaluated here are not intended to be exhaustive or to represent the likely results of a woodland creation programme; rather, they are intended to indicate the range of options that might be appropriate across the UK. It is important that these options are viewed alongside the information on possible climate change adaptation requirements in Section 4. For example, among the species included in the simulation for illustrative purposes is *Eucalyptus nitens*, which is just one of the possible exotic species listed in Table 6.5 (see 6.6, Chapter 6).

A number of approaches to low impact silviculture – continuous cover forestry (CCF) or alternative to clearfell (ACF) – are included within the options to reflect changing approaches to management, in part, as a response to climate change (options F, H, I). The GHG balance modelling of these options is less well understood than conventional rotational silvicultural systems. A detailed account of this aspect of the calculations is given in Morison *et al.* (2009). The three short rotation forestry (Eucalyptus) options (B1, B2, L) provide an initial evaluation of what may represent a new focus for forestry in the near future. The extent to which energy forestry will develop is uncertain, as are the levels of productivity that may be achieved. There is evidence (mostly anecdotal) that yield classes of up to 50 m³ ha⁻¹ year⁻¹ can be achieved for *Eucalyptus nitens* grown on a seven-year rotation, but the risk of frost damage is still considered as significant. The potential for short rotation forestry in the UK, including the use of native species such as ash and birch (option K) has been further explored by Hardcastle (2006).

The native woodland creation options (options D1 and D2) assume that the woodlands will be managed for biodiversity objectives and that there will be no abatement through fossil fuel substitution. This may well underestimate potential abatement in the longer term, particularly for option D1 (YC 4 native broadleaf woodland), as a future low carbon economy may place increasing emphasis on productive land covers that also deliver biodiversity benefits. This will certainly be the case if the woodfuel sector develops as outlined, for example, in England's Woodfuel Strategy, Scottish Forestry Strategy (SE, 2006), the UK Low Carbon Transition Plan (DECC, 2009b) and the UK Renewable Energy Strategy (DECC, 2009c).

For all options, abatement potential will be dependent on productivity (i.e. yield class) and, therefore, is subject to site conditions. The analysis presented here is thus not appropriate for application to specific projects for estimating abatement potential. It is also important to consider, as outlined by Broadmeadow and Matthews (2003) and in Chapter 6 of this report, that lower yielding sites are generally more suitable for developing woodland carbon reserves, particularly if located far from end-users or timber processing facilities. In contrast, higher yielding sites have the greatest abatement potential, whether through energy forestry or conventional approaches to management.

8.3.2 Abatement potential of different woodland creation options

The estimated abatement potential for each of the woodland creation options outlined in Table 8.4 is given in Table 8.5. For each option, potential abatement is presented as cumulative abatement that could be delivered in 2020, 2030, 2050 and 2100 (i.e. 10, 20, 40 and 90 years after first planting) for each hectare of woodland, assuming planting occurs in 2010. The breakdown between abatement through sequestration in trees, litter and soils and through fossil fuel substitution (both direct and indirect) is also given, although the split between the traded and non-traded sector (see 8.1.3 above and 8.4.2 below) is not shown. For clarity, the estimate of abatement potentials up to 2050 of different woodland creation options representative of the three principal objectives (see above), i.e. energy forestry, productive conifer/mixed forestry and low impact/multipurpose forestry is presented in Figure 8.5. It is clear that energy forestry options achieve their large cumulative abatement potential through the substitution benefits, while multi-purpose forestry achieves considerable abatement mainly through sequestration. Productive conifer/mixed forestry achieves abatement largely through sequestration in trees and soil, but substitution contributes significantly in all options.

Table 8.6 repeats Table 8.5 for total abatement, but excludes changes in soil carbon levels, as there is considerable uncertainty in the modelling of soil carbon, as outlined in 8.2 above.

It is clear from Table 8.5 that, in the short term (to 2020 and 2030) the energy forestry options represent the largest abatement potential, delivering up to 648 tCO₂ ha⁻¹ by 2030 (for YC 36 Eucalyptus: option B1). It should be noted that in common with other options that involve clearfell,

Table 8.5
Emissions abatement potential of different woodland creation options assumed to be planted in 2010. Sequestration: abatement through sequestration in biomass and soil carbon; substitution: abatement through direct and indirect fossil fuel substitution, with no distinction made between traded and non-traded sectors.

Option	Description	Source of abatement	Cumulative abatement potential (tCO$_2$ ha^{-1})			
			2020	2030	2050	2100
B1	YC 36 Eucalyptus	Sequestration[1]	94	180	−31	−29
		Substitution	234	468	1170	2575
		Total	328	648	1140	2546
B2	YC 20 Eucalyptus	Sequestration	86	167	84	130
		Substitution	130	260	650	1431
		Total	216	427	734	1561
C1	YC 6 broadleaf farm woodland	Sequestration	5	131	390	93
		Substitution	0	0	89	405
		Total	5	131	480	498
C2	YC 8 broadleaf farm woodland	Sequestration	5	143	477	143
		Substitution	0	0	115	542
		Total	5	143	592	685
D1	YC 4 native broadleaf woodland	Sequestration	32	180	517	854
		Substitution	0	0	0	0
		Total	32	180	517	854
D2	YC 4 native pine woodland	Sequestration	−4	30	144	543
		Substitution	0	0	0	0
		Total	−4	30	144	543
E1	YC 16 Douglas fir and Sitka spruce	Sequestration	75	254	487	336
		Substitution	0	0	160	698
		Total	75	254	647	1034
E2	YC 20 Douglas fir	Sequestration	73	276	573	388
		Substitution	0	87	284	1042
		Total	73	364	858	1430
F	YC 4/10/14 mixed woodland; ACF (selection)	Sequestration	56	194	343	337
		Substitution	0	16	133	417
		Total	56	210	476	753
G	YC 12 Sitka spruce and Douglas fir	Sequestration	67	244	477	373
		Substitution	0	0	123	525
		Total	67	244	600	899
H	YC 12 Sitka spruce; ACF (shelterwood)	Sequestration	13	131	324	273
		Substitution	0	0	122	454
		Total	13	131	446	728
I	YC 12 Sitka spruce/Douglas fir; ACF (selection)	Sequestration	46	191	316	207
		Substitution	0	0	152	511
		Total	46	191	468	718
J	YC 20 Short rotation willow coppice	Sequestration	47	17	−15	−25
		Substitution	59	177	353	766
		Total	106	193	338	740
K	YC 12 short rotation native species	Sequestration	92	49	56	45
		Substitution	0	141	281	704
		Total	92	189	338	749
L	YC 16 Eucalyptus (12-year rotation)	Sequestration	222	125	−68	−128
		Substitution	0	223	669	1561
		Total	222	348	601	1432

[1] Sequestration is the total additional carbon stored in biomass and soils in the year considered, less cumulative fossil fuel emissions resulting from management activity. Sequestration can therefore appear as negative in cases where the year considered immediately follows harvesting.

Table 8.6
Total abatement for the woodland creation options given in Table 8.5, but *excluding* emissions/sequestration in forest soils.

Option	Description	Cumulative abatement potential (tCO$_2$ ha^{-1})			
		2020	2030	2050	2100
B1	YC 36 Eucalyptus	318	626	1119	2519
B2	YC 20 Eucalyptus	184	359	628	1403
C1	YC 6 broadleaf farm woodland	13	129	446	526
C2	YC 8 broadleaf farm woodland	14	142	560	746
D1	YC 4 native broadleaf woodland	14	123	393	609
D2	YC 4 native pine woodland	−1	7	68	384
E1	YC 16 Douglas fir and Sitka spruce	74	243	632	1246
E2	YC 20 Douglas fir	71	334	882	1799
F	YC 4/10/14 mixed woodland; ACF (selection)	60	201	445	817
G	YC 12 Sitka spruce and Douglas fir	39	179	488	902
H	YC 12 Sitka spruce; ACF (shelterwood)	16	127	448	878
I	YC 12 Sitka spruce/Douglas fir; ACF (selection)	46	181	469	898
J	YC 20 Short rotation willow coppice	137	233	456	787
K	YC 12 short rotation native species	135	220	353	743
L	YC 16 Eucalyptus (12-year rotation)	257	401	675	1766

See Table 8.4 for key to abbreviations.

the energy forestry options can appear as negative sequestration abatement (i.e. net GHG emissions) if the time of reporting coincides with harvesting. This is because there has been no increase in biomass carbon stocks (with the exception of soil), but emissions associated with forest management have been accounted for. Other options deliver minimal abatement by 2020, even if instigated immediately. However, by 2030 and, particularly 2050 (Figure 8.5), the majority of options could provide significant abatement. The one exception is option D2, native pine woodland, which delivers minimal abatement up to 2050 because of the low yield class and low stocking density, reflecting that the principal objective of management is conservation of biodiversity.

Figure 8.5
Projected cumulative emissions abatement from 2010 to 2050 for different woodland creation options grouped by three objectives, showing the contribution from sequestration in the forest biomass and soil carbon components and from substitution by woodfuel and wood products for fossil fuel use and energy-intensive materials. Data from Table 8.5; woodland creation options are detailed in Table 8.4.

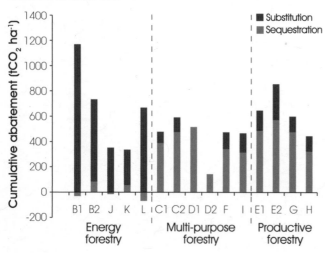

8.3.3 Case study of the potential abatement achieved by a 15-year woodland creation programme

The following case study illustrates the abatement potential that a 15-year, 10000 ha per year woodland creation programme could deliver and provides a comparison with abatement in other sectors and the possible contribution to national emissions reduction targets. The programme considers a narrow range of options, but inclusive of a broad range of forestry objectives:

- 1500 ha year^{-1} YC 36 Eucalyptus
- 1500 ha year^{-1} YC 20 Douglas fir
- 2000 ha year^{-1} YC 8 farm woodland
- 5000 ha year^{-1} YC 4 native broadleaf woodland.

Such a programme would provide a minimal contribution to emissions abatement by 2020 (Figure 8.6). This demonstrates that such programmes would not contribute significantly to the first three carbon budgets (and

Figure 8.6
Potential emissions abatement achievable by a woodland creation programme of 10 000 ha per year for 15 years. Time course of (a) cumulative and (b) annual carbon sequestration in woodland biomass and soils (light blue line) and the total abatement including emissions reductions through fossil fuel substitution (dark blue line).

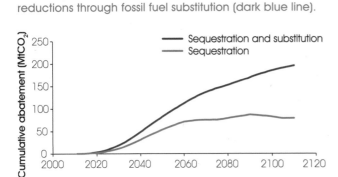

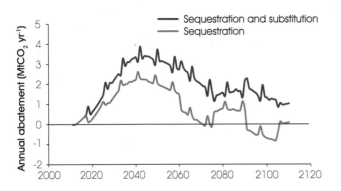

associated emissions reductions) outlined in the Climate Change Act (GB Parliament, 2008) and the UK's low carbon transition plan (DECC, 2009b). However, by the time that an 80% reduction in the UK's GHG emissions is required (i.e. 2050), such a woodland creation programme could have made a significant contribution (53 MtCO₂ cumulative abatement through sequestration alone). Moreover, if abatement through fossil fuel substitution is considered, this rises to 82 MtCO₂ with abatement of 3.7 MtCO₂ delivered in 2050 (Figure 8.6b). This is equivalent to over 2% of target UK emissions in 2050 (155 MtCO₂ per year: see 8.1 above). If this example based on a planting programme of 10 000 ha year^{-1} is scaled up to 25 000 ha year^{-1} (consistent with assumptions in the UK GHG inventory low emissions scenario (see 8.1.1 above)), the potential cumulative abatement by 2050 is 130 MtCO₂ through sequestration with a total abatement of 200 MtCO₂ when substitution benefits are included. It is therefore possible that, across the UK, woodland creation (and subsequent utilisation of wood products) could deliver additional emissions abatement equivalent to 5–10% of the UK's GHG emissions in the 2050s. The woodland created

would also represent an on-going source of renewable energy and sustainable timber products for future generations, continuing to contribute to a low carbon economy.

The selected options are optimistic in terms of productivity for both the energy forestry and conifer plantation options, assuming that site/species selection and silviculture will be appropriate. If a more conservative scenario is adopted with reduced yields (YC 20 Eucalyptus; YC 6 broadleaf farm woodland; YC 16 Douglas fir), abatement though sequestration in biomass and soil is only reduced to a small extent: from 11 to 10 MtCO₂ by 2030 and 53 to 50 MtCO₂ by 2050. However, the impact on total abatement (i.e. including substitution) is far more significant, with cumulative abatement falling from 82 to 67 MtCO₂ by 2050. The difference is even starker in the longer term, with cumulative abatement by 2100 falling from 185 to 144 MtCO₂ under the more conservative scenario. This again highlights that a greater contribution to emissions reduction can be achieved by focusing more intensive management on the more productive sites, and that lower yielding sites may be better managed as carbon (and biodiversity) reserves.

8.4 Cost-effectiveness of woodland creation

The cost-effectiveness of different abatement measures is clearly an important consideration. In establishing the capacity for abatement to be delivered by all sectors, the first report of the Committee on Climate Change (CCC, 2008) published a series of Marginal Abatement Cost (MAC) curves for the UK. For all sectors, abatement costing less than £100 per tonne CO₂ was considered as potentially cost-effective. As a contribution to the CCC's report, Moran et al. (2008) developed a marginal abatement cost curve for the agriculture, forestry and land management sector (AFLM). The abatement potential of two forestry options were considered in the analysis, both focussing on Sitka Spruce. This discussion expands on this MAC function for UK forestry, presenting a range of woodland creation options and including a breakdown of abatement in both traded and non-traded sectors. The analysis follows the majority of woodland creation options presented in 8.3 above, and is based on Crabtree et al. (2009). However, improvements to the CARBINE modelling system since the work of Crabtree concluded means that the analyses and results in 8.3 and 8.4 are not directly compatible.

8.4.1 Marginal abatement cost curves for forestry

A single afforestation option was considered Moran *et al.* (2008) focussing on YC 16 Sitka Spruce managed on a 49-year rotation. This afforestation measure resulted in an average rate of carbon sequestration in timber, soil and dead organic matter of 5 tC ha^{-1} year^{-1}. When lifetime cost-effectiveness (CE: expressed as a social metric) was considered, the measure was assessed as highly cost-effective (minus £7.12 to minus £1.82 per tonne CO_2), whether or not abatement through direct and indirect fossil fuel substitution was considered. The cost-effectiveness (for sequestration alone) was derived from a net present value (NPV) assessment of net costs of minus £6405 ha^{-1} divided by a lifetime abatement of 899 tCO$_2$ ha^{-1} (18.3 tCO$_2$ ha^{-1} year^{-1} for 49 years). The negative sign for CE indicates that the present value of timber revenue exceeded that of the costs and implies this type of forestry planting more than achieves the 3.5% social rate of return from timber output alone and thus has no net social cost. However, it should be noted that the analysis did not consider on-going management costs or guidance on declining discount rates after year 30 (HMT, 2008).

The approach of Moran *et al.* (2008) considered the extent of implementation of the afforestation measure through the adoption of two area-based thresholds: Maximum technical potential (MTP) is 'the absolute upper limit that might result from the highest technically feasible level of adoption or measure implementation'; while the central feasible potential (CFP) was defined as the 'adoption level most likely to emerge in the time scales and policy contexts under consideration'. CFP assumed a value of 50% of the MTP of 21500 ha per year additional to recent levels of woodland creation (8500 in 2006), based on a maximum level of woodland creation of 30000 ha per year. The abatement (CFP) from 10750 ha planting per year (starting in 2009) in 2022 was estimated at 0.98 MtCO$_2$, while the maximum abatement (MTP) that could be delivered by planting 21500 ha per year was 1.96 MtCO$_2$ in 2022. Moran *et al.* (2008) noted that the MTP (and consequently CFP) was a conservative estimate constrained by current policies.

As outlined above, the afforestation measure included a consideration of abatement arising from wood and timber products substituting for fossil fuels, both directly and indirectly. However, concerns over double counting resulted in the forestry option not being included in the CCC's MAC curve for the agriculture and land

management sector. The difficulties associated with accounting for direct and indirect fossil fuel substitution in both the traded and non-traded sectors are further explored in 8.1.3 above.

The analysis of Moran *et al.* (2008) also considered the abatement potential of reducing rotation length from 59 (as in GHG inventory projections: Thomson, 2009) to 49 years (i.e. similar to FMS-B in 8.2.4 above). This measure led to emissions from forest biomass and soil carbon of 0.29 MtCO$_2$ 2022 (central feasible potential: an additional 7100 ha harvested each year up to 2012 and 4200 ha per year between 2012 and 2022). However, when abatement through direct fossil fuel substitution was also included in the analysis, abatement of 1.1 MtCO$_2$ in 2022 was calculated at a cost-effectiveness of £12.1 per tCO$_2$. However, it was noted that this level of implementation of the measure was unsustainable in the long term. These apparent short-term opportunities for abatement should therefore be considered in the longer timeframe as presented in 8.2.4 above.

8.4.2 Costs for a range of woodland creation options

The net social cost of an afforestation option is taken as the sum of the establishment costs and the opportunity cost of the land used (i.e. net income associated with previous land use), less any revenue from wood or other forest product sales. This approximation understates the social cost because no account is taken of additional transaction costs involved in delivering the policy or of any ongoing forest management costs to woodland owners. In other respects, social costs are overstated because establishment costs may include an element of profit while ancillary public benefits (discussed in 8.4.3 below) are neglected in the analysis. It is assumed that policy measures could be designed to deliver the options and that output would not be diverted into other wood markets with different carbon characteristics which would affect cost effectiveness. Cost is derived as the equivalent annual cost (EAC) over one rotation. The EAC is estimated as the annuity equivalent at 3.5% to the net present value of the establishment and management costs plus the opportunity cost of land. It is assumed that subsequent rotations have the same cost structure as the initial rotation. The EAC derived for one rotation can therefore be applied to longer time periods than a rotation. For non-clearfell options, a 100-year life is used, on the assumption that cash flows beyond 100 years are beyond the planning horizon and have minimal present value. The costs used in calculating

EAC are given in Table 8.7, assuming that no stock fencing is used to reflect the nature of the sites considered most likely to be planted.

Revenue from timber sales is calculated as the present value (PV) over 100 years and converted to its equivalent annuity at 3.5% to give an equivalent annual revenue (EAR). This is then subtracted from the EAC (see above) to give a net cost per year. The EAR is derived over 100 years rather than over one rotation to facilitate comparisons between options. This approach also accommodates the likely variation in EAR between rotations that would be expected to result from the changing value of carbon over time. It should also be noted that changes in the value of carbon will influence the carbon substitution benefits of

electricity generation and the cost-effectiveness of relevant options, as discussed below.

Cost-effectiveness (CE) was calculated on a per hectare basis as the net cost per year divided by the abatement achieved on average per year over 100 years (in £ per tCO_2 per year). Thus a net cost of £500 ha^{-1} $year^{-1}$ divided by an average annual abatement of 15 tCO_2 per ha gives a CE of £36.6 per tCO_2.

As outlined in 8.1.3 above, emissions reductions that result from fossil fuel substitution in the traded sector (i.e. electricity generation) are not considered as abatement within evaluations of cost-effectiveness that are consistent with wider cross-sector studies. Here, the CO_2 emissions

Table 8.7
Costs assumed for forestry operations.

Option	Description	Establishment/restocking cost (£/ha)		Establishment and management cost (£ ha^{-1} $year^{-1}$)	Agricultural income foregone (£ ha^{-1} $year^{-1}$)	PV Agricultural income foregone (£ ha^{-1})	Equivalent annual cost (£ ha^{-1} $year^{-1}$) (EAC)
		With stock fencing	No stock fencing				
B1	SRF YC 36 energy forests	4400	2600	378	350	–2406	728
B2	SRF YC 20 energy forests	4400	2600	378	260	–1787	638
C1	YC 6 broadleaf farm woodland	6700	5400	202	480	–12839	682
D1	YC 4 native broadleaf woodland	5370	4070	147	260	–7190	407
D2	YC 4 native pine woodland	3580	2600	111	50	–1173	161
E1	YC 16 SS/DF	3580	2600	111	260	–6098	371
F	YC 4/10/14 mixed woodland: ACF (selection)	4400	3500	131	190	–5082	321
G	YC 12 SS/DF	3580	2600	111	160	–3753	271
H	YC 12 SS: ACF (shelterwood)	3580	2600	94	160	–4425	254
I	YC 12 SS/DF: ACF (selection)	3580	2600	94	160	–4425	254
J	SRC YC 20 willow	1310	1310	79	350	–5769	429
K	SRF YC 12 native species	5370	4070	353	260	–2995	613
L	SRF YC 16 energy forests	3580	2600	269	260	–2512	529

Note: Equivalent annual cost is the annual establishment and management cost plus the annuity derived from the PV of agricultural income foregone.

substituted for in the traded sector are reflected as an additional revenue stream based on the projected price of traded carbon (EUA: EU Allowance) as published by DECC, 2009a). In 2009 the central assumption is £21 per tCO_2, rising to £200 per tonne in 2050. The result of this approach is that although total abatement is reduced, the cost-effectiveness of abatement in the non-traded sector improves. To some extent, this reflects GHG emissions reduction through renewable energy reduction being rewarded through energy market instruments (ROCs) but the untraded carbon remains unrewarded. However, it could be argued that any inclusion of social values from the traded sector in the revenue stream distorts the basis of the cost-effectiveness calculation which is to identify the social cost per tonne of carbon (net of social revenue) of achieving a given reduction in net emissions. For this reason, Table 8.8 considers cost-effectiveness both with and without the value of the carbon substituted for in the traded sector being included.

8.4.3 Ancillary benefits

A range of ancillary benefits (and dis-benefits) may be associated with woodland creation. If these can be given a monetary valuation they should ideally be included in a social appraisal. Possible co-benefits include energy security, environmental gains and positive impacts on rural development (see Section 5). Moran et al. (2008) recognised that the exclusion of co-benefits was a

weakness in their analysis since woodlands may deliver sizeable public good impacts. Furthermore, significant policy synergies are achievable – for example emissions abatement and improvement of water quality to meet Water Framework Directive objectives (Nisbet et al., 2009). However, quantifying these is not straightforward, as outlined by Crabtree et al. (2009), in part because many of the environmental benefits are location specific. Some information is available in a UK context (Willis et al., 2003; Crabtree et al., 2003; Jaakko Poyry, 2006) and should be included in future evaluations, although it is not included in the evaluation presented here.

8.4.4 Cost-effectiveness of woodland options

Table 8.8 provides a ranking of the cost-effectiveness of the different woodland creation options based on a revenue calculation that includes a value for fossil fuel carbon emissions in electricity production that would be displaced. This produces negative CE values for SRC (option J), SRF (options B1, B2 and K) and the main conifer crops (i.e. socially desirable investments at no net cost). Native species and broadleaf options are less cost-effective but all give a CE below £100 per tCO_2 apart from SRF native species. This clearly highlights the potential for woodland creation to be employed as a cost effective approach to GHG emissions abatement. However, it is important to consider that woodland creation delivers abatement in the medium to longer term.

Table 8.8
Cost-effectiveness and average emissions abatement of woodland creation options over a 100-year period.

Option		Cost-effectiveness ($£/tCO_2$)	Cost-effectiveness ($£/tCO_2$) excluding traded carbon value	Abatement (tCO_2 ha^{-1} $year^{-1}$)
B1	SRF YC 36 energy forests	−60.8	24.8	15.1
J	SRC YC 20 willow	−50.3	58.6	3.7
L	SRF YC 16 energy forests	−45.3	41.3	8.4
B2	SRF YC 20 energy forests	−30.6	44.6	9.5
E1	YC 16 SS/DF	−17.3	−2.8	12.9
H	YC 12 SS: ACF (shelterwood)	−11.2	−0.1	9.7
G	YC 12 SS/DF	−9.6	5.3	9.1
I	YC 12 SS/DF: ACF (selection)	−4.7	8.1	9.1
F	YC 4/10/14 mixed broadleaf/conifer woodland: ACF (selection)	11.2	25.9	7.9
D2	YC 4 native pine woodland	21.1	21.1	7.0
K	SRF YC 12 native species	34.3	114.6	4.5
D1	YC 4 native broadleaf woodland	40.7	40.7	8.4
C1	YC 6 broadleaf farm woodland creation	72.7	75.8	5.2

See Table 8.4 for key to abbreviations.

Ideally, a marginal abatement cost curve for forestry would be developed that could be incorporated within wider evaluations of the delivery of Government's GHG emissions reductions commitments. Indeed, Crabtree *et al.* (2009) do quantify abatement potential for England at a cost of less than £100 per tonne CO_2 of 0.7 $MtCO_2$ in 2022 and 5.9 $MtCO_2$ in 2050 (assuming an additional 471000 ha of woodland is planted by 2050). However, it is difficult to establish a realistic value for either of the two area-based thresholds MTP or CFP. The estimates of cost-effectiveness presented in Table 8.8 do, however, provide a basis for developing a woodland creation programme to meet climate change objectives, including identifying where grant aid or other financial incentives might be required to achieve such a programme.

8.5 Conclusions: the potential of UK forestry to contribute to emissions abatement

It is clear that the forestry sector can make a significant contribution to emissions reduction commitments. If enhanced woodland creation and appropriate forest management measures were implemented as a matter of urgency, total emissions abatement delivered by the sector could reach 15 $MtCO_2$ annually by the 2050s (Figure 8.4b). This level of abatement would equate to about 10% of total GHG emissions from the UK if recent emissions reductions commitments are achieved. Planting a total of 23200 ha year[-1] over the next 40 years would provide nearly 1 million additional hectares of woodland that would be required to achieve this level of abatement. Including the BAU level of woodland creation, this would represent a 33% increase in woodland area bringing total woodland cover to approximately 3.8 million hectares. Although this would clearly represent a major change in, and challenge to, the forestry sector it only represents a 4% change in land use. Indeed, the resulting forest cover of 16% would still be well below the European average. However, given the degree of change in the landscape, it would be important to ensure that the strong regulatory framework for woodland creation in the UK is maintained to prevent inappropriate woodland creation. Given the wide variation in cost-effectiveness measures reported above for new planting options, it will be important to ensure that the 'right' planting options are exercised in both the private and public forest estate.

Opportunities for forest management measures in the UK to contribute to GHG abatement are more limited than for woodland creation. This observation differs from that of the IPCC (Nabuurs *et al.*, 2007), at least in part because of the relatively slow growth rate of UK forests compared with those in some other parts of the world. The low level of abatement that could be delivered by forest management – in absolute terms – also reflects the limited extent of woodlands in the UK. Importantly, forest management abatement measures are more difficult to interpret because much of the abatement is delivered outside the forestry sector through direct and indirect fossil fuel substitution. If an holistic view of abatement is not taken, supported by appropriate approaches to carbon accounting, there is a risk that measures will be implemented to maximise forest and soil carbon stocks that limit the delivery of abatement. Modest additional abatement can be delivered by optimising forest management for timber production, which will also provide raw materials for a future low carbon society.

8.6 Research priorities

- Development of the UK's LULUCF GHG inventory is required to reflect emissions from, and uptake by, forests that result from changes in management practice. This development should be coupled with improved reporting of forest carbon stocks through the National Forest Inventory.
- Improved understanding of changes in forest soil carbon stocks, based on empirical evidence, is required to underpin accounting models of forest carbon balance.
- A comprehensive evaluation of life cycle analyses for a wide range of wood products compared to alternative materials is required to better demonstrate the role of forest management and product displacement (indirect fossil fuel substitution) in GHG emissions abatement.
- The economic value of ancillary benefits of woodland creation (biodiversity, water quality, recreation, soil protection) needs to be incorporated within cost-effectiveness assessments, cost-benefit analyses and marginal abatement cost curve analysis for the forestry sector.
- An operational decision support system should be developed to downscale national level assessments of abatement potential through changes in forest management to aid the implementation of appropriate abatement measures.
- Development of carbon accounting models for new forest species and new silvicultural systems that take into account the possible effects of the changing climate and possible adaptation measures is required.

References

BROADMEADOW, M. and MATTHEWS, R. (2003). *Forest, carbon and climate change: the UK contribution.* Forestry Commission Information Note 48. Forestry Commission, Edinburgh.

COMMITTEE ON CLIMATE CHANGE (2008). *Building a low-carbon economy: the UK's contribution to tackling climate change.* The First Report of the Committee on Climate Change, December 2008. The Stationery Office, London.

CRABTREE, R., PEARCE, D. and WILLIS, K. (2003). *Economic analysis of forestry policy in England.* Final report to the Department for Environment, Food and Rural Affairs and H.M. Treasury. CJC Consulting, Oxford.

CRABTREE, R., MATTHEWS, R.W., HARRIS, D. AND RANDLE, T.J. (2009). *Analysis of policy instruments for reducing greenhouse gas emissions from agriculture, forestry and land management - forestry options.* Final contract report to Defra and the Forestry Commission. ADAS, Abingdon.

DEPARTMENT OF ENERGY AND CLIMATE CHANGE (2009a). *The UK low carbon transition plan: national strategy for climate and energy.* The Stationery Office, London.

DEPARTMENT OF ENERGY AND CLIMATE CHANGE (2009b). *The UK renewable energy strategy.* The Stationery Office, London.

DEPARTMENT OF ENERGY AND CLIMATE CHANGE (2009c). *Carbon valuation in UK policy appraisal: a revised approach.* Department of Energy and Climate Change, London.

DEWAR, R.C. (1990). A model of carbon storage in forests and forest products. *Tree Physiology* 6, 417–428.

DEWAR, R.C. (1991). Analytical model of carbon storage in trees, soils, and wood products of managed forests. *Tree Physiology* 8, 239–258.

DEWAR, R.C. and CANNELL, M.G.R. (1992). Carbon sequestration in the trees, products and soils of forest plantations: an analysis using UK examples. *Tree Physiology* 11, 49–71.

EDWARDS, P.N. and CHRISTIE, J.M. (1981). *Yield models for forest management.* Forestry Commission Booklet 48. HMSO, London.

FORESTRY COMMISSION (2001a). *National inventory of woodland and trees: England.* Inventory report. Forestry Commission, Edinburgh.

FORESTRY COMMISSION (2001b). *National inventory of woodland and trees: Scotland.* Inventory report. Forestry Commission, Edinburgh.

FORESTRY COMMISSION (2002). *National inventory of woodland and trees: Wales.* Inventory report. Forestry Commission, Edinburgh.

FORESTRY COMMISSION (2007). *A woodfuel strategy for England.* Forestry Commission England, Bristol.

FORESTRY COMMISSION (2009). *Climate Change Action Plan 2009–11.* Forestry Commission Scotland, Edinburgh.

FORSTER, D. and LEVY, P. (2008). *Policy options development and appraisal for reducing GHG emissions in Wales.* Report to Welsh Assembly government. AEA Energy and Environment, Harwell.

GREAT BRITAIN PARLIAMENT (2008). *Climate Change Act 2008.* The Stationery Office, London.

HARDCASTLE, P. (2006). *A review of the potential impacts of short rotation forestry.* Final report to the Forestry Commission. LTS International, Edinburgh.

HM TREASURY (2008). *The green book: appraisal and evaluation in central government.* Treasury guidance. The Stationery Office, London.

HM TREASURY (2009). *Budget 2009: Building Britain's future.* Economic and fiscal strategy report and financial statement and budget report. The Stationery Office, London.

IPCC (2006). *Guidelines for National Greenhouse Gas Inventories, 2006.* Prepared by the National Greenhouse Gas Inventories Programme. IPCC, Japan.

JAAKKO POYRY (2006). *Woodland and forest sector in England. A mapping study carried out on behalf of the English Forest Industries Partnership.* Jaakko Poyry Consulting, Cheam.

KURZ, W.A., DYMOND, C.C., WHITE, T.M., STINSON, G., SHAW, C.H., RAMPLEY, G.J, SMYTH, C., SIMPSON, B.N., NEILSON, E.T., TROFYMOW, J.A., METSARANTA, J. and APPS, M.J. (2009). CBM-CFS3: a model of carbon-dynamics in forestry and land-use change implementing IPCC standards. *Ecological Modelling* 220, 480–504.

LAVERS, G. M. (1983). *The strength properties of timber.* 3rd edn. Building Research Establishment Report. HMSO, London.

MATTHEWS, G.A.R. (1993). *The carbon content of trees.* Forestry Commission Technical Paper 4. Forestry Commission, Edinburgh.

MATTHEWS, R.W. (1994). Towards a methodology for the evaluation of the carbon budget of forests. In: Kanninen, M. (ed.) *Carbon balance of the world's forested ecosystems: towards a global assessment.* Proceedings of a workshop held by the Intergovernmental Panel on Climate Change AFOS, Joensuu, Finland, 11–15 May 1992. Painatuskeskus, Helsinki. pp. 105–14.

MATTHEWS, R.W. (1996) The influence of carbon budget methodology on assessments of the impacts of forest management on the carbon balance. In: Apps, M.J. and Price, D.T. (eds) *Forest ecosystems, forest management and the global carbon cycle.* NATO ASI Series I, Vol. 40. Springer-Verlag, Berlin. pp. 233–243.

MATTHEWS, R.W. and DUCKWORTH, R.R. (2005). BSORT: a model of tree and stand biomass development and production in Great Britain. In: Imbabi, M.S. and Mitchell, C.P. (eds) *Proceedings of World Renewable Energy Congress,* 22–27 May, Aberdeen, UK. Elsevier, Oxford.

MATTHEWS, R.W., ROBERTSON, K., MARLAND, G. and MARLAND, E. (2007). Carbon in wood products and product substitution. In: Freer-Smith, P.H., Broadmeadow, M.S.J. and Lynch, J.M. (eds) *Forestry and climate change.* CABI, Wallingford. pp. 91–104.

MOHREN, G.M.J. and KLEIN GOLDEWIJK, C.G.M. (1990). *CO₂FIX: a dynamic model of the CO₂-fixation in forest stands.* "De Dorschkamp", Report 624. Research Institute for Forestry and Urban Ecology, Wageningen.

MOHREN, G.M.J., GARZA CALIGARIS, J.F., MASERA, O., KANNINEN, M., KARJALAINEN, T. and NABUURS, G-J. (1999). *CO₂FIX for Windows: a dynamic model of the CO₂ fixation in forest stands.* Institute for Forestry and Nature Research (The Netherlands), Instituto de Ecologia (UNAM, Mexico), Centro Agronomico Tropical de Investigacion y Ensenanza (Costa Rica) and European Forest Institute (Finland).

MORAN, D., MACLEOD, M., WALL, E., EORY, V., PAJOT, G., MATTHEWS, R., MCVITTIE, A., BARNES, A., REES, B., MOXEY, A., WILLIAMS, A. and SMITH, P. (2008). *UK Marginal abatement cost curves for the agriculture and land use, land-use change and forestry sectors out to 2022, with qualitative analysis of options to 2050.* Final Report to the Committee on Climate Change. Scottish Agricultural College, Edinburgh.

MORISON, J.I.L., MATTHEWS, R.W., PERKS, M., RANDLE, T.J., VANGUELOVA, E., WHITE, M.E. and YAMULKI, S. (2009). *The Carbon and GHG balance of UK forests - a review.* Forest Research, Farnham.

NABUURS, G.J. (1996). Significance of wood products in forest sector carbon balances. In: Apps, M.J. and Price, D.T. (eds) *Forest ecosystems, forest management and the global carbon cycle.* NATO ASI Series I, Vol. 40. Springer-Verlag, Berlin.

NABUURS, G.J., MASERA, O., ANDRASKO, K., BENITEZ-PONCE, P., BOER, R., DUTSCHKE, M., ELSIDDIG, E., FORD-ROBERTSON, J., FRUMHOFF, P., KARJALAINEN, T., KRANKINA, O., KURZ, W.A., MATSUMOTO, M., OYHANTCABAL, W., RAVINDRANATH, N.H., SANZ SANCHEZ, M.J. and ZHANG, X. (2007). Forestry. In: Metz, B., Davidson, O.R., Bosch, P.R., Dave, R. and Meyer, L.A. (eds) *Climate change 2007: mitigation.* Contribution of Working Group III to the Fourth Assessment. Report of the Intergovernmental Panel on Climate Change. Cambridge University Press, Cambridge.

NISBET, T.R., SILGRAM, M., SHAH, N., BROADMEADOW, S.B. and MORROW, K. (2009). *Woodlands and the Water Framework Directive.* Final contract report to the Environment Agency and Forestry Commission. Environment Agency, Bristol.

ROBERTSON, K., FORD-ROBERTSON, J., MATTHEWS, R.W. and MILNE, R. (2003). Evaluation of the C-FLOW and CARBINE carbon accounting models. In: Milne, R. (ed.) *UK emissions by sources and removals by sinks due to land use, land use change and forestry activities.* Report to DEFRA (contract EPG1/1/160). Centre for Ecology and Hydrology, Edinburgh.

SCOTTISH EXECUTIVE (2006). Scottish forestry strategy. Forestry Commission Scotland, Edinburgh.

THOMPSON, D.A. and MATTHEWS, R.W. (1989a). *The storage of carbon in trees and timber.* Forestry Commission Research Information Note 160. Forestry Commission, Edinburgh.

THOMPSON, D.A. and MATTHEWS, R.W. (1989b). CO₂ in trees and timber lowers greenhouse effect. *Forestry and British Timber* 18 (October), 19, 21, 24.

THOMSON, A. (ed.) (2009). *Inventory and projections of UK emissions by sources and removals by sinks due to land use, land use change and forestry.* Annual contract report (GA01088) to Defra. Centre for Ecology and Hydrology, Edinburgh.

TIPPER, R., CARR, R., RHODES, A., INKINEN, A., MEE, S. and DAVIS, G. (2004). The UK's forest: a neglected resource for the low carbon economy. *Scottish Forestry* 58, 8–19.

WILLIS, K.G., GARROD, G., SCARPA, R., POWE, N., LOVETT, A., BATEMAN, I.J., HANLEY, N. and MACMILLAN, D.C. (2003). *The social and environmental benefits of forests in Great Britain. Phase 2 report.* Centre for Research in Environmental Appraisal and Management, University of Newcastle-Upon-Tyne, Newcastle. Report for the Forestry Commission, Edinburgh.

SECTION 4
ADAPTATION

Chapter 9 - The adaptation of UK forests and woodlands to climate change
Chapter 10 - Woodlands helping society to adapt

THE ADAPTATION OF UK FORESTS AND WOODLANDS TO CLIMATE CHANGE

Chapter

9

K. J. Kirby, C. P. Quine and N. D. Brown

Key Findings

There is considerable uncertainty about how trees, woods and forests will respond to climate change, but change they will. New cultural landscapes will develop in response to the new conditions that are not simple transpositions of those that currently occur at lower elevations, or further south in Europe or that have occurred under warm periods in the past. Indirect responses as consequence of changes in other land management/policy will be as important as direct impacts.

Action to start to change the extent, composition and structure of our woodland is needed in order to avoid future serious limitation of goods and services from our forests and potentially also wildlife losses. A move towards planned rather than reactive adaptation is desirable, given the long response rates of trees and forests. We need to increase the resistance and resilience of existing woodland. Within existing woods an overall increase in management intervention is likely to be needed, even in semi-natural woodland, to modify the biological and ecological response of forests to climate change in order to maintain and increase benefits to society. This will be a challenge as much broadleaved woodland is currently unmanaged, costs of management can be high and owners may require professional advice and support. A proportion of woods should however, be left as minimum intervention areas, in order to assess the degree of 'passive adaptation'.

The majority of woods are likely to be treated as high forest in different forms. Whereas clearfell systems have predominated in the past, in future continuous cover forestry approaches may become more advantageous, because they are thought to be more wind-firm; maintain a more even carbon storage; show lower soil carbon losses during harvesting; and maintain higher humidity levels. However, the evidence that they will deliver these benefits needs strengthening. There may be less need for coppice systems to maintain southern or thermophilic elements of the woodland system, although coppice may still be desirable for light-demanding species and young regrowth for bird species needing dense shrub layers. The silvicultural system *per se* is however, less important than the structures that it creates and their resilience and robustness in relation to climate change.

Adaptation measures on the ground need to be set in clear national and regional frameworks, and based upon regular updating of the climate change projections and their specific implications for forests and trees. There needs to be ongoing iteration between:

*   national targets and aspirations, reflected in forestry strategies and other guidance;
*   regional and other sub-country level frameworks that indicate the priorities for different types of forestry activity and the balance between forestry and other land use;
*   local projects and regulation which determine what work actually goes ahead.

Practical adaptation measures need to be tailored to the different types of woods, woodland owners and their objectives.

Adaptation will involve increasing the tree and woodland cover to develop new habitat networks for biodiversity and for other purposes. New afforestation must be developed sensitively with full recognition of the potential implications for biodiversity, agriculture, water harvesting, housing and infrastructure development, alongside the other associated costs and benefits.

In order to continue to meet demands for timber, fuel, and some ecosystem services, we may need to introduce new provenances and new species, although research is needed to establish which are fit for purpose and likely to survive. The nature conservation community needs to be clearer as to:

- what it is trying to conserve in a changing environment;
- whether the past emphasis on use of native species and local provenances is still valid;
- where might species and provenances from the near continent be better suited to future conditions, or provide refuge for rare and threatened species.

Many social values of woodland are related to how accessible they are. In future, increased emphasis may be placed on accessibility without reliance on cars, and hence on the benefits of urban and peri-urban woodland. There will also be a need to help people understand the changed appearance of some landscapes.

There must be adequate monitoring of forest and woodland states and processes to assess and adjust the use of adaptive management; improved decision-making processes will be needed to cope with the assessment of risk, and the inherent uncertainties.

In this chapter, we consider the adaptation of the UK's tree and woodland cover to make it more resilient to climate change over the next 50–100 years. Resilience implies that the future tree and woodland cover recovers quickly from climate change impacts and the ecosystem services provided are maintained across the landscape.

The UK woodland resource is frequently split into semi-natural stands and plantations, ancient and recent woodland (Spencer and Kirby, 1992; Goldberg *et al.*, 2007). All are within the scope of this chapter. These distinctions have underpinned forestry policy in the recent past, but may become less useful and clear-cut in the longer term, for example as species' distributions change and, if production forests are managed more as mixed species stands with varied structure, as they are re-stocked by natural regeneration ('close to nature' forests, see Chapter 6, Forest Management Alternative, FMA2).

9.1 What is 'an adapted forest cover' for the UK?

An adapted forest cover is one that is resilient under changing environmental conditions and continues to meet society's needs for goods and services. Adaptation results first from the biological and ecological response of trees and woods to changes in their environment (Lindner *et al.*, 2008). Forests adapt as the environment changes the presence or absence of species and their abundance. There can be longer term changes as species themselves adapt to new environmental pressures.

In the past, forests and trees have responded to climate change through changes in their range and distribution. In the post-glacial period, the British landscape went from tundra to pine–birch forests, to mainly mixed broadleaved forests (Godwin, 1975). However, there may be lags in species movement into areas that have become environmentally suitable (Svenning and Skov, 2005, 2007); mature trees may survive in an area long after the conditions for regeneration have ceased to be suitable, for example the small-leaved lime in northern England (Pigott and Huntley, 1981). The current composition may be limited by past environmental or historical factors: the English Channel appears to have limited the spread to Britain of some species common on the near Continent. Under conditions of rapid environmental change, the biological response does not necessarily produce forests that are optimally matched to the current environment.

Biological and ecological adaptation to climate change alone may not produce the kinds of woods and forests that society wants. Our needs and demands are likely to include the provision (Defra, 2007; Forestry Commission, 2001; Forest Service, 2006; Scottish Executive, 2006) of:

- carbon sequestration
- conservation of biodiversity
- environmental services such as soil and water protection, improvement of air quality
- forest products such as timber, fuel
- employment in forestry and forest-related industries
- recreation, attractive landscapes, cultural and historic features, and other contributions to people's quality of life.

In the UK, with limited land area and substantial human population, there are no significant moves towards major zonation of woodland for a single use. While the balance of services provided may vary, both nationally and between individual woods, the majority of woodland is likely to have to provide a range of services (Hunter, 1999).

Adaptation measures are often easier to implement in managed forests such as plantations than in natural forests (Nabuurs et al., 2007) because the tree species composition and forest structure are more under managerial control (European Commission, 2007). An increase in management intervention is therefore likely, even in semi-natural woodland, to modify the biological response of forests to climate change, and in order to maintain benefits to society. Reductions in the risk of detrimental changes, such as loss of productivity, woodland cover, or

species richness will also be dependent on management intervention. Management should not be aimed at adaptation to some specific, predicted climate regime, but towards developing the forest's capacity for adaptation to continuing climate change. There should also be a suite of minimum-intervention forests where natural processes predominate (Peterken, 2000); understanding what happens in the absence of intervention is a valuable guide to management elsewhere.

Vegetation dominated by long-lived species may be more vulnerable to increasing climate variability (Notaro, 2008). However, the long life of some forest species means that they may be relatively tolerant of wide variations in annual weather conditions. For example, some veteran oak trees in Windsor Great Park started their growth in the Little Ice Age (17th century), but have survived, so far, the increasing frequency of hot summers. The projected future climate impacts on trees and woods as they are now, need to be balanced against the risks from trying to change our trees and woods to meet the projected future climate conditions. Introducing species more tolerant of the expected higher summer temperatures in 2050, may not be worthwhile if they cannot tolerate current winter temperatures or late spring frosts (see also Chapters 4 and 5).

The current age distribution of our trees strongly influences their future potential to provide both timber and biodiversity. The conifer crops to be harvested in 2050 and the mature broadleaf stands in the second half of the 21st century will be largely those that are already growing – their area cannot be increased. The level of wood fibre production after 2050 will depend on how existing crops are managed, but also on how much new productive woodland is created between now and then. The future area of ancient woodland and the numbers of veteran trees are also largely constrained by what exists now, although other aspects can be influenced by the management approach adopted. The extent of open space and young growth in ancient semi-natural woods over the next 50 years will depend largely on active management. Thereafter, natural processes may create gaps more regularly as the current stands (mostly around 60–90 years old) start to mature (Hopkins and Kirby, 2007; Kirby et al., 2005). Whether the broadleaved woods being created now are of timber quality when they are mature will depend on how they are treated over the next century.

Adaptation measures over the next two decades are needed to avoid future bottlenecks in the provision of goods and services from our forests and potential wildlife losses. While some adaptation measures cannot have much impact on our tree and forest cover for some decades, if we do not take action now, then much greater efforts will be needed subsequently.

9.2 Developing adaptation strategies and actions for UK forests

Several principles and priorities for climate change adaptation have been suggested, both in general terms and specifically for forest and tree cover (Mitchell *et al.,* 2007; Hopkins *et al.,* 2007; Smithers *et al.,* 2008; Nabuurs *et al.,* 2007; Millar *et al.,* 2007; Lindner *et al.,* 2008; Forestry Commission (a), in press). They can be summarised as: creating resistance and promoting resilience to change (see below); monitoring change and accepting landscape change.

Creating resistance to change. The longer that the current tree and woodland cover can be maintained as productive forests or rich wildlife sites, the more time that there is for other adaptation measures to be brought in.

- Resistance to change can be improved by reducing the impacts of other stressors on the systems, such as pests and diseases, pollutants, over-grazing and development pressures. Reducing deer pressure in woods, for example, allows more flowering and seed setting of species such as primroses, so increasing the potential for populations to survive drought years (Rackham, 1999).
- Management practices, such as rotation length, coupe size, tree species composition and canopy cover can be modified to favour retention of current production, habitat conditions, features or species (Humphrey, 2005).
- Resistance is likely to be higher in large blocks of woodland because they contain more internal variety of structure and are less affected by adverse edge effects, for example increased water loss or spray drift from adjacent farmland (Herbst *et al.,* 2007; Gove *et al.,* 2007); species populations within them tend to be larger and hence less susceptible to random extinction.
- Sites, species and features most vulnerable to threat need to be identified, as has been suggested for different groups of plants (Gran Canaria Group, 2006).
- Potential refugia need to be identified where the direct impacts of climate change may be less than in the surrounding region. Gorge oakwoods in north Wales may provide refuges for Atlantic bryophytes, sensitive

to reduced humidity, as they appear to have done when much of the rest of the woodland was actively managed as coppice (Edwards, 1986).

Resistance is unlikely to be absolute however, and once critical thresholds are passed, change may be rapid and catastrophic.

Promoting resilience to change. We should seek to adapt the current tree and woodland extent, location, structure and composition towards those that will be more suitable for future conditions. For example, because of disease risks, alternatives to Corsican pine should be encouraged, where previously it was the favoured productive species.

Measures that have been suggested to increase resilience include:

- contingency planning for outbreaks of new pests or major new disturbance regimes (e.g. increased fire risk);
- encouraging a variety of species that can occupy the same functional space within the forest ecosystem, as has happened naturally at Lady Park Wood (Peterken and Mountford, 2005);
- increasing regeneration rate to allow more potential for selective pressures to work on seedlings;
- greater diversity of planting material, both at the species and population genetic level.

A disadvantage of higher resilience is that, much of the time, delivery of services from the forest may be sub-optimal. Economic analysis techniques are needed to judge the relative costs and benefits of occasional catastrophic losses of services versus regular, but sub-optimal delivery.

Monitoring both the processes taking place and the outcomes in order to:

- track whether climate change and its impacts are as expected;
- identify where and what forms of adaptation are successful or unsuccessful, as the case may be;
- provide a context that allows appropriate responses to rare, catastrophic events such as the 1987 storm or emerging issues such as 'acute oak decline' syndrome;
- validate models of species and ecosystem responses to climate change (e.g. Berry *et al.,* 2002; Sykes *et al.,* 1996; Giesecke *et al.,* 2007; Thuiller *et al.,* 2002), in order to improve future projections.

However, it can be difficult to separate the climate signal

from other changes affecting monitoring results (Kirby *et al.,* 2005).

Accepting that new cultural landscapes will develop in response to the new climate conditions: The forests that develop over the next century will not be simple transpositions of those that currently occur further south in Europe (or at lower elevations), nor will they necessarily be like those that have occurred under similar climates in the past. Non-analogous assemblages (Huntley 1990; Keith *et al.,* in press) will form because:

- the new climates are not the same as those currently in southern Europe, for example there may be changes in storm frequency or severity;
- the 'starting point' (including the impact of landscape history on the composition and distribution of our forests) is different to that which led to the evolution of the southern European landscape;
- species respond individually to climate change, not as assemblages or communities (e.g. Berry *et al.,* 2002; Kirby *et al.,* 2005; Hill *et al.,* 1999);
- species distributions are affected by other factors (rates of spread, competition between species, herbivory, predation and facilitation) that interact with the direct climate impact (e.g. Svenning and Skov, 2005; Beale *et al.,* 2008).

9.3 National level challenges

9.3.1 Shifts in major forest zones

Much of the country is likely to remain within the broad temperate forest zone (Lindner *et al.,* 2008), although opportunities for Mediterranean-type species will increase and boreal species may come under more stress (see Boxes 9.1 and 9.2). Temperate broadleaved woodland may change its detailed composition and structure but retain a similar overall appearance (see Box 9.3). There may be shifts in which climatic factors limit particular species distributions. For example, hyper-Atlantic bryophytes (Ratcliffe, 1968) at the southern and eastern edges of their range may decline because of hotter summers, but in the north this effect may be offset by increased winter rainfall. Small-leaved lime regeneration may increase because of hotter summers, leading to infilling of its distribution in the south, but range expansion in the north (Pigott and Huntley, 1981). While it is possible to suggest possible adaptation measures, their application in practice will depend on an assessment of the benefits and costs that will accrue on any individual site.

BOX 9.1 Mediterranean treescapes in southern England?

In Spain, holm oak (*Quercus ilex*) is spreading to higher elevations, replacing heather and beech woodland (Penuelas and Boada, 2003). By analogy, on south-facing slopes in southern England, the vegetation may develop a 'Mediterranean' character. Possible adaptation responses might be:

- pines become more important for conifer production than firs and spruces (subject to diseases such as red band needle blight);
- walnut and sweet chestnut are favoured for broadleaved production;
- more need to plan for fire because of hotter conditions but also more open grassy woods;
- increased value placed on shade trees in rural and urban settings;
- active management for some open-space species is less important because they depend less on open areas providing warm microclimates;
- maintenance of shade and internal woodland humidity becomes more important;
- acceptance of southern tree species, e.g. holm oak, that are already established locally as part of our future wildlife.

9.3.2 Movement of individual species

The rate at which woods and forests adapt depends on species' ability to track climate change either by moving northward, or upward, or on to cooler or wetter aspects (such as north-facing slopes). Some species spread rapidly, such as grey squirrels, deer species or rosebay willowherb. Rapid spread also occurs through human assistance: trees planted outside their past natural range, the deliberate and accidental spread of species (including diseases) on cars, boots, among logs or other plant or soil material moved about the country.

For other species, the potential for movement to keep ahead of climate change is uncertain. Barriers to species colonisation, such as the English Channel, or the availability of suitable sites or soils and associated species within the preferred climate zone may slow the response to climate change. For example, the climate space for nuthatch (*Sitta europaea*) may increase to the north and west (Harrison *et al.,* 2001), but the availability of habitat in the form of old trees and behavioural factors, such as the perceptual

BOX 9.2 Future shifts in boreal forests?

The boreal forests in Britain are in the more mountainous regions so there is potential for some movement of forest upwards, widely reported elsewhere (Hawkins et al., 2008), as well as to the north east, as has occurred in the past (Gear and Huntley, 1991). At lower altitudes, broadleaved trees may spread into pine woodland. Some boreal forest species are potentially sensitive to climate change, e.g. capercaillie (Harrison et al., 2001), although forest management and predation may be more immediately critical. Possible adaptation responses include:

- review of where the natural elevation and latitude limits are for growth of different trees;
- assess potential impact of emerging pests and pathogens (e.g. the impact of red band needle blight on native pine);
- reduce deer browsing, which is limiting forest spread both altitudinally and latitudinally;
- accept changing composition of southern or lower altitude native pine stands through spread of broadleaved species;
- review the implications of forest spread for montane and northern open habitat assemblages.

threshold at which a bird is willing to disperse to a distant wood (Alderman et al., 2004) may restrict its actual spread. Failure of species to spread into newly available climate space, for whatever reason, may allow the existing species to survive longer in sub-optimal climate space.

As species disperse in response to climate change, they must establish in competition with existing ones. Animal communities typically respond quickly to environmental change: there can be rapid taxonomic turnover and ecological rearrangement of the fauna (Wing and Harrington, 2001). However, fossil evidence suggests that plant communities may exhibit considerable 'inertia': pre-existing vegetation has a competitive advantage over new arrivals because it can monopolise resources and shade out invaders. There is typically a substantial time-lag between the arrival of new species and any significant change in the structure and composition of vegetation. This type of two-phase sequence of invasion was exhibited by Eastern hemlock (*Tsuga canadensis* (L.) Carrière) over the last 2500 years in Wisconsin (Parshall, 2002). Community inertia may also explain why beech, which seems to have

been present in England since around 9000 BP (Rackham, 2003), did not become abundant in the pollen record until about 3000 years ago. Extreme events such as droughts or fire, which cause significant mortality, may trigger community turnover.

Modelling of species spread from southern European refugia in the current post-glacial period suggests that rates of 50–100 m per year are needed to explain current distributions of some major tree species, but some trees appear not to occupy their full climatic range (Svenning and Skov, 2005, 2007). However, Banuelos et al., 2004 considered that the eastern range edge of holly in Denmark has shifted about 100 km within half a century (2000 m per year), possibly due to increasingly mild winter temperatures; a recent study of tree-line shifts in north America suggests tree migration rates of 100 km a century (1000 m per year) (Woodall et al., 2009). Estimates for some herbaceous plants and invertebrates are much lower, only a few metres a year (Rackham, 2003). However, unless there have been significant changes in their dispersal ability over the last two millennia, there must be alternative 'rare long-distance' mechanisms that enabled them to reach Britain. For example, wild boar may be significant as dispersers of seeds of ancient woodland indicators on their coats or feet (Schmidt et al., 2004). For some poor dispersers, there is a case for direct translocation as a human analogue of such rare long-distance dispersal events. Translocations within range are already a target in some Species Action Plans (DoE, 1995).

The adaptation strategy needs to consider:

- For how long should current species be maintained at the expense of allowing species of the future to establish and spread?
- Where is it appropriate to assist the dispersal process through active and deliberate human intervention and on what scale?

9.3.3 Interaction with other land uses

Any change in woodland cover to increase the supply of ecosystem services will not occur uniformly across the country and will be the product of interaction with other land-use policies and management choices. Historically, woodland remained abundant where there were reasonably close markets for the products (e.g. peri-urban forests or close to sources of iron); the expansion during the 20th century focused on land of little value for modern farming (Rackham, 2003; Smout, 2003). These produced

BOX 9.3 Re-sorting of temperate broadleaved woodland types

Across much of the UK, there are likely to be shifts in the main tree species. Oak was often favoured in planting programmes, where ash was the more natural dominant. Current trends are for ash to re-assert itself. Minor trees typical of southern and more continental woods may increase, for example lime and field maple; there may be an increased mixed deciduous component in woods formerly dominated by beech in the south, whereas beech may continue to spread into oakwoods in the north and west. In the uplands, oak and birch may grow and regenerate more vigorously at higher altitudes, as at Wistman's Wood on Dartmoor over the last century (Proctor *et al.*, 1980). In floodplains and other wet situations, changes in the water regime may favour or act against alders and willows. Shifts in the distribution and abundance of associated flora and fauna will occur, in part directly from climate change, but also influenced by management in and around the woods (Flemming and Svenning, 2004; Mitchell *et al.*, 2007).

Possible adaptation responses include:

* reassessment of productive potential of broadleaved species
* review of impact of emerging pests and pathogens, e.g. gypsy moth, acute oak decline
* accept changing distributions and assemblages of trees in broadleaved woods including near-continental species as part of future wildlife
* review the balance of management in woodland; there may be less need for coppice systems to maintain southern/thermophilic species, although coppice may still be desirable for light-demanding species and for birds needing dense shrub layers
* accept expansion of oak and birch woodland to higher altitudes than commonly found at present, assuming reductions on grazing pressures allow this to happen
* accept changing field layer compositions and associated faunal changes
* review the balance of open and wooded landscapes in both upland and lowlands.

Wood-pasture and parkland

The UK is believed to have a particularly high density of veteran trees, often associated with wood-pasture and parkland. These trees are increasingly vulnerable to extreme drought and storms, new or invigorated pests and pathogens, and their loss would cause loss of associated saproxylic invertebrates, lichens and fungi. Possible adaptation responses include:

* continued management of the individual trees to prolong their life
* establishment of new generations of trees where these are currently lacking
* speeding up development of 'veteran tree' features to allow colonisation of younger trees by specialist species
* precaution against increased fire risk (often these trees are surrounded by grass and bracken)
* the development of new open-grown trees in fields, hedges and other locations to spread the future overall population and distribution of veteran trees.

a heterogeneous woodland cover with distinct well- and poorly-wooded regions (e.g. Spencer and Kirby, 1992; Forestry Commission, 2003). Similar pressures will shape future land-use patterns.

Agriculture will remain the priority land use in the 21st century, because of increasing concerns about food security. There may be pressure to convert forests to farming (as is happening in New Zealand; Stevenson and Mason, 2008), where woodland occupies land seen as particularly productive. New opportunities for forests may emerge on sites that are too drought prone for

un-irrigated farming, too wet in winter on floodplains or poorly drained soils; too degraded/polluted for economical food production (urban brown-field sites), or too remote. Local demand for wood or fuel may shift the balance towards clusters of woods around new markets; but conversely, there is pressure locally to clear forests to restore open habitats such as heathland or bog for biodiversity reasons (Forestry Commission Scotland, 2009; Forestry Commission, 2009). The carbon-sequestration consequences and sustainability of such decisions will be increasingly critical.

The role of woodland in regulating water flows will become more important, favouring development of increased woodland cover in upper catchments (except those on peat soils because of concerns about net carbon loss) (Nisbet and Broadmeadow, 2004; IFRMRC, 2008; Woodland Trust, 2008; see also Chapter 10). Large-scale woodland expansion may, however, not be suitable where water yields are already low and trees would have higher evapotranspiration rates than current crops.

Trees and woodland around settlements and cities are likely to be encouraged as improving quality of life (shade and shelter may reduce costs of air conditioning and heating), for recreation, and to a lesser extent as sources of local wood products (Britt and Johnston, 2008; developed further in Chapter 10). The numerous trees outside woodland in rural areas (Forestry Commission, 2003) may increase in value as shade for livestock. However, non-woodland trees can be more vulnerable to climate change because of their exposed situation.

There will be continued pressure on land for urban and infrastructure development that will impact on trees and forests, although these may trigger changes to planning guidance (e.g. in England Planning Policy Statement 9; ODPM, 2005) that may help reduce direct loss and increase compensatory planting where losses do occur.

A more integrated approach to land use is highly desirable since many ecosystem service flows depend on the interaction between wooded and open elements of the landscape. New combinations of land use, for example, agro forestry (Morgan-Davies *et al.*, 2007) or re-wilded areas (www.wildland-network.org.uk; Taylor 2005) may further blur past distinctions between forest and open land.

The above pressures and priorities need to be translated into new woodland or improved woodland management on the ground, through iteration between:

- national targets and aspirations, reflected in, for example, national forestry strategies (e.g. Scottish Executive, 2006), other guidance (e.g. Planning Policy Statement 9 in England emphasises the need for woodland protection), and Biodiversity Action Plan targets;
- regional and other sub-country level frameworks that indicate where the priorities for different types of forestry activity lie;
- local projects and regulation which determine what work actually goes ahead.

9.4 Regional and landscape-level adaptation

'Adaptation' measures may need to be different in southeast England from those in northwest Scotland, because the landscapes, the woods, and what is expected of them differ. Adaptation recommendations for managers have been made for woods in different parts of the UK (Ray, 2008a,b; Broadmeadow, 2002a,b). Similar attempts have been made to apply the biodiversity adaptation principles for England (Hopkins *et al.*, 2007; Smithers *et al.*, 2008) to local levels (Natural England, 2009). However, even within a single landscape, the critical factors may vary: changes in winter rainfall might be important for valley bottoms, whereas summer drought could be critical on adjacent south-facing slopes. In Catalonia, Jump *et al.*, (2006) found that beech growth was more limited by drought at the species' southern limits, but not where it occurred at higher altitudes. A priority for future research is identifying the extent to which it is possible and useful to refine impact and response data to sub-national levels.

At the regional level, the coupling between forests and land-use management and the local markets for forest products need to be planned in an integrated manner. This may be key for the success of forest adaptation. For instance, the development of large heating units based on wood chips or pellets, e.g. in schools, hospitals and other public buildings, should be coupled to the development of adequate management of woodlands and woodland creation in their vicinity. This guarantees the sustainable supply of wood at low transport cost to the combined heat and power (CHP) unit and secures a market for local forestry. It may also lead to greater social acceptability of intensive forestry alternatives.

9.4.1 Woodland products

Timber and wood products are traded globally, but for certain types of product and producer adaptation at the sub-national level may still be important to provide the right type, age and size of material locally. There needs to sufficient 'available' woodland to ensure long-term supply of raw material within economic transport distance. The resource must be accessible with no significant conflicts with other objectives that would limit its use. Mechanisms for linking small producers with local markets need to be better developed, for example the web-based service initiated by the Sylva Foundation (www.myforest.org.uk).

A more balanced age range across the landscape would ensure that there is consistency of supply year-to-year; at present there is often an imbalance in age classes because of periods of high exploitation (such as the two World Wars), followed by neglect and subsequent peaks of woodland creation (Forestry Commission, 2003; Mason, 2007). The wood produced should be of a size and quality suitable for a range of products. Historically, markets have often changed during the course of a rotation (particularly for slower-growing broadleaves), as exemplified by the decline in ship-building timbers, the rise and fall of mining timber, or of poplar for matchwood. However, quality and uniformity of product tend to be valued across a range of different uses. In addition, the size and distribution of stands for harvesting should be such that this can be done efficiently. The trend in the 20th century was towards large-scale harvesting, but a shift towards smaller-scale working, such as the promotion of continuous cover forestry in state forests in Wales, would have implications for harvesting technology and practice (for more detailed consideration of forest products see Chapter 7).

9.4.2 Biodiversity

The characteristics of landscapes which will retain or develop high biodiversity under climate change have been summarised by Hopkins et al., (2007). They include:

- variation in topography, particularly slope, aspect and height. Lenoir et al., (2008) report an upward shift in the optimum elevation of forest plant species by around 29 m per decade;
- diversity in soils and water regimes;
- numerous semi-natural land-cover types, which may provide the conditions that will allow a wide range of species to move through the landscape (Watts et al., 2007; Watts, 2006);
- diverse and structurally-varied vegetation.

The underlying robustness derived from topography, soil or water regimes, can be modelled if the above is true: for example, biodiversity in Snowdonia should be more robust than that in the East Anglian plain under climate change scenarios. Manipulation of land cover and vegetation structure can further improve landscape resilience; in a predominantly forested landscape creating different stand structures increases microclimate variation between and within stands and glades (Morecroft et al., 1998). In a largely open landscape encouraging small woods, tree lines and scattered trees, e.g. along river corridors, field boundaries and breaks of slope has a similar effect.

Such manipulation should allow more species to be able to spread through at least the local landscape, to take advantage of differences in microclimate conditions. Models of this landscape 'permeability' have been used to explore the best places to put new woodland to facilitate species movement (Watts, 2006; Watts et al., 2007). Further work is needed to validate the underlying assumptions in these connectivity models; for example, as the cover of suitable habitat increases, the benefit of deliberately targeting the location of additional habitat declines (Pearson and Dawson, 2005).

There is a tension between maintaining separate, dispersed populations of species to reduce the risk of localised extinction from catastrophic events and promoting networks to create opportunities for migrations and adaptation to change (Nabuurs et al., 2007). In the UK, the balance of advantage is seen as being more towards promoting networks, because pests and pathogens are not particularly limited in their dispersal, at the scale of UK landscapes.

Some of the adaptation measures for biodiversity involve increases in woodland cover – woodland creation. Other woodland creation schemes (to provide carbon sequestration, improve water supply, produce wood fibre or fuel) may not have biodiversity as a prime objective, but are likely to have some negative impact on biodiversity of open ground. Some 20th century afforestation schemes have developed into valued new cultural landscapes, with their own distinctive nature conservation values; others have not; and some are targeted for modification under open habitat restoration programmes. The history of past conflicts (e.g. Symonds, 1936; Nature Conservancy Council, 1986; Tompkins, 1989) colours reactions to afforestation proposals, so that new woodland creation must therefore be developed sensitively with full recognition of the potential implications for biodiversity, alongside the other associated costs and benefits.

Triage at the landscape level may help guide action for woodland species:

- In landscapes with little inherent robustness and low permeability, action concentrated on and immediately around individual sites may be more cost-effective than work on improving the landscape matrix, where more effort is needed to make a significant difference.
- Landscapes with intermediate vulnerability or permeability are a high priority for creating new woodland (or other habitats) to improve the landscape permeability,

since there is the potential to make a large difference with relatively little effort.

- In landscapes with low vulnerability or high permeability, additional woodland or habitat creation may make little difference to the landscape's adaptive capability for the majority of woodland species.

9.4.3 Social values of woodland

Many social values of woodland are related to how accessible they are. In future, increased emphasis may be placed on accessibility without reliance on cars. Adaptation might therefore be considered in terms of the match between tree and woodland distribution and population density: how much woodland is physically accessible via public transport, footpaths, for example (McKernan, 2007; Woodland Trust, 2004); and the degree to which it is actually accessed when cultural and behavioural factors are included (Burgess, 1996). The importance of local accessibility increases for short visits while for more remote, but attractive forest areas, the available accommodation and the range of activities are the important factors. Forest design, structure and relative openness are important, but the composition, in terms of tree species, tends to be less so (e.g. Coles and Bussey, 2000). There is some favouring of broadleaves over conifers by forest visitors, but this may be outweighed by the age and or size of the trees. In contrast, forest owners and managers are likely to consider the ease of maintenance or robustness to anti-social behaviour of the trees and woodlands.

A second element of social adaptation is helping people prepare for the changes in landscape appearance that will undoubtedly occur. For example, if beech becomes less common in the Chilterns; if oak spreads through some native pinewoods; or the balance of woodland and open land changes, the reasons for this need to be explained if unnecessary pressure to resist these changes is to be avoided. Both rapid increases and decreases in tree and woodland cover can generate active local opposition, which diverts resources from other adaptation measures.

9.4.4 Dealing with pests, pathogens and other disturbances

A landscape-level approach can help in planning for future potential climate change-related threats, whether these are abiotic or biotic. Some examples of such action are: deer management groups; organised control/eradication of potential hosts to new diseases (Rhododendron as a host of *Phytophthora ramorum* and *P. kernoviae* and timber

movement restrictions to reduce spread of *Dendroctonus micans*). The impact of new diseases on large-scale plantations is often emphasised because of the economic consequences, but native species in semi-natural stands are not immune, as shown by Dutch elm disease, alder dieback, and current concerns over 'acute oak decline'. Across Europe, oak is considered, like pine and spruce, to be potentially vulnerable to major diseases (Lindner *et al.*, 2008), including outbreaks of pests and pathogens not currently found in the UK.

There is the potential for 'new' species to become invasive: these may be recent arrivals (oak processionary moth) but could also be species that have been long established in parks and gardens. For example there is some concern that *Robinia pseudacacia* might become a more aggressive invader of both woodland and non-wooded habitats under a warmer climate.

Increased risks of fire and possibly severe storm damage should be considered at the landscape level. Lessons can be learnt from the responses to past landscape-scale disturbances, such as the effects of the 1987 storm in southeast England (Kirby and Buckley, 1994; Grayson, 1989). There may also be interactions between climate change effects and pollutants, such as nitrogen deposition, that increase risk. For example, increased carbon dioxide and increased nitrogen may initially lead to increased tree growth but higher summer temperatures may combine with traffic pollution to increase levels of ozone, leading to increased damage to trees (see Chapter 3).

9.5 Adaptation measures at site level

Individual owners and managers can reduce the impact of climate change on the ability of their woods to deliver their desired range of objectives through altering the silvicultural system, the structure of the forest within a system and the main crop species used (Lindner *et al.*, 2008; Forestry Commission (a) in press; see also Section 3). The silvicultural system *per se* is, however, less important than the structures that it creates and their resilience and robustness in relation to future needs and conditions.

It would be possible to develop adaptation prescriptions based on the Forest Management Alternatives which were described in Section 3 and the majority of woods are likely to be treated as high forest in different forms (FMAs 1–4). Whereas clearfell systems have predominated in the past, continuous cover forestry approaches are increasingly promoted because they maintain a more even carbon storage, show lower soil carbon losses during harvesting, and maintain more even humidity levels. Mixed-age structured woods may be more resilient in the longer term. However, moving from even-aged to uneven-aged structures often involves short- to medium-term costs in production, biodiversity or social acceptance during the transition. Where the balance of advantage lies will often involve a range of site-specific factors.

For example, woods might be made more structurally diverse by reducing coupe size and encouraging continuous cover forestry. This may make the woods less susceptible to extreme windthrow, but increase harvesting costs and make the wood look more uniform in distant views. Shade-bearing woodland plants might benefit but the habitat available for species associated with open woodland might decline. Other examples of the trade-offs that have to be considered are that coppice and pollard systems maintain cultural continuity and past genetic variation by prolonging the life span of individual trees, involve very little soil disturbance to achieve regeneration, but reduce the potential for genetic change between generations. Dense natural regeneration provides more potential for natural selection to operate, but may require more intervention to achieve (such as fencing to remove grazing and scarification to improve establishment). Reducing rotation lengths will result in loss of potential old growth development, but permits more rapid testing of genetic material and thus may increase adaptation to the emerging new conditions (Hubert and Cottrell, 2007).

Young stands are usually faster growing than old ones, and so may sequester carbon more rapidly. However, old growth stands also build up carbon in the form of slowly decomposing organic matter in litter and soil (Luyssaert et al., 2008; Zhou et al., 2006) and are highly valued for biodiversity. More varied woodland, both in species composition and structures, means that there may always be some stands present that are vulnerable to particular threats which will result in chronic low-level disturbance. However, high diversity should make it easier to contain major disturbances because only a minority of stands are ever at the susceptible stage at any one time. This approach is already practised with respect to wind-hazard management (Gardiner and Quine, 2000).

Forestry and woodland management makes only limited use of external inputs of fertiliser, pesticides and other agrochemicals, and therefore do not have a large

burden of the embedded carbon costs involved in their manufacture. However, the scope for further reduction of external inputs should be considered, for example by careful matching of species to site type (Pyatt et al. 2001); or stand manipulation to control competing ground vegetation. Where high production is needed, the use of nitrogen-fixing species as part of the crop mix might be appropriate, although the impact on biodiversity needs to be considered. Alder is the only native tree to fix nitrogen, but some non-native trees and shrubs are being tried (Hemery, 2001).

9.6 Adaptation in tree species choice

Climate change will affect the survival and growth patterns of tree species, with consequences both for semi-natural woodland composition and for production patterns and growth potential. The impact and the need for adaptive action will depend on whether desirable species are increasing or decreasing; whether they are major or minor parts of forest systems; and whether there is a net change in species occurrence and/or abundance overall. Where species are at the boundaries of acceptable growth (e.g. Norway spruce in eastern England), alternatives need to be sought; and even where the species remains within its tolerances, different provenances may be required (Broadmeadow and Ray, 2005; Ray, 2001). From a biodiversity perspective, species currently native only in southern Britain (e.g. beech), should become accepted in northern Britain; species from the near continent (e.g. sycamore) accepted in southern Britain, as part of the re-sorting of species that is likely to follow climate change (see Box 9.3).

As the progress of climate change becomes clearer, an even wider range of species may need to be considered (Table 9.1). Most of the species in this Table are already grown in Britain, at least in collections. From a biodiversity and landscape perspective, more emphasis should be placed on the broadleaved species as these are more likely to produce stands that are close in ecological and visual terms to current semi-natural woodland over most of the country. Further work is however needed on this approach and which particular species to include.

Uncertainties over future growth and potential threats to particular species (e.g. Brown and Webber, 2008) has led to favouring the use of mixtures of species and provenances at a variety of scales as an 'insurance mechanism' (Broadmeadow and Ray, 2005). Lady Park Wood in the Wye Valley can be considered as a natural microcosm of how, over decades, different components of a mixed stand may be hit by different disturbances, but the site remains wooded (Peterken and Mountford, 1995). One approach may be to introduce to woods relatively small amounts of novel species and provenances (Lindner et al., 2008) that may prove useful in the future. Where native tree species are involved, this could affect the genetic diversity that has developed in the UK during the current post-glacial period, although as the environment changes such diversity will change anyway. In addition, many UK populations are exposed to long-distance wind pollen transport and to mixing with 'non-native' material from parks and gardens, for example. A general insistence on local provenance in native species planting may no longer be tenable, although it should remain a consideration, particularly where minor or insect-pollinated species are concerned. However, introductions of species and provenances have risks, notably the risk that the material will spread in woods or open ground where it is not wanted. Therefore, risk assessments for tree species need to be developed.

Table 9.1
Continental European tree species not native to the UK that warrant consideration as 'alternative species' in developing climate change adaptation strategies.

Broadleaf species	Conifer species
Acer monspessulanum	Abies alba
Acer opalus	Abies borisii-regis
Alnus cordata	Abies cephalonica
Castanea sativa	Abies cilicica
Celtis australis	Abies pinsapo
Fagus orientalis	Picea omorika
Fraxinus angustifolia	Pinus brutia
Juglans regia	Pinus pinaster
Ostrya carpinifolia	Pinus pinea
Platanus orientalis	Pinus peuce
Populus alba	
Quercus faginea	
Quercus ilex	
Quercus pyrenaica	
Quercus pubescens	

Table provided by Bill Mason, Richard Jinks and Mark Broadmeadow.

9.7 Advice and regulation adaptation

The foregoing sections have implications for how woods are managed and regulated. Some of what is required will be new. However, much will be a challenging extension (because of the major uncertainties about climate change and its impacts) of the current paradigm of sustainable forest management (Forestry Commission (b) in press).

Practical adaptation measures need to be tailored to the different types of woods, woodland owners and their objectives. Developing and implementing adaptation measures for coniferous production forests is likely to be easier than for the broadleaved/semi-natural resource. The former are already managed more actively and the ownership is more concentrated among the public forest estate and large management companies. There is also a need to revisit conservation designations and guidance: to develop site objectives and designation practices that can cope with more dynamic environments both for wooded and non-wooded habitats; and review approaches and attitudes to non-native species from the near continent. Perhaps as big a challenge as adapting to the changes in the physical climate will be responding to developments in policy, regulation and public attitudes. Such changes are hardly likely to be less frequent than over the last 60 years (see also Chapter 5).

9.8 Research priorities

Given the uncertainty identified above, there are some pressing research needs:

- Development of databases and knowledge on how different species are expected to respond to climate change (e.g. Ecological Site Classification, Climate Envelope Modelling, see Chapter 4), matched by studies on how their populations and distributions are actually changing (including new provenance/species trials).
- Improved understanding as to which factors will become limiting for which species at a regional level; and how climate change factors will change disturbance regimes of wind, fire, pests and pathogens.
- Improved understanding of how climate change factors will change disturbance regimes of wind, fire, pests and pathogens.
- Exploration of the scope for and limits of 'technical fixes' such as species translocations, genetically improved trees, and so on.

- Improved modelling of how climate impacts on other land uses and societal behaviour will impact on trees, woods and forestry, combined with development of appropriate decision-making methods that can deal with uncertainty and integration of different societal values.
- Improved monitoring and modelling of the degree to which more varied composition and structure does improve resilience; and studies of how to measure the economic value of changes in forest system resilience.
- Improved understanding of appropriate decision making methods, including methods of dealing with uncertainty and the integration of multiple societal values.
- Developing practical ways of applying research results to effect change across landscapes that integrate areas of high social and productive land use with those where conservation has a higher priority (*cf.* a revitalised Man and the Biosphere programme?).

9.9 Conclusions

Adaptation needs to be an ongoing process, with continuing testing of orthodoxies and re-calibration of experiences, which have often been based on a static view of the natural and social environments, a focus on preservation of past structures, communities, systems or markets. Equally, views on and understanding of climate change and its impacts will evolve. Incentives, controls, education and knowledge transfer need to be kept in line with progress on adaptive measures. Ultimately it will be a case of accepting the impacts and changes we can make little difference to; concentrating our efforts on those which can be changed; and having the wisdom to separate the two.

References

ALDERMAN, J., McCOLLIN, D., HINSLEY, S., BELLAMY, P., PICTON, P. and CROCKETT, R. (2004). Simulating population viability in fragmented woodland: nuthatch (*Sitta europaea* L.) population survival in a poorly wooded landscape in eastern England. In: Smithers, R. (ed.) *Landscape ecology of trees and forests,* IALE, UK, pp. 76–83.

BANUELOS, M.J., KOLLMANN, J., HARTVIG, P. and QUEVEDO, M. (2004). Modelling the distribution of *Ilex aquifolium* at the north-eastern edge of its geographic range. *Nordic Journal of Botany* 23, 129–142.

BEALE, C.M., LENNON, J.J. and GIMONA, A. (2008). Opening the climate envelope reveals no macroscale associations with climate in European birds. *Proceedings of the National Academy of Science* 105, 14908–14912.

BERRY, P.M., DAWSON, T.P., HARRISON, P.A. and PEARSON, R.G. (2002). Modelling potential impacts of climate change on the bioclimatic envelope of species in Britain and Ireland. *Global Ecology and Biogeography* 11, 453–462.

BRITT, C. and JOHNSTON, M. (2008). *Trees in Towns II – A new survey of urban trees in England and their condition and management.* Department for Communities and Local Government, London.

BROADMEADOW, M. (2002a). *A review of climate change implications for trees and woodland in the East of England.* Forest Research, Farnham. (unpublished).

BROADMEADOW, M. (ed) (2002b). *Climate change: impacts on UK forests*. Forestry Commission Bulletin 125. Forestry Commission, Edinburgh.

BROADMEADOW, M. and RAY, D. (2005). *Climate change and British woodland.* Forestry Commission Information Note 69. Forestry Commission, Edinburgh.

BROWN, A. and WEBBER, J. (2008). *Red band needle blight of conifers in Britain*. Research Note 002, Forestry Commission, Edinburgh.

BURGESS, J. (1996). Focussing on fear: the use of focus groups in a project for the Community Forest Unit, Countryside Commission. *Area* 28, 130–135.

COLES, R.W. and BUSSEY, S.C. (2000). Urban forest landscapes in the UK – progressing the social agenda. *Landscape and Urban Planning* 52, 181–188.

DEPARTMENT OF THE ENVIRONMENT (1995). *Biodiversity: the UK Steering Group report (Volume 2: Action plans).* HMSO, London.

DEFRA (2007). *A strategy for England's trees, woods and forests*. Department for Environment, Food and Rural Affairs, London.

EDWARDS, M.E. (1986). Disturbance histories of four Snowdonian woodlands and their relation to Atlantic bryophyte distributions. *Biological Conservation* 37, 301–320.

EUROPEAN COMMISSION (2007). *Adapting to climate change in Europe – options for EU action*. European Commission Communication 354.

FLEMMING, S. and SVENNING, J. (2004). Potential impact of climate change on the distribution of forest herbs in Europe. *Ecography* 27, 366–380.

FLOOD RISK MANAGEMENT RESEARCH CONSORTIUM (2008). *Impacts of upland management on flood risk: multi scale modelling methodology and results from the Pontbren experiment*. FRMRC Research Report UR 16.

FOREST SERVICE (2006). *Northern Ireland forestry – a strategy for sustainability and growth*. Forest Service, Northern Ireland, Belfast.

FORESTRY COMMISSION (2001). *Woodlands for Wales*. Forestry Commission Wales, Aberystwyth.

FORESTRY COMMISSION (2003). *National inventory of woodland and trees*. Forestry Commission, Edinburgh.

FORESTRY COMMISSION (2009a). *Forests and climate change guidelines.* Consultation draft , July 2009. Forestry Commission, Edinburgh.

FORESTRY COMMISSION (2009b). *The UK Forestry Standard.* Consultation draft, July 2009. Forestry Commission, Edinburgh.

FORESTRY COMMISSION (2009). *Restoring and expanding open habitats from woods and forests in England: a consultation*. Forestry Commission England, Bristol.

FORESTRY COMMISSION SCOTLAND (2009). *Control of woodland removal: the Scottish government's policy*. Forestry Commission Scotland, Edinburgh.

GARDINER, B.A. and QUINE, C.P. (2000). Management of forests to reduce the risk of abiotic damage – a review with particular reference to the effects of strong winds. *Forest Ecology and Management* **135**, 261–277.

GEAR, A.J. and HUNTLEY, B. (1991). Rapid changes in the range limits of Scots pine 4000 years ago. *Science* **251**, 544–547.

GESSLER, A., KEITELL, C., KEUZWIESER, J., MATYSSEK, R., SEILER, W. and RENNENBERG, H. (2007). Potential risk for European beech (*Fagus sylvatica L.*) in a changing climate. *Trees* **21**, 1–11.

GIESECKE, T., HICKLER, T., KUNKEL, T., SYKES, M.T. and BRADSHAW, R.H.W. (2007). Towards an understanding of the Holocene distribution of *Fagus sylvatica* L. *Journal of Biogeography* **34**, 118–131.

GODWIN, H. (1975). *History of the British flora*. Cambridge University Press, Cambridge.

GOLDBERG, E.A., KIRBY, K.J., HALL, J.E. and LATHAM, J. (2007). The ancient woodland concept as a practical conservation tool in Great Britain. *Journal of Nature Conservation* **15**, 109–119.

GOVE, B., POWER, S.A., BUCKLEY, G.P. and GHAZOUL, J. (2007). Effects of herbicide spray drift and fertilizer overspread on selected species of woodland ground flora: comparison between short-term and long-term impact assessments and field surveys. *Journal of Applied Ecology* **44**, 374–384.

GRAN CANARIA GROUP (2006). *The Gran Canaria Declaration II on climate change and plants*. Jardin Botanico 'Viera y Clavijo' and Botanic Gardens Conservation International, Gran Canaria, Spain.

GRAYSON, A.J. (1989). *The 1987 storm: impact and responses*. Forestry Commission Bulletin 87. HMSO, London.

HARRISON, P.A., BERRY, P.M. and DAWSON, T.P. (2001).

Climate change and nature conservation in Britain and Ireland: modelling natural resource responses to climate change (the MONARCH project). UKCIP, Oxford.

HAWKINS, B., SHARROCK, S. and HAVENS, K. (2008). *Plants and climate change: which future?* Botanic Gardens Conservation International, Richmond.

HEMERY, G.E. (2001). Growing walnut in mixed stands. *Quarterly Journal of Forestry* **95**, 31–36.

HERBST, M., ROBERTS, J.M., ROSIER, P.T.W., TAYLOR, M.E. and GOWING, D.J. (2007). Edge effects and forest water use: a field study in a mixed deciduous woodland. *Forest Ecology and Management* **250**, 176–186.

HILL, J.K., THOMAS, C.D. and HUNTLEY, B. (1999). Climate and habitat availability determine 20th century change in a butterfly's range margin. *Proceedings of the Royal Society of London B Series* **266**, 1197–1206.

HOPKINS, J.J. and KIRBY, K.J. (2007). Ecological change in British broadleaved woodland since 1947. *Ibis* **149**(Suppl. 2), 29–40.

HOPKINS, J.J., ALLISON, H.M., WALMSLEY, C.A., GAYWOOD, M. and THURGATE, G. (2007). *Conserving biodiversity in a changing climate: guidance on building capacity to adapt*. DEFRA, London.

HUBERT, J. and COTTRELL, J. (2007). *The role of forest genetic resources in helping British forests respond to the effects of climate change*. Forestry Commission Information Note 86. Forestry Commission, Edinburgh.

HUMPHREY, J.W. (2005). Benefits to biodiversity from developing old-growth conditions in British upland spruce plantations: a review and recommendations. *Forestry* **78**, 33–53.

HUNTLEY, B. (1990). European post-glacial forests: compositional changes in response to climatic change. *Journal of Vegetation Science* **1**, 507–518.

HUNTLEY, B. (1991). How plants respond to climate change: individualism and the consequences for plant communities. *Annals of Botany* **67**, 15–22.

HUNTER, Jr, M.L. (ed.) (1999). *Maintaining biodiversity in forested ecosystems*. Cambridge University Press, Cambridge.

JUMP, A., HUNT, J. and PENUELAS, J. (2006). Rapid climate change-related growth decline at the southern range edge of *Fagus sylvatica*. *Global Change Biology* **12**, 2163–2174.

KEITH, S.A., NEWTON, A.C., HERBERT, R.J.H., MORECROFT, M.D. and BEALEY, C.E. (2009). Non-analogous community formation in response to climate change. *Journal of Nature Conservation*. (in press)

KIRBY, K.J. and BUCKLEY, G.P. (eds) (1994). *Ecological responses to the 1987 Great Storm in the woods of south-east England*. Science Report No. 23. English Nature, Peterborough.

KIRBY, K.J., SMART, S.M., BLACK, H.I.J., BUNCE, R.G.H., CORNEY, P.M. AND SMITHERS, R.J. (2005). *Long term ecological change in British woodland (1971–2001).* Research Report 653. English Nature, Peterborough.

LENOIR, J., GEGOUT, J.C., MARQUET, P.A., DE RUFFRAY, P. and BRISSE, H. (2008). A significant upward shift in plant species optimum elevation during the twentieth century. *Science* 320, 1768–1771.

LINDNER, M., GARCIA-GONZALO, J., KOLSTROM, M., GREEN, T., MAROSCHEK, M., SEIDL, R., LEXER, M.J., NETHERER, S., SCHOPF, A., KEREMER, A., DELZOIN, S., BARBATI, A., MARCHETTI, M. and CORONA, P. (2008). *Impacts of climate change on European forests and options for adaptation.* European Commission (D-G for Agriculture and Rural Development), Brussels.

LUYSSAERT, S., DETLEF SCHULZE, E., BORNER, A., KNOHL, A., HESSENMOLLER, D., CIAIS B.P., GRACE, J. (2008). Old-growth forests as global carbon sinks. *Nature* 455, 213–215.

MASON, W. L. (2007). Changes in the management of British forests between 1945 and 2000 and possible future trends. *Ibis* 149(Suppl. 2), 41–52.

McKERNAN, P. and GROSE, M. (2007). *An analysis of accessible natural greenspace provision in the South-east.* Forestry Commission, Farnham.

MILLAR, C.L., STEPHENSON, N.L. and STEPHENS, S.L. (2007). Climate change and forests of the future: managing in the face of uncertainty. *Ecological Applications 17,* 2145–2151.

MITCHELL, R.J., MORECROFT, M.D., ACREMAN, M., CRICK, H.Q.P., FROST, M., HARLEY, M., MACLEAN, I.M.D, MOUNTFORD, O., PIPER, J., PONTIER, H., REHFISCH, M.M., ROSS, L.C., SMITHERS, R.J., STOTT, A., WALMSLEY, C.A., WATTS, O. and WILSON, E. (2007). *England Biodiversity Strategy – towards adaptation to climate change.* Final report to DEFRA for Contract CR0327, London.

MORGAN-DAVIES, C., WATERHOUSE, A., POLLOCK, M.L. and HOLLAND, J.P. (2007). Integrating hill sheep production and newly established native woodland: achieving sustainability through multiple land-use in Scotland. *International Journal of Agricultural Sustainability* 6, 133–147.

MORECROFT, M.D., TAYLOR, M.E. and OLIVER, H.R. (1998). Air and soil microclimates of deciduous woodland compared to an open site. *Agricultural and Forest Meteorology* 90, 141–156.

NABUURS, G.J., MASERA, O., ANDRASKO, K., BENITEZ-PONCE, P., BOER, R., DUTSCHKE, M., ELSIDDIG, E., FORD-ROBERTSON, J., FRUMHOFF, P., KARJALAINEN, T., KRANKINA, O., KURZ, W.A., MATSUMOTO, M.,

OYHANTCABAL, W., RAVINDRANATH, N.H., SANZ SANCHEZ, M.J., ZHANG, X. (2007). Forestry. In: Metz, B., Davidson, O.R., Bosch, P.R., Dave, R., Meyer L.A. (eds) *Climate Change 2007: mitigation.* Contribution of Working Group III to the Fourth Assessment Report of the Intergovernmental Panel on Climate Change. Cambridge University Press, Cambridge. pp. 541–584.

NATURAL ENGLAND (2009). *Responding to the impacts of climate change on the natural environment: the Cumbria High Fells.* Research Report 115. Natural England, Peterborough.

NATURE CONSERVANCY COUNCIL (1986). *Nature conservation and afforestation.* Nature Conservancy Council, Peterborough.

NISBET, T. and BROADMEADOW, S. (2004). *A guide to using woodland for sediment control.* Forest Research, Farnham.

NOTARO, M. (2008). Response of the mean global vegetation distribution to inter-annual climate variability. *Climate Dynamics* 30, 845–854.

ODPM (2005). *Biodiversity and geological conservation.* Planning Policy Statement 9. HMSO, London.

PARSHALL, T. (2002). Late Holocene stand-scale invasion by hemlock (*Tsuga canadensis*) at its western range limit. *Ecology* 83, 1386–1398.

PEARSON, R.G. and DAWSON, T.P. (2005). Long-distance plant dispersal and habitat fragmentation: identifying conservation targets for spatial landscape planning under climate change. *Biological Conservation* 123, 389–401.

PENUELAS, J. and BOADA, M. (2003). A global-induced shift in the Montseny mountains (NE Spain) *Global Change Biology* 9, 131–140.

PETERKEN, G.F. (2000). *Natural forest reserves.* Research Report 384. English Nature, Peterborough.

PETERKEN, G.F. and MOUNTFORD, E. (1995). Lady Park Wood Reserve – the first half century. *British Wildlife* 6, 204–213.

PETERKEN, G.F. and MOUNTFORD, E. (2005). Natural woodland reserves - 60 years of trying at Lady Park Wood. *British Wildlife* 17, 7–16.

PIGOTT, C.D. and HUNTLEY, J.P. (1981). Factors controlling the distribution of *Tilia cordata* at the northern limits of its geographical range. III. Nature and causes of seed sterility. *New Phytologist* 87, 817–839.

PROCTOR, M.C.F., SPOONER, G.M. and SPOONER, M.F. (1980). Changes in Wistman's Wood, Dartmoor: photographic and other evidence. *Transactions of the Devon Association for the Advancement of Science* 112, 43–79.

PYATT, D.G., RAY, D. and FLETCHER, J. (2001). *An ecological site classification for forestry in Great Britain.* Forestry Commission Bulletin 124. Forestry Commission,

Edinburgh.

RACKHAM, O. (1999). The woods 30 years on: where have the primroses gone? *Nature in Cambridgeshire* **41**, 73–87.

RACKHAM, O. (2003). *Ancient woodland* (revised edn). Castlepoint Press, Dalbeattie.

RATCLIFFE, D.A. (1968). An ecological account of Atlantic bryopytes in the British Isles. *New Phytologist* **67**, 365–439.

RAY, D. (2008a) *Impacts of climate change on forestry in Wales*. Forestry Commission Research Note 301. Forestry Commission Wales, Aberystwyth.

RAY, D. (2008b). *Impacts of climate change on forestry in Scotland – a synopsis of spatial modelling research*. Forestry Commission Research Note 101. Forestry Commission Scotland, Edinburgh.

SCHMIDT, M., SOMMER, K., KRIEBITZSCH, W-U., ELLENBERG, H. and VON OHEIMB, G. (2004). Dispersal of vascular plants by game in northern Germany. Part I: Roe deer (*Capreolus capreolus*) and wild boar (*Sus scrofa*). *European Journal of Forest Research* **123**, 167–176.

SCOTTISH EXECUTIVE (2006). *Scottish forestry strategy*. Forestry Commission, Edinburgh.

SMITHERS, R.J., COWAN, C., HARLEY, M., HOPKINS, J.J., PONTIER, H. and WATTS, O. (2008). *England Biodiversity Strategy climate change adaptation principles*. DEFRA, London.

SMOUT, T.C. (ed) (2003). *People and woods in Scotland – a history*. Edinburgh University Press, Edinburgh.

SPENCER, J.W. and KIRBY, K.J. (1992). An inventory of ancient woodland for England and Wales. *Biological Conservation* **62**, 77–93.

STEVENSON, H. and MASON, E, (2008). In the face of deforestation. *New Zealand Journal of Forestry* **53**, 40–41

SVENNING, J.C. and SKOV, F. (2005). The relative roles of environment and history as controls of tree species composition and richness in Europe. *Journal of Biogeography* **32**, 1019–1033.

SVENNING, J.C. and SKOV, F. (2007). Could the tree diversity pattern in Europe be generated by postglacial dispersal limitation? *Ecology Letters* **10**, 453–460.

SYKES, M.T., PRENTICE, C. and CRAMER, W. (1996). A bioclimatic model for the potential distributions of north European tree species under present and future climates. *Journal of Biogeography* **23**, 203–233.

SYMONDS, H.H. (1936). *Afforestation in the Lake District*. Dent, London.

TAYLOR, P. (2005). *Beyond conservation – a wildland strategy*. Earthscan and BANC, London.

THUILLER, W., LAVOREL, S., ARAUJO, M, SYKES, M. and PRENTICE, I. (2002). Climate change threats to plant diversity in Europe. *Proceedings of the National Academy of Science* **102**, 8245–8250.

TOMPKINS, S. (1989). *Forestry in crisis: the battle for the hills*. Christopher Helm, London.

WATTS, K. (2006). British forest landscapes: the legacy of woodland fragmentation. *Quarterly Journal of Forestry* **100**, 273–279.

WATTS, K., RAY, D., QUINE, C.P., HUMPHREY J. and GRIFFITHS, M. (2007). *Evaluating biodiversity in fragmented landscapes: applications of landscape ecology tools*. Forestry Commission Information Note 85. Forestry Commission, Edinburgh.

WING, S.L. and HARRINGTON, G.J. (2001). Floral response to rapid warming in the earliest Eocene and implications for concurrent faunal change. *Paleobiology* **27**, 539–563.

WOODALL, C.W., OSWALT, C.M., WESTFALL, J.A., PERRY, C.H., NELSON, M.D. and FINLEY, A.O. (2009). An indicator of tree migration in forests of the eastern United States. *Forest Ecology and Management* **257**, 1434–1444.

WOODLAND TRUST (2004). *Space for people*. Woodland Trust, Grantham.

WOODLAND TRUST (2008). Woodland actions for biodiversity and their role in water management. Woodland Trust, Grantham. Online at: www.woodland-trust.org.uk/pdf/woodswater26_03–08.pdf

ZHOU, G., LIU, S., LI, A., ZHANG, D., TANG, Z., ZHOU, C., YAN, J. and MO, J. (2006). Old-growth forests can accumulate carbon in soils. *Science* **314**, 1417.

WOODLANDS HELPING SOCIETY TO ADAPT

J. F. Handley and S. E. Gill

Chapter

10

Key Findings

In a changing climate, tree and woodland cover in and around urban areas becomes increasingly important for managing temperatures, surface water and air quality. Large tree canopies are the most beneficial, and guidelines should be followed by all concerned parties to ensure that we continue to maintain and plant such trees in urban areas and to overcome perceived risks including subsidence and windthrow. Care should be taken to select tree species which will not contribute to urban ozone pollution and, where water stress is likely to be a problem, planting should focus on more drought-tolerant species.

It is crucial that we have a thorough understanding of the current pattern of tree cover in urban areas, to target where we need to maintain and increase cover. Tree and woodland creation in urban areas should then have two key aims: to manage temperatures and to manage surface water.

The role of trees and woodlands in managing temperatures is clear. Tree and woodland planting should be targeted to: (1) places where people live (especially the most vulnerable members of society) which currently have low tree cover; (2) places where people gather (such as town and local centres), which currently have low tree cover. There is a need for clearer guidelines to encourage the establishment of tree cover in town centres and high density residential areas. Such guidelines need to cover the perceived risks from subsidence and windthrow.

Trees and woodlands have a role to play in managing surface water in urban areas. Their interception of rainfall could be significant. Planting here could be targeted to soils with higher infiltration rates and areas with a history of surface water flooding.

In a changing climate, woodland cover in rural areas will also have a role in helping society, as well as other species, to adapt. Its main role is in managing water resources and reducing flooding. The impact of woodland on water resources will become increasingly important and care will be needed in woodland siting and design to reduce potential losses of water. Its role in managing flooding is more complicated. Large-scale planting in catchments is neither feasible nor likely to control extreme floods, and could indeed have a significant adverse impact on water resources. Targeted woodland creation to manage flooding and improve water status is more effective and could be very beneficial. This would include: planting woodland on (or downslope of) soils vulnerable to erosion and structural damage, woodland buffers along watercourses, planting on derelict and disused land and as a priority, re-creation of carefully designed and located floodplain and riparian woodland.

The overall climate change picture for the UK is summarised in Chapter 2 of this report. Regional studies of climate change impacts (including potential benefits and opportunities) have been carried out throughout the UK as part of the UK Climate Impacts Programme (Box 10.1).

The detailed picture varies from region to region, reflecting changing socio-economic patterns and the strong climate gradient from south east to north west across the UK.

This chapter explores whether trees and woodland, particularly in and around urban areas, can help to moderate the societal impacts of climate change and realise potential 'benefits'. Climate change is expected to bring with it a more outdoor lifestyle and to increase the variety and intensity of visitor use in recreational landscapes (Box 10.1). Managed forests with their ready-made access network are well placed to help meet such demand close to urban centres (e.g. Delamere Forest in northwest England) and to deflect pressure from more fragile landscapes (e.g. Grizedale Forest in the English Lake District).

The degree of urbanisation is especially significant given the concentration of people and property in urban areas, together with the way in which urbanisation itself influences the local climate (Wilby, 2007; Gill et al., 2007). Climate change adaptation strategies need to operate at a series of interlocking spatial scales, an approach exemplified by Shaw et al., 2007, which identifies three levels of inquiry: conurbation or catchment scale, neighbourhood scale and building scale.

Trees and woodland have a potential contribution at each level of scale, and particularly in and around urban areas, in moderating the societal impacts of climate change and realising potential 'benefits'. The ensuing review takes a multi-scalar approach and focuses on the role of trees and woodland with regard to two key impacts of climate change: higher temperatures and changes to the hydrological cycle. It then explores how these roles can be realised in practice, in spite of climate-related hazards.

BOX 10.1 Key threats and opportunities of climate change impacts in the UK

The most widely recognised problems include:

- an increase in the risk of riverine and coastal flooding and erosion
- increased pressure on drainage systems
- a potential increase in winter storm damage
- habitat loss
- summer water shortages, low stream flows and water quality problems
- increased risk of subsidence in subsidence prone areas
- increasing thermal discomfort in buildings and health problems in summer.

Common benefits include:

- a longer growing season and enhanced crop yields
- less cold weather, transport disruption
- reduced demand for winter heating
- fewer cold-related illnesses and deaths.

Opportunities include:

- agricultural diversification and the potential to grow new crops
- an increase in tourism and leisure pursuits
- a shift to more outdoor-oriented lifestyles.

Source: Regional Scoping Studies, West and Gawith, 2005

10.1 Managing high temperatures

Heat waves are expected to increase in frequency and severity in a warmer world (IPCC, 2007; Meehl and Tebaldi, 2004) and urban heat islands will exacerbate the effects of regional warming by increasing summer temperatures relative to outlying districts (Wilby, 2003). Heat stress is likely to increase morbidity and mortality both directly and indirectly through cardiovascular and respiratory disease, which may be further exacerbated by an interaction between temperature stress and air pollution (Davis and Topping, 2008). The severe heat wave in southern and central Europe in 2003 which extended from June to mid-August may have caused 35 000 excess deaths, especially among the elderly (Kosatsky, 2005). Average summer temperatures (June to August) exceeded the long-term mean by up to five standard deviations, but this extreme event is well within the range anticipated by climate models for the 21st century (Stott *et al.,* 2004). Renaud and Rebetz (2009) compared below-canopy and open-site air temperatures at 14 forest sites in Switzerland over 11 days during the August 2003 heat wave. Maximum mean temperatures were significantly cooler (up to 5.2°C) under the canopy, with broadleaved and mixed forests containing beech being particularly effective. They commented on the potential role of forests to provide cool shelter, especially in urban areas, 'where forested parks could provide an important source of relief during heat waves' (Renaud and Rebetz, 2009, p. 873).

In fact, the importance of trees and woodland for the urban microclimate has long been recognised. Oke (1989) notes that tree cover in urban areas frequently exceeds that of the surrounding peri-urban environment and that the trees collectively constitute an 'urban forest'. He reviews the influence of the urban forest on micrometeorology at a range of scales, noting the:

> 'radiative, aerodynamic, thermal and moisture properties of trees that so clearly set them apart from other urban materials and surfaces in terms of their exchanges of heat, mass and momentum with the atmosphere. Their resulting ability to produce shade, coolness, shelter, moisture and air filtration

makes them flexible tools for environmental design' (Oke, 1989, p. 335).

Indeed, provision of vegetation, particularly large broadleaved trees is proposed as one of the more effective strategies for maintaining human comfort during high temperature episodes in urban areas (Watkins *et al.,* 2007). Bioclimatic design at the city scale requires an understanding of both the moderating role of vegetation (the green infrastructure) and the need to maintain ventilation and cool air drainage (Oke, 1989; Eliasson, 2000). European cities such as Stuttgart have long been planned with such considerations in mind (Hough, 2004) and it is important that planting design facilitates, rather than obstructs, ventilation. The current state of the art in bioclimatic planning and design is exemplified by the city of Berlin (Berlin Digital Environmental Atlas, 2009; www. stadtentwicklung.berlin.de/umwelt/umweltatlas/edua_ index.shtml).

Besides providing shade, urban greenspace contributes to cooling through evapotranspiration. Modelling studies in Greater Manchester have shown significant differences for surface temperature between different urban morphology types, depending on the amount of greenspace present (Figure 10.1). At the neighbourhood level, a 10% decrease in urban green results in an increased maximum surface

Figure 10.1
Urban morphology types (UMTs) plotted in order of maximum surface temperature for 1961–1990 along with evaporating fraction (Gill, 2006).

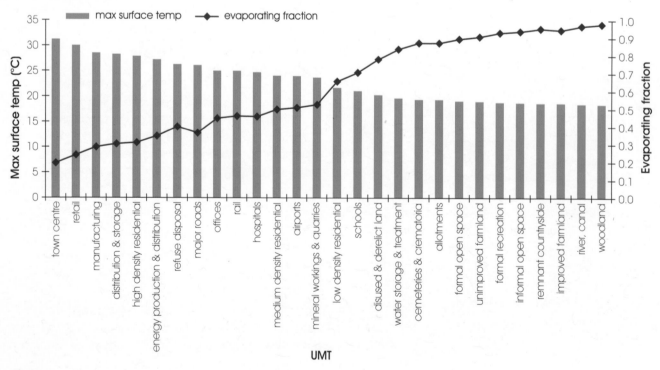

temperature of 7°C in high density residential areas and 8.2°C in town centres (compared with the 1961–1990 current form case) under the UKCIP02 2080s high emissions scenario (Gill et al., 2007). By contrast, adding 10% green cover keeps maximum surface temperatures at or below the 1961–1990 baseline up to, but not including, the 2080s high emissions scenario (Figure 10.2). There are large differences in tree cover with changes in residential density; in Greater Manchester for example, average tree cover varies from 27% (low density) through 13% (medium density) to 7% (high density). The high density areas include wards where socio-economic disadvantage and ill-health are concentrated and active greening programmes will be needed to ensure that residents are not further disadvantaged by climate change (Tame, 2006).

Surface temperature is just one among a number of parameters that determine human comfort; these include air temperature, radiant temperature field, direct solar radiation, air speed and humidity (Watkins et al., 2007). Studies in Hungary (Gulyás et al., 2006) and Germany (Mayer and Höppe, 1987) show a strong correlation between radiation modifications and changes in thermal stress, focusing on the role of trees; especially large canopy deciduous trees in the public realm. We have seen that a shift to a more outdoor-oriented lifestyle is likely to accompany climate change in the UK (Box 10.1). Wilson et al. (2008) have endorsed this and emphasised the role of well designed (and well treed) public open spaces in building adaptive capacity. Parks, with mature trees, will provide cool lacunae in an increasingly inhospitable urban environment with microclimatic benefits which can extend beyond, into surrounding urban neighbourhoods (Oke, 1989; Spronken-Smith and Oke, 1998). It is however vital to ensure that appropriate water supplies are available to sustain urban greenspace, and its evaporative cooling function, in periods of heat stress which will inevitably coincide with periods of potential soil water deficit (Gill et al., 2007 and Watkins et al., 2007). Drought tolerance could become an increasingly important consideration in tree species selection for planting schemes.

The wish to escape from uncomfortably hot, poorly ventilated offices, shops and public buildings will create the need to adapt public space (Wilson et al., 2008). The UK building stock is not well adapted to a warmer climate and problems of thermal comfort are already being experienced in London (Hacker and Holmes, 2007). Rather than install air conditioning (which drives up energy use and therefore GHG emissions), these authors advocate the use of advanced passive features, including shade. Retrofitting buildings is difficult and therefore broadleaved trees may have an important contribution to make (Huang et al., 1987; Simpson, 2002). Indeed in the USA, tree planting programmes have been devised with the express intention to reduce or avoid peak energy demand for cooling (Akbari, 2002; McPherson and Simpson, 2003).

10.2 Responding to change in the hydrological cycle

'Future Water', the UK government's water strategy for England states:

Figure 10.2
Maximum surface temperature in the town centre UMT with current form and plus or minus 10% green cover (black dashed line shows 1961–1990 current form temperature) (Gill, 2006).

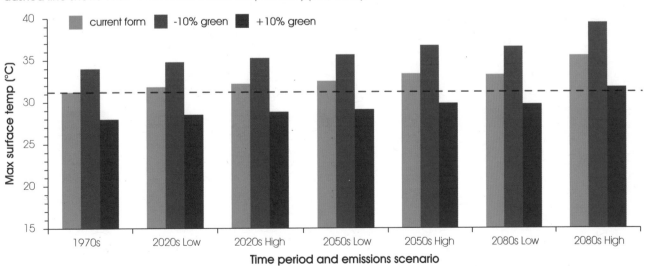

'Climate change is already a major pressure. With predictions for the UK of rising temperatures, wetter winters, drier summers, more intense rainfall events and greater climate variability, we can expect to experience higher water demand, more widespread water stress with increased risk of drought, more water quality problems, as well as more extreme downpours with a greater risk of flooding' (Defra, 2008).

It is clear that climate change impacts on society will be powerfully mediated through change in the hydrological cycle. Water shortage in summer is unlikely to be compensated by excess rainfall in winter which will, in turn, increase flood risk, especially in urban areas. The Foresight Report on Future Flooding (Evans *et al.*, 2003) sought to quantify the economic consequences of increasing flood risk. Their work was updated and summarised by Sir Michael Pitt in his review of the June/July floods of 2007 (Pitt, 2008). The new analysis indicates the potential for even warmer and wetter winters, together with summers that are also warmer but not quite so dry as previously predicted. The increased intensity of rainfall, in both winter and summer, increases the risk from intra-urban (i.e. surface water) flooding in urban areas.

Woodland is potentially beneficial in helping society to adapt to climate change because of its ability to intercept rainfall, some of which is evaporated back to the atmosphere before it reaches the forest floor by stemflow and throughfall. Within woodland, the infiltration of that water is more effective than under alternative types of land cover and forest soils tend to be deeper, and therefore contain a greater storage capacity. It follows that woodland may help to moderate peak flows of water in high rainfall events, while sustaining infiltration to aquifers and baseflow in rivers during periods of drought. Thus, commentators such as Seppälä (2007), when writing about forestry and climate change are able to assert that 'forests maintain

much of the water supply, and trees make a contribution to water management and hence reduce the threat of flooding and erosion'. The extent to which such claims can be sustained in the face of scientific evidence, and their relevance in a UK context is, however, very much open to debate (Newson and Calder, 1989; McCulloch and Robinson, 1993; Calder and Aylward, 2006; Calder, 2007). The position is well summarised by Calder (2007) in a paper which evaluates forest benefits against water costs and argues that forest management programmes need to be set in the context of long-term sustainable land and water management.

Some of the adverse impacts of forests on the water cycle identified by Calder (2007) are actually a function of poor forest management and in the UK much has been learnt about how problems of enhanced runoff and sediment yield can be avoided by good husbandry (Forestry Commission, 2003a).

Here, three issues concerning the role of trees and woodland in helping society to adapt to climate change are considered:

1. The impact of trees and woodland on water supply
2. The impact of trees and woodland in moderating flooding
3. The impact of trees and woodland in managing surface water runoff within urban areas.

10.2.1 The impact of trees and woodland on water supply

We have seen that sustaining the quantity and quality of water supply is likely to become a critical issue for society in a changing climate; indeed it already presents a severe challenge in the south and east of England. Water use by trees and woodland and its implications for water supply

Table 10.1
Typical range of annual evaporation losses (mm) for different land covers receiving 1000 mm annual rainfall (Nisbet, 2005).

Land cover	Transpiration	Interception	Total evaporation
Conifers	300–350	250–450	550–800
Broadleaves	300–390	100–250	400–640
Grass	400–600	–	400–600
Heather	200–420	160–190	360–610
Bracken	400–600	200	600–800
Arable*	370–430	–	370–430

* Assuming no irrigation

has been very effectively reviewed by Nisbet (2005). Table 10.1 provides comparative data on annual evaporation losses between different types of forest cover and a range of alternative land covers.

While interception loss tends to be greater for woodland, transpiration rates are somewhat reduced due, among other factors, to more effective stomatal control (Roberts, 1983). The multifarious influences of climate, geology, forest management, design, scale and land cover make it difficult to generalise about the effects of forestry on water resources. Nevertheless, some important distinctions can be drawn between the likely impact of conifers and broadleaves in the uplands and lowlands, respectively (Nisbet, 2005), as shown in Table 10.2. Climate change, with enhanced temperatures in both summer and winter throughout Britain, will increase evaporation, and this, together with seasonal changes in rainfall, could exert a strong influence on forest water use and water yields

(Nisbet, 2002). The drive to plant more forest energy crops for renewable fuel could further increase the threat to water supplies (Calder et al., 2009).

It follows from this discussion that the role of trees and woodland in the management of water resources is likely to become more important in the future as the combination of rising water demand and the likelihood of drier summers generates even greater pressure on water resources. Application of the precautionary principle suggests that the extensive planting of woodland, especially of conifers would require very careful evaluation from the perspective of water yield, particularly over significant aquifers in the English lowlands.

10.2.2 The impact of trees and woodland in moderating flooding

We have seen that in principle, woodland should be

Table 10.2
Impact of woodland on water yield.

Conifer woodland/upland	Broadleaved woodland/upland
<ul><li>Evidence available from major catchment studies in Wales (Plynlimon), England (Coalburn) and Scotland (Balquhidder).</li><li>For every 10% covered by mature forest Calder and Newson (1979), suggest a 1.5–2.0% reduction in water yield.</li><li>More recent evidence suggests impact of well designed, mixed age forest may be somewhat less on whole forest rotation and impact of forest may decline with time (e.g. Hudson et al., 1997)</li><li>Difficult to identify a response to felling of between 20 and 30% of forested catchment.</li></ul>	<ul><li>No, or very limited, research evidence available.</li><li>For secondary (scrub) woodland colonising upland landscapes surmised that the light leaved species involved (e.g. birch and rowan) unlikely to increase interception significantly above moorland (e.g. heather).</li></ul>
Conifer woodland/lowland	**Broadleaved woodland/lowland**
<ul><li>Interception and transpiration loss becomes proportionately more significant as rainfall is reduced.</li><li>Long-term recharge rate reduced by 75% under pine compared with grassland over sandstone in English midlands (Calder et al., 2003).</li><li>For years of average annual rainfall no recharge at all beneath pine forest on this midlands' site (Calder et al., 2003)</li><li>Spruce in Netherlands reduces water recharge by 79% compared with arable land (Van der Salm et al., 2006).</li></ul>	<ul><li>Long-term recharge rate reduced by about half (48%) beneath oak compared with grassland on sandstone in English midlands (Calder et al., 2003).</li><li>Oak in Netherlands reduces water recharge by 64% compared with arable land (Van der Salm et al., 2006)</li><li>Drainage water from beech over chalk in Hampshire estimated to be 13% greater than grassland during 18 month measurement period (Roberts et al., 2001).</li><li>However, grass drainage could exceed that of woodland on chalk in very wet years due to much higher woodland interception loss (Roberts et al., 2001)</li></ul>

effective at reducing overland flow, and therefore peak discharge from catchments, during high rainfall events. Soils under natural forests tend to be relatively porous with high infiltration rates and consequently low rates of surface runoff, and generally exhibit low rates of erosion (Calder and Aylward, 2006).

In a recent European study, Serrano-Muela *et al.* (2008) provide an elegant demonstration of these properties from an undisturbed forest catchment (San Salvador) in the Spanish Pyrenees, which is contrasted with neighbouring deforested catchments. The forest cover significantly moderates flood response until late spring when following recharge of the soil profile and a heavy rainfall event, the flood peak is similar to that of the deforested catchment. Thus, mature forests reduce the number of floods but do not significantly alter the hydrological impacts of extreme rainfall. It is clear that as the severity of the flood increases, the impact of land use change appears to be reduced (Calder and Aylward, 2006). Similarly, the impact of forests on peak flows becomes harder to detect as the geographical scale of inquiry is increased. Robinson *et al.* (2003) studied 28 river basins across Europe sampling a wide range of managed forest types, climates and ground conditions. They concluded that:

'For all the forest types studied the changes to extreme flows will be diluted at the larger basin scale, where forest management is phased across a catchment, or only a part of the basin is forested. Overall, the results from these studies conducted under realistic forest management procedures have shown that the potential for forests to reduce peak and low flows is much less than has often been widely claimed' (Robinson *et al.*, 2003, p. 96).

These findings are broadly in line with a major review of the impacts of rural land use and management on flood generation commissioned by Defra (O'Connell *et al.*, 2004) which assessed and critiqued the available literature. Their conclusions have been neatly summarised by Heath *et al.* (2008):

- The past 50 years have seen a significant intensification of agriculture, with anecdotal evidence that this has had an effect on flood peaks.
- There is evidence from small-scale manipulation experiments that land use/management has a significant effect on runoff at local scales.
- There is very limited evidence that the effects of land use/management can be distinguished at catchment

scales in the face of climate variability.
- There is evidence that surface flow can be reduced by local land management, but effects on flood peaks may depend on spatial and temporal integration to catchment scales.
- It is not possible (at least yet) to rely on rainfall-runoff modelling to predict impacts of land use/management changes.

It seems that while large-scale woodland creation could not be justified on the grounds of flood control alone, there are specific situations where carefully designed woodland planting could be beneficial. These are discussed below.

Planting woodland buffers on compacted upland pastures

High stocking rates in upland pastures have resulted in soil compaction and reduced infiltration. Planting of woodland buffers normal to the direction of flow significantly boosts infiltration and can reduce overland transport of sediment (Carroll *et al.*, 2004; Ellis *et al.*, 2006; Marshall *et al.*, 2009). Recent modelling studies suggest that these effects could reduce local peak flows by 13–48% (Jackson *et al.*, 2008).

Riparian planting along stream sides

The planting of riparian woodland can also attenuate flood peaks by increasing hydraulic roughness and reducing wave velocity (Anderson *et al.*, 2006). It acts in a similar way to floodplain woodland but on a smaller scale. Thomas and Nisbet (2006) showed that the formation of large woody debris dams within stream channels could significantly lengthen local peak flow response times. Planting riparian woodland can also help flood control by reducing bank erosion, sediment delivery and the siltation of flood channels (Nisbet *et al.*, 2004). Other water benefits of riparian woodland are cited as reducing diffuse pollution through the capture of nutrients and pesticides draining from the adjacent land, and alleviating thermal stress to fish by shading (Calder *et al.*, 2008; Parrott and Holbrook, 2006).

Woodland establishment on disused and derelict land in the urban and peri-urban environment

Runoff modelling under climate change scenarios (Gill, 2006) suggests that derelict land is potentially significant in managing surface water flows in the urban and peri-urban environment. Such areas can act as a source or sink for surface water depending on their origin, condition, management and, when reclaimed, the nature of the after-use. Certain categories of disued and derelict land, e.g. colliery spoil, are often highly compacted (Moffat and McNeil, 1994); runoff from such areas is extremely rapid and can contribute to surface water flooding of residential areas. Afforestation not only relieves compaction, so improving infiltration and soil storage, but reduces runoff through interception and transpiration. The potential for woodland on derelict and disued land has been reviewed by Perry and Handley (2000) who identified a wide range of opportunities, including closed landfill sites. Bending and Moffat (1997) have shown that tree growth can be compatible with landfill integrity and here, woodland establishment could be particularly beneficial in reducing the potential for contaminative excess runoff.

Flood plain forests to increase storage and attenuate flow

The recreation of floodplain woodland to delay the progression of floods may offer the greatest potential for forestry to assist flood control. The potential of floodplain woodland to attenuate flood peaks, as well as delivering associated benefits such as water quality, biodiversity, fisheries, recreation and landscape have been highlighted by Kerr and Nisbet, 1996. The flood attenuation principle relies on the hydraulic roughness created by woody debris dams within stream channels and the physical presence of trees on the flood plain. The net effect is to reduce flood velocities, enhance out of bank flow, and increase water storage on the floodplain so, potentially, moderating the downstream flood impact. Hydraulic modelling studies in southwest England indicate that planting woodland across the flood plain could have a marked effect on flood flows (Thomas and Nisbet, 2006). Implementation will need to address concerns such as flooding of local properties due to back-up of flood water and blockage of downstream structures by woody debris. Moreover, the attenuation of flood-flows in one sub-catchment will change the synchrony of the drainage network (which could be either problematic or highly beneficial) and therefore the location of floodplain forest needs to be carefully designed so as to reduce flood risk, rather than accentuate it.

10.2.3 The impact of trees and woodland in managing surface runoff within urban areas

Urbanisation of a catchment fundamentally changes its hydrological character, due to extensive surface sealing with reduced infiltration and enhanced surface runoff; much of that via the sewer network (Bridgman

et al., 1995). The effect of progressive urbanisation and deforestation on simulated flood frequency curves for a naturally wooded watershed is shown in Figure 10.3 (Wissmar *et al.,* 2004).

Figure 10.3
Simulated flood-frequency curves (m³ s⁻¹) for Maplewood Creek. Flood-frequency curves indicate that annual flood discharge rates for 1991 and 1998 exceed pre-settlement discharge rates. Symbols: open triangles, the flood-frequency curve for 1998; open squares, the 1991 flood-frequency curve; solid circles, the flood-frequency curve for pre-settlement conditions. Pre-settlement conditions represent fully forested cover and no impervious surfaces.

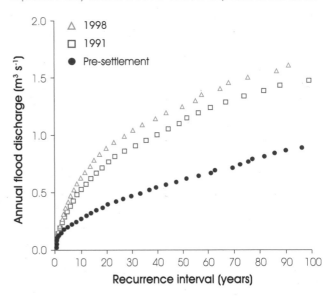

With kind permission from Springer Science + Business Media: Environmental Management, Effects of changing forest and impervious land covers on discharge characteristics of watersheds, 34, 2004, 91–98, Wissmar, R.C., Timm, R.K. and Logsdon, M.G., Figure 2.

The profound implications of urbanisation for catchment behaviour are well illustrated by Bronstert *et al.,* (2002) who simulate the effect of two contrasting flood events on the Lein catchment (Germany). As shown in Figure 10.4, for high precipitation intensities, an increase in settlement leads to markedly higher peak flows and flood volumes, whereas the influence of urbanisation on advective rainfall events with lower precipitation intensities and higher antecedent soil moisture is much smaller. In this example, the increase in stream runoff for the convection event is mainly due to an increase in sewer overflow in settlement areas and infiltration-excess overload flow on farmland. This type of short duration, high intensity rainfall is likely to become more frequent with climate change, and the Pitt review (Pitt, 2008) highlights the potential for enhanced intra-urban or surface water flooding. However, the implications of changing patterns of rainfall on urban

Figure 10.4
Simulation of two flood events in the Lein catchment with a return period of approximately three years for present conditions and two urbanisation scenarios (10% and 50% increase in settlement area): (a) convective storm event; (b) advective storm event (Bronstert *et al.*, 2002).

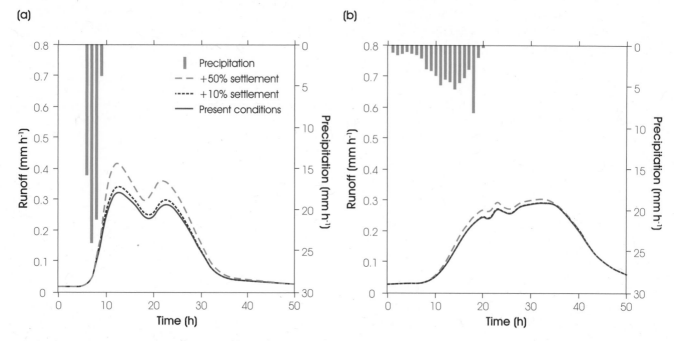

With kind permission from John Wiley & Sons Ltd: Hydrological Processes, Effects of climate and land-use change on storm runoff generation: present knowledge and modelling capabilities, 16, 2002, 509–529, Bronstert, A., Niehoff, D. and Bürger, G.

drainage are even more pervasive, with significant consequences for water quality and waste-water treatment (Balmforth, 2002).

In the USA, conservation of native woodland is considered to be one of the most effective measures for countering the quantitative and qualitative impacts of urbanisation on river systems (Booth *et al.*, 2002). Indeed Matteo *et al.* (2006) drawing on the results of modelling studies, advocate the creation of woodland buffers along streams and highways with a view to handling adverse conditions such as large storms, non-point source pollution and flooding. Those characteristics of tree cover which are potentially problematic for water supply (notably rainfall interception) could be beneficial in managing flood flows in the urban environment. In the urban ecosystem, canopy rainfall interception changes the urban runoff process by reducing the flow rate and shifting the runoff concentration time via water storage on the canopy surface (Sanders, 1986). The high ventilation of the built environment could be expected to enhance woodland interception losses beyond the typical values of 25–45% for conifers and 10–25% for broadleaves (Nisbet, 2005). A decrease in storm runoff volume reduces flooding hazard, surface pollutant wash-off and pollutant loading of the runoff – all key features of summer storm impacts on urbanised areas

in Britain. Having studied in detail rainfall interception by mature open-grown trees in Davis, California, Xiao and his co-workers (2000, p. 782) conclude that interception losses may be even 'higher in places that have frequent summer rainfall and warm, sunny conditions'. These properties are important because urban runoff reduction ultimately reduces expenditure on urban runoff control and waste-water treatment. These financial benefits have been quantified for settlements in California (Xiao *et al.*, 2002) and the research findings used to inform the design and management of the urban forest resource.

By contrast with the USA, the capacity of trees in the urban environment (i.e. the urban forest) to store and also to infiltrate water (Bartens *et al.*, 2008) tends not to have been recognised as part of sustainable urban drainage systems within the UK. Modelling in the Greater Manchester area (Gill, 2006; Gill *et al.*, 2007) suggests that increasing or decreasing the tree cover has an effect on runoff, especially on more porous soils. While under the type of extreme rainfall event that is envisaged in a changed climate, the capacity of urban trees to counter increased runoff is exceeded (Gill *et al.*, 2007), it does not follow that urban trees do not have a significant role to play within the storm water chain, especially under less extreme conditions. Research findings from the USA would suggest within the storm water chain, especially under less extreme conditions. Research findings from the USA would suggest that broadleaved trees in British cities are already playing an important role in moderating runoff and protecting water quality. This role is likely to grow in importance, especially in summer, when Britain's broadleaved deciduous trees are best placed to intercept rainfall in intermittent summer storms.

10.3 Managing trees and woodlands to optimise benefits to society and reduce climate-related hazards

The vision in Defra's Strategy for England's Trees, Woods and Forests begins: 'It is 2050, and England's trees, woods and forests are helping us to cope with the continuing challenge of climate change …' (Defra, 2007, p. 10). The climate change adaptation roles of trees and woodlands, discussed in the previous sections, will determine, to some extent, where we want our trees and woodlands to be in order to optimise the benefits to society. Trees and woodlands in rural areas have some role to play in helping society to adapt, in particular through strategic planting to manage riverine flood risk, water resources and quality, as well as helping other species to adapt and

realising opportunity in visitor landscapes. However, the evidence set out above strongly makes a case for trees and woodland to be located where people are most concentrated; in the peri-urban and urban environments. It is here, where, along with the other adaptation roles, they can contribute significantly to managing high temperatures and reducing pressure on drainage systems.

In contrast to the USA, where tools to characterise the structure, function and economic benefits of urban forests have been developed, and assessments have been undertaken for many major cities (e.g. American Forests, 2009; Nowak, 2008), little is known in the UK, and indeed in Europe (Konijnendijk, 2003), about this urban forest. The 'Trees in towns' surveys highlight how little local authorities know about the urban tree populations in their district (Britt and Johnston, 2008; Land Use Consultants, 1993). Of the sites surveyed for 'Trees in towns II', the mean canopy cover was 8.2%. While town size had no effect on this, it does vary within urban areas according to land use; from 3.6% in industrial and high density residential areas to 22.8% in low density residential areas. A characterisation of Greater Manchester found that medium-density residential areas are especially important as they account for 37% of the 'urbanised' area, such that just under 30% of all the tree cover found in the urban areas occur here (Figure 10.5) (Gill et al., 2008; Gill, 2006).

Figure 10.5
Percentage of all surface covered by trees across 'urbanised' Greater Manchester (from data generated for Gill, 2006).

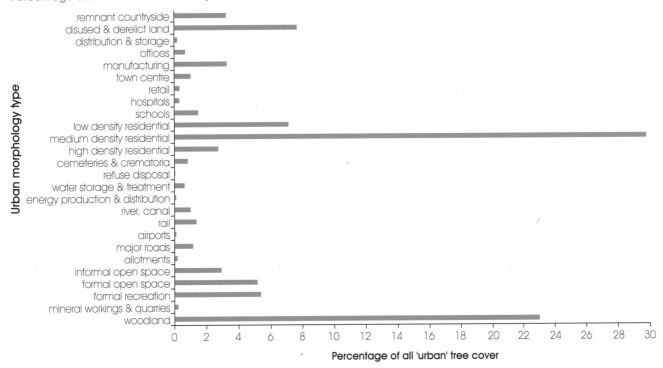

It is imperative that the trees and woodlands in urban areas are strategically planned and managed. Many local authorities lack basic information about the nature and extent of trees and woodlands in their district and, while a substantial number have produced tree strategies, these are often seriously deficient in terms of their content and detail (Britt and Johnston, 2008). 'Trees in towns II' calls for local authorities to develop and implement a comprehensive tree strategy, and that it is beneficial to think beyond trees to the wider context of urban green space and the environment (Britt and Johnston, 2008). This chimes well with recent moves towards green infrastructure planning, with trees and woodlands being significant components (e.g. North West Green Infrastructure Think Tank, 2008; Kambites and Owen, 2006).

There is clearly a need to bring tree cover into the most built up parts of our urban areas, including town centres and high-density residential areas. In addition, the most vulnerable members of society often live in areas with the lowest tree cover (e.g. Tame, 2006; Pauleit et al., 2005). However, the Trees and Design Action Group has highlighted the huge gap between aspirations for more and larger trees in the urban realm and practical considerations which are creating a landscape devoid of large tree species. They are championing a 'new culture of collaborative working that places trees and their requirements at the forefront of the decision-making process' (Trees and Design Action Group, 2008, p. 1). This echoes a previous call by the Royal Commission on Environmental Pollution for the natural environment to be at the heart of urban design and management (RCEP, 2007). Indeed, certain perceived tree hazards could be compounded further by climate change. Below, we explore three of these: building and/or infrastructure subsidence, air quality (ozone precursors) and windthrow.

10.3.1 Building and/or infrastructure subsidence

On shrink-swell clay soils in particular, changes in soil moisture content result in dimensional changes in the soil (Percival, 2004). Soil moisture content varies with season, and trees can add to this change (Roberts et al., 2006). If the dimensional changes in the soil occur below the foundation level of buildings, this can result in subsidence damage. While both the intensity and frequency of shrink-swell soil hazards may increase with climate change, the spatial extent is unlikely to change (Forster and Culshaw, 2004).

A street tree survey in London revealed that about 5%

of all trees removed in the previous five years were the result of subsidence claims, with some boroughs reporting losses of 10–40% (London Assembly, 2007). However, the perceived threat of subsidence may be much greater than the actual threat (GLA, 2005). Biddle (1998) has suggested that, while tree roots are involved in at least 80% of subsidence claims on shrinkable clay soils, even on clay soils the risk of a tree causing damage is less than 1%. In addition to the removal of trees, subsidence fears may also lead to the planting of smaller tree species (London Assembly, 2007). Given the importance of trees in the urban environment, a proper understanding is required of the mechanism of damage, how this can be prevented, and appropriate remedies if damage occurs (TDAG, 2008; Biddle, 1998).

10.3.2 Trees and air quality

As a result of their greater leaf areas and of the air turbulence created by their structure, trees and woodlands take up more gaseous pollutants, aerosols and particulates than shorter vegetation. While this can increase nitrogen and sulphur interception with effects on water quality (see preceding discussion), it can also improve air quality in urban areas (Beckett et al., 2000). A number of UK studies have identified small potential decreases in urban air concentrations of NO_2, SO_2 and O_3 (e.g Broadmeadow and Freer-Smith, 1996) however, because of the known adverse effects of particulate pollution on human health, attention has recently focused on the uptake of particles by urban greenspace. Health benefits arising from improved urban air quality have been included as one of a number of the economic benefits of urban greenspace in the UK (Willis and Osman, 2005).

While all trees and woodland can improve air quality through the deposition of ozone, nitrogen dioxide, carbon monoxide, and nitric acid, certain tree species emit volatile organic compounds (VOCs), such as isoprene and monoterpenes. This can contribute to the formation of secondary pollutants such as ozone due to the reaction of VOCs and nitrogen oxides in the presence of sunlight (see Chapter 3). Donovan et al. (2005) have published a model which considers both pollutant uptake and VOC emissions and identifies the potential of different tree species to improve urban air quality.

10.3.3 Windthrow in storms

How wind patterns are likely to alter with climate change is poorly understood (Hulme et al., 2002). Due to inconsistencies between models and the physical representation within them, the UKCIP02 climate scenarios were unable to attach any confidence level to the projections for wind speed and urged for caution in interpreting changes in wind speed (Hulme *et al.,* 2002). The UKCIP02 scenarios suggested little change in average spring and autumn wind speed and stronger winter winds in southern and central Britain.

In London, the vast majority of trees that have been removed in the last five years have been for health and safety reasons, including trees that have been damaged by storms and pose a risk from falling branches (Trees and Design Action Group, 2008). However, given that there is no robust understanding of wind speed changes with climate change, it may be most appropriate to follow existing guidance on planting and managing trees in urban environments.

10.4 Research priorities

- As the UK climate changes trees in cities and in urban greenspace will become increasingly important in managing temperatures, surface water and air quality. In the UK (and Europe) decision support systems are required to integrate understanding and to characterise the structure, function and economic benefits of urban and peri-urban trees and woodlands.
- It is important that trees and woodlands in urban areas are strategically planned and managed. Most UK local authorities lack the basic information on the nature and extent of trees and woodlands in their districts. This information gap needs to be addresses urgently and urban and peri-urban trees and woodlands should be included in national forest inventories.
- The interactions between interception of precipitation by trees, urban tree effects on soil infiltration and sustainable urban drainage need to be better understood. Information is required to identify the optimum tree and woodland component for the planning and design of green infrastructure, and planting along urban watercourses, on derelict and disused land (urban and peri-urban) requires more systematic consideration. The role of woodlands and forests in flood management requires further research but forests have an important role in the management of water resources.

References

AKBARI, H. (2002). Shade trees reduce building energy use and CO_2 emissions from power plants. *Environmental Pollution* **116**, 119–126.

ANDERSON, B.G., RUTHERFORD, I.D. and WESTERN, A.W. (2006). An analysis of the influence of riparian vegetation on the propagation of flood waves. *Environmental Modelling and Software* 21, 1290–1296.

AMERICAN FORESTS (2009). CITYgreen. Online at: www.americanforests.org/productsandpubs/citygreen/ (accessed 7 May, 2009).

ANDERSON, H.R., DERWENT, R.G. and STEDMAN, J. (2002). Air pollution and climate change. In: *Health effects of climate change*. Department of Health, London.

BALMFORTH, D. (2002). Climate change and SUDS (Scottish Hydraulics Study Group) In: Mansell, M.G. (ed.) *Rural and urban hydrology*. Thomas Telford, London, p. 384.

BARTENS, J., DAY, S.D., HARRIS, J.R., DOVE, J.E. and WYNN, T.M. (2008). Can urban tree roots improve infiltration through compacted subsoils for stormwater management? *Journal of Environmental Quality* 37, 2048–2057.

BECKETT, K.P., FREER-SMITH P.H. and TAYLOR, G. (2000). Particulate pollution capture by urban trees: effect of species and wind speed. *Global Change Biology* 6, 995–1003

BENDING, N.A.D. and MOFFAT, A.J. (1997). *Tree establishment on landfill sites: research and updated guidance*. Report to DETR. Forestry Commission, Edinburgh.

BERLIN DIGITAL ENVIRONMENTAL ATLAS (2009). Online at: www.stadtentwicklung.berlin.de/umwelt/umweltatlas/edua_index.shtml (accessed 7 May 2009).

BIDDLE, P.G. (1998). *Tree root damage to buildings – volume 1: causes, diagnosis and remedy*. Willowmead, Wantage.

BOOTH, D.B., HARTLEY, D. and JACKSON, R. (2002). Forest cover, impervious-surface area and the mitigation of stormwater impacts. *Journal of the American Water Resources Association* 38, 835–845.

BRIDGMAN, H.A., WARNER, R.F. and DODSON, J. (1995). *Urban biophysical environments*. Oxford University Press, Oxford.

BRITT, C. and JOHNSTON, M. (2008). *Trees in towns II – A new survey of urban trees in England and their condition and management*. Department for Communities and Local Government, London.

BROADMEADOW, M.S.J. and FREER-SMITH, P.H (1996). *Urban woodland and the benefits for local air quality*. Research for amenity trees No. 5. The Stationery Office, London.

BRONSTERT, A., NIEHOFF, D. and BÜRGER, G. (2002). Effects of climate and land-use change on storm runoff generation: present knowledge and modelling capabilities.

Hydrological Processes **16**, 509–529.

CALDER, I.R. (2007). Forests and water – ensuring forest benefits outweigh water costs. *Forest Ecology and Management* **251**, 110–120.

CALDER, I.R. and NEWSON, M.D. (1979). Land use and upland water resources in Britain – a strategic look. *Water Resources Bulletin* **16**, 1628–1639.

CALDER, I.R., REID, I., NISBET, T.R. and GREEN, J.C. (2003). Impact of lowland forests in England on water resources – application of the HYLUC model. *Water Resources Research* **39**, 1319–1328.

CALDER, I.R. and AYLWARD, B. (2006). Forests and floods: moving to an evidence-based approach to watershed and integrated flood management. *Water International* **31**, 87–99.

CALDER, I.R., HARRISON, J., NISBET, T.R. and SMITHERS, R.J. (2008). *Woodland actions for biodiversity and their role in water management*. Woodland Trust, Grantham, Lincolnshire.

CALDER, I.R., NISBET, T.R. and HARRISON, J. (2009). An evaluation of the impacts of energy tree plantations on water resources in the UK under present and future UKCIP02 climate scenarios. *Water Resources Research* **45**, W00A17. doi:10.1029/2007WR006657

CARROLL, Z.L., BIRD, S.B., EMMETT, B.A., REYNOLDS, B. and SINCLAIR, F.L. (2004). Can tree shelterbelts on agricultural land reduce flood risk? *Soil Use and Management* **20**, 357–359.

DAVIS, D.L. and TOPPING, J.C. JR. (2008). Potential effects of weather extremes and climate change on human health. In: MacCracken, M.C., Moore, F. and Topping, J.C. Jr. (eds) *Sudden and disruptive climate change: exploring the real risks and how we can avoid them*. Earthscan, London. pp. 39–42.

DEFRA (2007). *A strategy for England's trees, woods and forests*. Department for Environment, Food and Rural Affairs, London.

DEFRA (2008). *Future water: the Government's water strategy for England*. Cm. 7319. The Stationery Office, London.

DONOVAN, R.G., STEWART, H.E., OWEN, S.M., MACKENZIE, A.R., HEWITT, C.N. (2005). Development and application of an urban tree air quality score for photochemical pollution episodes using the Birmingham, UK, area as a case study. *Environmental Science and Technology* **39**, 6730–6738.

ELIASSON, I. (2000). The use of climate knowledge in urban planning. *Landscape and Urban Planning* **48**, 31–44.

ELLIS, T.W., LEGNÉDOIS, S., HAIRSINE, P.B. and TONGWAY, D.J. (2006). Capture of overland flow by a tree belt on a pastured hillslope in south-eastern Australia. *Australian Journal of Soil Research* **44**, 117–125.

EVANS, E.P., THORNE, C.R., SAUL, A., ASHLEY, R., SAYERS, P.N., WATKINSON, A., PENNING-ROWSELL, E.C. and HALL, J.W. (2003). *Future flooding: an analysis of future risks of flooding and coastal erosion for the UK between 2030 and 2100*. Office of Science and Technology, Department of Trade and Industry, London.

FORESTRY COMMISSION (2003). *Forests and water guidelines*. 4th edn. Forestry Commission, Edinburgh.

FORSTER, A. and CULSHAW, M. (2004). Implications of climate change for hazardous ground conditions in the UK. *Geology Today* **20**, 61–66.

GILL, S.E. (2006). *Climate change and urban greenspace*. PhD thesis, University of Manchester.

GILL, S.E., HANDLEY, J.F., ENNOS, A.R. and PAULEIT, S. (2007). Adapting cities for climate change: the role of the green infrastructure. *Built Environment* **33**, 115–133.

GILL, S.E., HANDLEY, J.F., ENNOS, A.R., PAULEIT, S., THEURAY, N. and LINDLEY, S.J. (2008). Characterising the urban environment of UK cities and towns: a template for landscape planning. *Landscape and Urban Planning* **87**, 210–222.

GREATER LONDON AUTHORITY (GLA). (2005). *Connecting Londoners with trees and woodlands: a Tree and Woodland Framework for London*. GLA, London.

GULYÁS, A., UNGER, J. and MATZARAKIS, A. (2006). Assessment of the microclimatic and human comfort conditions in a complex urban environment: modelling and measurements. *Building and Environment* **41**, 1713–1722.

HACKER, J.N. and HOLMES, M.J. (2007). Thermal comfort: climate change and the environmental design of buildings in the United Kingdom. *Built Environment* **33**, 97–114.

HEATH, T., HAYCOCK, N., EDEN, A., HEMSWORTH, M. and WALKER, A. (2008). *Evidence based review: does land management attenuate runoff?* Report to the National Trust, Haycock Associates, Pershore, Worcestershire.

HOUGH, M. (2004). *Cities and natural process: a basis for sustainability*. 2nd edn. Routledge, London.

HUANG, Y.J., AKBARI, H., TAHA, H. and ROSENFELD, A.H. (1987). The potential of vegetation in reducing summer cooling loads in residential building. *Journal of Climate and Applied Meteorology* **26**, 1103–1116.

HUDSON, J.A., CRANE, S.B. and BLACKIE, J.R. (1997). The Plynlimon water balance 1969–1995: the impact of forest and moorland vegetation on evaporation and streamflow in upland catchments. *Hydrology and Earth Sciences* **1**, 409–427.

HULME, M., JENKINS, G., LU, X., TURNPENNY, J., MITCHELL, T., JONES, R., LOWE, J., MURPHY, J., HASSELL, D., BOORMAN, P., MCDONALD, R. and HILL, S. (2002). *Climate change scenarios for the United Kingdom*. The UKCIP02 Scientific Report. Tyndall Centre for Climate Change Research, School of Environmental Sciences, University of East Anglia, Norwich.

IPCC (2007). Climate change 2007: Impacts, adaptation and vulnerability. In: Perry, M.L., Canziani, O.F., Palutikof, J.P., Van der Linden, P.J. and Hanson, C.E. (eds) *Contribution of Working Group II to the Fourth Assessment Report of the Intergovernmental Panel on Climate Change*. Cambridge University Press, Cambridge.

JACKSON, B.M., WHEATER, H.S., MCINTYRE, N.R., CHELL, J., FRANCIS, O.J., FROGBROOK, Z., MARSHALL, M., REYNOLDS, B. and SOLLOWAY, I. (2008). The impact of upland land management on flooding: insights from a multiscale experimental and modelling programme. *Journal of Flood Risk and Management* 1, 71–80.

KAMBITES, C. and OWEN, S. (2006). Renewed prospects for green infrastructure planning in the UK. *Planning, Practice and Research* 21, 483–496.

KERR, G. and NISBET, T.R. (1996). *The restoration of floodplain woodlands in lowland Britain: a scoping study and recommendations for research*. Environment Agency, Bristol.

KONIJNENDIJK, C.C. (2003). A decade of urban forestry in Europe. *Forest Policy and Economics* 5, 173–186.

KOSATSKY, T. (2005). The 2003 European heatwave. *European Surveillance* 10, 148–149.

LAND USE CONSULTANTS (1993). *Trees in towns*. Department of the Environment, London.

LONDON ASSEMBLY (2007). *Chainsaw massacre: a review of London's street trees*. Greater London Authority, London.

MARSHALL, M.R., FRANCIS, O.J., FROGBROOK, Z.L., JACKSON, B.M., MCINTYRE, N., REYNOLDS, B., SOLLOWAY, I., WHEATER, H.J. and CHELL, J. (2009). The impact of upland land management on flooding: results from an improved pasture hillslope. *Hydrological Processes* 23, 464–475.

MATTEO, M., RANDHIR, T. and BLONIARZ, D. (2006). Watershed-scale impacts of forest buffers on water quality and runoff in urbanizing environment. *Journal of Water Resource Planning and Management* 132, 144–152.

MAYER, H and HÖPPE, P. (1987). Thermal comfort of man in different urban environments. *Theoretical and Applied Climatology* 38, 43–49.

McCULLOCH, J.S.G. and ROBINSON, M. (1993). History of forest hydrology. *Journal of Hydrology* 150, 189–216.

McPHERSON, E.G. and SIMPSON, J.R. (2003). Potential energy savings in buildings by an urban tree planting programme in California. *Urban Forestry and Urban Greening* 2, 73–86.

MEEHL, G.A. and TEBALDI, C. (2004). More intense, more frequent and longer lasting heatwaves in the 21st Century.

Science 305, 994–997.

MOFFAT, A.J. and MCNEIL, J. D. (1994). Reclaiming disturbed land for forestry. *Forestry Commission Bulletin 110.* HMSO, London.

NEWSON, M.D. and CALDER, I.R. (1989). Forests and water resources: problems of prediction on a regional scale. *Philosophical Transactions of the Royal Society of London B Series* 324, 283–298.

NISBET, T.R. (2002). Implications of climate change: soil and water. In: Broadmeadow, M. (ed.) *Climate change: impacts on UK forests. Forestry Commission Bulletin 125*. Forestry Commission, Edinburgh. pp. 53–68.

NISBET, T.R., ORR, H.G. and BROADMEADOW, S. (2004). *Evaluating the role of woodlands in managing soil erosion and sedimentation within river catchments: Bassenthwaite Lake study*. Report to the Forestry Commission, Forest Research, Farnham.

NISBET, T. (2005). *Water use by trees*. Forestry Commission Information Note 65. Forestry Commission, Edinburgh.

NORTH WEST GREEN INFRASTRUCTURE THINK TANK (2008). *North West Green Infrastructure Guide*. Online at: www.greeninfrastructurenw.co.uk/resources/GIguide.pdf

NOWAK, D.J. (2008). Assessing urban forest structure: summary and conclusions. *Arboriculture and Urban Forestry* 34, 391–392.

O'CONNELL, P.E., BEVEN, K.J., CARNEY, J.N., CLEMENTS, R.O., EWEN, J., FOWLER, H., HARRIS, G.L., HOLLIS, J., MORRIS, J., O'DONNELL, G.M., PACKMAN, J.C., PARKIN, A., QUINN, P.F., ROSE, S.C., SHEPHERD, M. and TELLIER, S. (2004). *Review of impacts of rural land use and management on flood generation: Impact study report*. R&D Technical Report FD2114/TR, DEFRA, London.

OKE, T.R. (1989). The micrometeorology of the urban forest. *Philosophical Transactions of the Royal Society of London B Series* 324, 335–349.

PARROTT, J. and HOLBROOK, J. (2006). *Natural Heritage Trends: riparian woodlands in Scotland – 2006*. Scottish Natural Heritage Commissioned Report No. 204 (ROAME No. FOINB02a) SNH, Edinburgh.

PAULEIT, S., ENNOS, R. and GOLDING, Y. (2005). Modelling the environmental impacts of urban land use and land cover change – a study in Merseyside, UK. *Landscape and Urban Planning* 71, 295–310.

PERCIVAL, G. (2004). Tree roots and buildings. In: Hitchmough, J. and Fieldhouse, K. (eds) *Plant user handbook: a guide to effective specifying*. Blackwell Science, Oxford. pp. 113–127.

PERRY, D. and HANDLEY, J.F. (2000). The potential for woodland on urban and industrial wasteland in England and Wales. *Forestry Commission Technical Paper 29*. Forestry Commission, Edinburgh.

PITT, M. (2008). *The Pitt Review: Learning lessons from the 2007 floods.* An independent review by Sir Michael Pitt. Cabinet Office, London.

RENAUD, V. and REBETZ, M. (2009). Comparison between open-site and below-canopy climatic conditions in Switzerland during the exceptionally hot summer of 2003. *Agricultural and Forest Meteorology* 149, 873–880.

ROBERTS, J.M. (1983). Forest transpiration: a conservative hydrological process? *Journal of Hydrology* 66, 133–141.

ROBERTS, J.M., ROSIER, P.T.W. and SMITH, D.M. (2001). *Effects of afforestation on chalk groundwater resources.* Centre for Ecology and Hydrology Report to the Department for Environment, Food and Rural Affairs (DEFRA). Centre for Ecology and Hydrology, Wallingford.

ROBERTS, J., JACKSON, N., SMITH, M. (2006). *Tree roots in the built environment.* Research for Amenity Trees No. 8. Department for Communities and Local Government. The Stationery Office, London.

ROBINSON, M., COGNARD-PLANCQ, A.L., COSANDEY, C., DAVID, J., DURAND, P., FÜHRER, H.-W., HALL, R., HENDRIQUES, M.O., MARC, V., McCARTH, R., McDONNELL, M., MARTIN, C., NISBET, T., O'DEA, P., RODGERS, M. and ZOLLINER, A. (2003). Studies of the impact of forests on peak flows and baseflows: a European perspective. *Forestry Ecology and Management* 186, 85–97.

ROYAL COMMISSION ON ENVIRONMENTAL POLLUTION (RCEP) (2007). *The urban environment.* HMSO, London.

SANDERS, R.A. (1986). Urban vegetation impacts on the hydrology of Dayton, Ohio. *Urban Ecology* 9, 361–376.

SEPPÄLÄ, R. (2007). Global forest sector: trends, threats and opportunities. In: Freer-Smith, P.H., Broadmeadow, M.S.J. and Lynch, J.M. (eds) *Forestry and climate change.* CABI, Wallingford. pp. 25–30.

SERRANO-MUELA, M.P., LANA-RENAULT, N., NADAL-ROMERO, E., REGÜÉS, D., LATRON, J., MARTI-BONO, C., and GARCIA-RUIZ, J.M. (2008). Forests and their hydrological effects in Mediterranean mountains. *Mountain Research and Development* 28, 279–285.

SHAW, R., COLLEY, M. and CONNELL, R. (2007). *Climate change adaptation by design: a guide for sustainable communities.* Town and Country Planning Association, London.

SIMPSON, J.R. (2002). Improved estimates of tree-shade effects on residential energy use. *Energy and Buildings* 34, 1067–1076.

SPRONKEN-SMITH, R.A. and OKE, T.R. (1998). The thermal regime of urban parks in two cities with different summer climates. *International Journal of Remote Sensing* 19, 2085–2104.

STOTT, P.A., STONE, D.A. and ALLEN, M.R. (2004). Human contribution to the European heatwave of 2003. *Nature* 432, 610–614.

TAME, I.D. (2006). *Developing an intervention plan to challenge the environmental inequity of urban trees.* MPlan thesis, University of Manchester.

THOMAS, H. and NISBET, T.R. (2006). An assessment of the impact of floodplain woodland on flood flows. *Water and Environment Journal* 21, 114–126.

TREES and DESIGN ACTION GROUP (TDAG) (2008). *No trees, no future – trees in the urban realm.* TDAG, London.

VAN DER SALM, C., DENIER VAN DER GON, H., WIEGGERS, R., BLECKER, A.B. and VAN DEN TOORN, A. (2006). The effect of afforestation on water recharge and nitrogen leaching in the Netherlands. *Forest Ecology and Management* 221, 170–182.

WATKINS, R., PALMER, J. and KOLOKOTRONI, M. (2007). Increased temperature and intensification of the urban heat island: implications for human comfort and urban design. *Built Environment* 33, 85–96.

WEST, C.C. and GAWITH, M.J. (eds) (2005). *Measuring progress: preparing for climate change through the UK Climate Impacts Programme.* UKCIP, Oxford.

WILBY, R.L. (2003). Past and projected trends in London's urban heat island. *Weather* 58, 251–260.

WILBY, R.L. (2007). A review of climate change impacts on the built environment. *Built Environment* 33, 31–45.

WILLIS, K. and OSMAN L. (2005). Economic benefits of accessible green spaces for physical and mental health. CJC consulting report to the Forestry Commission. Online at: www.forestry.gov.uk

WILSON, E., NICOL, F., NANAYAKKARA, L. and UEBERJAHN-TRITTA, A. (2008). Public urban open space and human thermal comfort: the implications of alternative climate change and socio-economic scenarios. *Journal of Environmental Policy and Planning* 10, 31–45.

WISSMAR, R.C., TIMM, R.K. and LOGSDON, M.G. (2004). Effects of changing forest and impervious land covers on discharge characteristics of watersheds. *Environmental Management* 34, 91–98.

XIAO, Q., McPHERSON, E.G., USTIN, S.L., GRISMER, M.E. and SIMPSON, J.R. (2000). Winter rainfall interception by two mature open-grown trees in Davis, California. *Hydrological Processes* 14, 763–784.

XIAO, Q. and McPHERSON, E.G. (2002). Rainfall interception by Santa Monica's municipal urban forest. *Urban Ecosystems* 6, 291–302.

SECTION 5
SUSTAINABLE DEVELOPMENT

0°   1°   2°   3°   4°

FORESTRY, CLIMATE CHANGE AND SUSTAINABLE DEVELOPMENT

Chapter

11

P. Snowdon

Key Findings

Forests and timber offer many sustainable credentials through their capacity to generate economic, social and environmental benefits. The contribution of forests to tackling climate change must not be seen in isolation from other benefits provided by sustainable forest management. Integrated approaches are needed, since managing forests with the main objective of reducing net carbon emissions may imply trade-offs with other socially-desirable objectives of forestry. Local and regional conditions and knowledge are required in developing forest management solutions that help to tackle climate change while meeting the needs of sustainable development.

Sustainable development[1] is one of the principal objectives of the UK Government. It is vital, therefore, that policies and actions in the forestry sector on climate change mitigation and adaptation contribute to the objectives of sustainable development, and to the principle of sustainable forest management.

At the same time, the imperative of tackling climate change means that changes may be required in the coming years both to the balance of forest policy objectives and to the management practices that underpin them. The introduction of a Climate Change Guideline supporting the UK Forest Standard (see 1.5.1, Chapter 1) demonstrates the importance of this issue.

The Government's approach to sustainable development is set out in its strategy, Securing the Future, published in 2005. The UK and devolved administrations have developed a common conceptual framework for sustainable development. This is shown in Figure 11.1. The framework is supported by separate strategies for each administration that reflect their priorities and specific needs.

The framework identifies two outcomes; first, '*living within environmental limits*' and, second, '*ensuring a strong, healthy and just society*'. Arguably, the development of forest policy and practice in recent decades has done much to support these outcomes through, for example, the contribution of woodlands to biodiversity, recreation and amenity and through an increasing focus on woodland creation in and around towns and cities (see Forestry Commission, 2009). The forestry strategies for England, Scotland, Wales and Northern Ireland, and the environmental and social objectives of the UK Forestry Standard, show this contribution at a strategic level. Both of these outcomes can be consistent with managing forests for climate change. For example, protecting forests as environmental assets is integral to the analysis of climate change impacts and adaptation as shown in Section 2 and Section 4 respectively, while the social contribution of woodlands is made clear in the examination in Chapters 10 and 13 of how woodlands help society to adapt to climate change.

The sustainable development framework identifies three actions for achieving the outcomes.

- **Creating a sustainable economy.** This action envisages an economy that attaches a full value to the natural environment (including full pricing of carbon) and to the benefits that it provides for people. Economic prosperity is seen within the framework as a way of achieving target outcomes above rather than an end in itself.
- **Promoting good governance.** This action seeks to

[1] A widely-used definition of sustainable development is 'development which meets the needs of the present without compromising the ability of future generations to meet their own needs' (World Commission on Environment and Development, 1987).

Figure 11.1
The UK's shared framework for sustainable development.

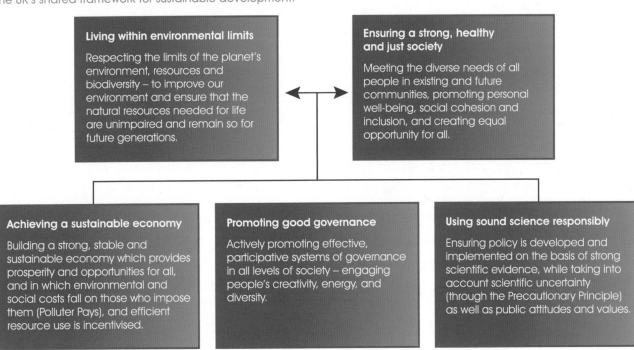

Living within environmental limits

Respecting the limits of the planet's environment, resources and biodiversity – to improve our environment and ensure that the natural resources needed for life are unimpaired and remain so for future generations.

Ensuring a strong, healthy and just society

Meeting the diverse needs of all people in existing and future communities, promoting personal well-being, social cohesion and inclusion, and creating equal opportunity for all.

Achieving a sustainable economy

Building a strong, stable and sustainable economy which provides prosperity and opportunities for all, and in which environmental and social costs fall on those who impose them (Polluter Pays), and efficient resource use is incentivised.

Promoting good governance

Actively promoting effective, participative systems of governance in all levels of society – engaging people's creativity, energy, and diversity.

Using sound science responsibly

Ensuring policy is developed and implemented on the basis of strong scientific evidence, while taking into account scientific uncertainty (through the Precautionary Principle) as well as public attitudes and values.

(Further details on the framework are available online at: www.defra.gov.uk/sustainable/government/publications/uk-strategy/index.htm).

incorporate the knowledge and aspirations of different stakeholders in policy development and to put in place clear principles to guide decision-making.

- **Using sound science responsibly.** This action relates closely to this Assessment report. Scientific evidence should be used responsibly in a rounded way, taking account of uncertainties, and considering public attitudes and values.

11.1 Sustainable forest management

The importance of managing forests in a 'sustainable' way was formally recognised with the adoption of the Statement of Forest Principles at the United Nations Conference on Environment and Development (the Earth Summit) in 1993 (see 1.5.1, Chapter 1). At a European level, the EU, its member states and other European countries have made high-level commitments to sustainable forest management through the Ministerial Conferences on the Protection of Forests in Europe (MCPFE).

Sustainable forest management thus requires multiple objectives to be considered in an integrated way. This is recognised in Article 2.1 (a,b) of the Kyoto Protocol in

which signatories agreed various ways of considering potential impacts of mitigation options and of establishing common approaches to promoting sustainable development through forestry actions. The issue is also underlined in the IPCC's 4th Assessment Report by Nabuurs *et al.* (2007) who stress the importance of understanding the many functions of forest ecosystems and the effects of human activities, and of not treating different socioeconomic and environmental outputs in isolation:

'Important environmental, social, and economic ancillary benefits can be gained by considering forestry mitigation options as an element of the broad land management plans, pursuing sustainable development paths, involving local people and stakeholders and developing adequate policy frameworks' (p. 574).

A further implication of sustainable forest management is the need for local and regional institutions and people to play an effective role in shaping management practices. In the UK, this has been recognised in forest planning and consultation procedures, and in processes for the disbursement of planting and management grants through the Rural Development Programmes in England, Scotland, Wales and Northern Ireland. Guidelines under the UKFS also highlight the importance of regional and local biogeographical conditions in determining appropriate management practices. The importance of spatial differences emerges strongly in the analysis of climate change impacts in Section 2, mitigation in Section 3 and adaptation in Section 4. The UKCP09 projections illustrate that the impacts of climate change will vary substantially across different parts of the UK. Forest management practices will, therefore, need to be carefully tailored in future to reflect the suitability of different species and management regimes to different locations (see 1.6, Chapter 1 and Section 2).

11.2 Implications of climate change mitigation and adaptation for sustainable forest management

It is of critical importance for policy to evaluate whether different actions to achieve climate change mitigation and adaptation are compatible with sustainable forest management. High-level, collaborative work, such as by the Collaborative Partnership on Forests (CPF) (2008), has advocated sustainable forest management as the appropriate framework for actions on climate change in all types of forest.

> 'It [sustainable forest management] can be applied to forests in which wood production takes place, including planted forests, as well as to protected forests and to degraded forests in need of restoration' (p. vii).

A particular strength of sustainable forest management is that it stresses an adaptive approach, through which forest management practices can change as conditions change. This will be particularly important if climatic changes result in significant alterations to growing conditions and to the suitability of different species and management practices (Section 2). In keeping with an integrated approach, mitigation and adaptation practices should not be seen as necessarily being mutually exclusive. In many cases, forestry actions (e.g. planting on floodplains) bring benefits both for mitigation and adaptation.

As noted above, properly managed forests have many properties which are consistent with sustainable development. This is shown in their capacity to generate economic, social and environmental benefits. Firm evidence of the magnitude of these benefits was provided by Willis *et al.* (2003). However, many of these benefits

(e.g. carbon sequestration, biodiversity conservation) are not typically rewarded by the market. Adequately reflecting these benefits in financial incentives would produce more efficient forest management in a broad economic sense. Work by Moxey (2009) has begun to examine the incentives that are currently in place in the forestry sector (see 1.5.1, Chapter 1) and how further analysis could ascertain whether changes to the suite of policy instruments are needed in future.

A summary of the effects in terms of sustainable development of different forestry actions on climate change mitigation and adaptation is shown in Table 11.1. This Table illustrates potential effects rather than providing definitive statements. Building on evidence from earlier chapters (particularly Chapters 6, 8 and 9,), Table 11.1 shows that the high-level actions on climate change identified for forestry have many synergies with the objectives of sustainable development. However, trade-offs may also need to be faced in some instances if other forestry objectives (such as timber production, or biodiversity conservation) are not to be unduly compromised by the pursuit of climate change objectives. For example, managing forests solely for their carbon mitigation potential is unlikely to be always consistent with managing them for biodiversity conservation. The precise effects of any action depend on the type of woodland, the characteristics of the site and the specific management activities employed.

Two important points arise from this consideration of multiple outcomes. First, 'conventional' planted conifer forests tend to be the most cost-effective in securing carbon abatement whereas native woodlands are in general the most highly valued for biodiversity benefits. (These two woodland types equate broadly to the Forest Management Alternatives described in Section 3 as 'Intensive even-aged forestry' and the 'Close-to-nature forestry', respectively.) Cost-effectiveness here means that forests offer an opportunity to reduce net emissions at a lower relative cost than other options (which includes other forestry management options, other land-use changes, and emission reductions from, say, industry or transport). Second, any assessment of cost and benefits from using forests as mitigation options should take account of the whole forest life cycle, including the use of timber after harvesting: the 'carbon footprint' of forest management.

However, a fuller understanding of the scale of synergies and trade-offs between mitigative and adaptive actions and the other benefits provided by forestry requires further

Table 11.1
Sustainable development implications of forestry actions for mitigation and adaptation.

Action	Economic	Social	Environmental
Planting new woodlands	Depends on any displacement of other land uses, on rotation lengths and on opportunities for carbon payments. Financial opportunities through generating carbon credits. Employment creation (if replacing less labour-intense activity).	Tree planting to improve urban temperatures and surface water conditions provides favourable living and working environments.	Depends on the 'forest management alternative'. Benefits highest in native woodlands but sequestration levels will tend to be lower. UK Forest Standard to support appropriate planting (e.g. avoiding deep peat soils).
Protect and manage existing forests	Increased long-term employment in managed woodlands, both direct and indirect.	Amenity and recreation values arise. Forests offer a resource for anticipated higher visitor numbers in a warmer climate.	Protection of forest carbon stocks may reduce sequestration rates. Biodiversity and landscape benefits. Protects watersheds and soils. Adaptation measures protect the ecosystem service functions of woodlands.
Use wood for energy	Income for woodland owners. Employment opportunities. Provision of renewable energy. Increased local income.	Potential competition with other land-uses.	Reduced use of fossil fuels. Short rotation plantations may reduce environmental values (depending on previous land-use). Loss of deadwood habitat. Less carbon locked up in soils.
Replace other construction/ manufacturing materials with wood	Potential economic diversification. Income for woodland owners and timber suppliers.	Potential competition with other land-uses.	Avoided GHG emissions associated with the manafacture and use of those materials replaced by wood.
Plan to adapt to a changing climate	Reduced economic damage from extreme weather events. Reduce risk of pest outbreak.	Enhanced living and working environments. Protection against environmental hazards. Reduced impacts of climate change, particularly on urban populations.	Enhanced habitat networks. Impacts from possible use of non-local provenance species.

Adapted from Nabuurs *et al.* (2007).

research and analysis. Research is needed to strengthen scientific understanding in this area and then to apply economic analysis to this understanding of trade-offs and synergies. Research and analysis of this type will provide important evidence in helping decision-makers to re-evaluate current policy and practice so that mitigation and adaptation are firmly embedded in sustainable forest management alongside the other benefits that forests provide.

11.3 Research priorities

- Forestry actions can, in many cases, bring benefits in terms of both mitigation and adaptation. These benefits must be properly incorporated into the concept and practice of sustainable forest management. Integrated approaches are needed.
- Local and regional institutions and knowledge are required in developing forest management solutions that

help to tackle climate change while meeting the needs of sustainable development.

• Further scientific and economic analysis is required to understand the nature and scale of synergies and trade-offs between forestry mitigative and adaptive actions and other outputs from forestry.

• Further effort is needed to re-design policy incentives so that adequate reward is given to the provision of non-market benefits, including those relating to the mitigative and adaptive functions of forests.

References

COLLABORATIVE PARTNERSHIP ON FORESTS (2008). *Strategic framework for forests and climate change: a CPF proposal.* Online at: www.fao.org/forestry/media/16639/1/0/ See also other CPF publications online at Collaborative Partnership on Forests website at: www.fao.org/forestry/cpf/en

FORESTRY COMMISSION. (2009). *Study of the Forestry Commission Estate in England.* Forestry Commission, Edinburgh. Online at: www.forestry.gov.uk/england-estatestudy

MOXEY, A. (2009). *Scoping study to review forestry policy instruments in the UK.* Report to the Forestry Commission, Edinburgh.

NABUURS, G.J., MASERA, O., ANDRASKO, K., BENITEZ-PONCE, P., BOER, R., DUTSCHKE, M., ELSIDDIG, E., FORD-ROBERTSON, J., FRUMHOFF, P., KARJALAINEN, T., KRANKINA, O., KURZ, W.A., MATSUMOTO, M., OYHANTCABAL, W., RAVINDRANATH, N.H., SANZ SANCHEZ, M.J. and ZHANG, X. (2007). Forestry. In: Metz, B., Davidson, O.R., Bosch, P.R., Dave, R. and Meyer, L. A. (eds) *Climate change 2007: mitigation.* Contribution of Working Group III to the Fourth Assessment Report of the Intergovernmental Panel on Climate Change. Cambridge University Press, Cambridge. pp. 541–584.

WILLIS, K.G., GARROD, G., SCARPA, R., POWE, N., LOVETT, A., BATEMAN, I.J., HANLEY, N. and MACMILLAN, D.C. (2003). *The social and environmental benefits of forests in Great Britain. Phase 2 report.* Centre for Research in Environmental Appraisal and Management, University of Newcastle-Upon-Tyne, Newcastle. Report for the Forestry Commission, Edinburgh.

WORLD COMMISSION ON ENVIRONMENT AND DEVELOPMENT. (1987). *Our common future.* Oxford University Press, Oxford.

FORESTRY AND CLIMATE CHANGE: A SOCIO-ECONOMIC PERSPECTIVE

M. Nijnik, J. Bebbington, B. Slee and G. Pajot

Chapter

12

Key Findings

Economic analysis reveals that forestry projects involving carbon capture and storage have the potential to postpone climate change, reduce net emissions, while allowing time for adaptation and technological innovation. Implementation requires policy measures to be cost-effective, ecologically sustainable and socially desirable. Appropriate public investment, economic incentives and institutional and governance capacities are required to bring about such projects.

There is evidence in support of cost-effective woodland creation programmes on marginal land where opportunity costs are lowest. The choice of location for forestry development, and the choice of management regimes to be applied, are important factors in determining economic costs. Overall, new tree planting is deemed to be economically viable either when SRC and SRF are established for bio-energy, or when afforestation provides environmental and/or social co-benefits.

Adaptation and mitigation activities are linked together, and the knowledge built up in the UK and beyond should be used to facilitate more successful mitigation–adaptation interactions in the forestry/land use sectors in the wider context of sustainable development and promoting rural livelihoods.

Problems with the inclusion of carbon credits from forestry into regulatory emission trading schemes arise as a result of the perception that forestry sinks are temporary and from issues such as 'leakages', double-counting and high transaction costs associated with measuring, assessing and monitoring of carbon. Opportunities to increase the cost-efficiency of climate change mitigation will arise if solutions to these problems are found.

This chapter provides an economic perspective of forestry and climate change in the UK. A growing body of literature has developed on this subject, particularly in Europe and overseas.

This literature suggests that the mitigative and adaptive roles of forests can be enhanced by both new planting and forest management (IPCC, 2007). However, forests' roles will be mediated and shaped by market signals, policy frameworks, and governance approaches as well as by attitudes and behavioural patterns. All will be considered in this chapter.

Major studies have been carried out in recent years into economic aspects of climate change generally. Foremost among these has been the Stern Review (Stern 2006) which placed scientific observations and policy choices in an economic framework, and did much to increase awareness of the costs of failing to take adequate action now on climate change.

12.1 The economics of carbon sequestration and storage through forestry

From an economic perspective, it is important to compare the cost-effectiveness of different approaches to GHG abatement. This requires data on the cost, for each abatement activity, of removing a tonne of CO_2 equivalent. From an economic viewpoint, it makes sense to choose those mitigation options with relatively low costs, as then, GHG reduction targets can be met at a lower overall cost to the economy[1]. Since forest management involves costs and benefits which extend over time, discounting is used

[1] Indeed, some sources may face negative costs for carbon mitigation: investments in household energy efficiency being one example.

to calculate the net present value of forestry management options per tonne of carbon equivalent abated. In particular, economists have focused on the marginal costs of carbon abatement through forest management and forest creation: how much does it cost, in net terms, to sequester one more tonne of CO_2? We would expect these marginal costs to vary across forest management options, and to vary spatially for a given option (because, for instance, of variations in growing conditions and in the price of land).

Various studies have examined the cost-effectiveness of forestry as a carbon sink, relative to other mitigation options (Crabtree, 1997; Newell and Stavins, 2000; Stavins and Richard, 2005; Nijnik 2005; Enkvist et al., 2007; Nijnik and Bizikova, 2008; Moran et al., 2008). Such work is vital in assessing the economic feasibility of forestry in tackling climate change. These studies have identified substantial variability in marginal costs in different countries and in different settings. A meta-analysis of 68 studies (Van Kooten et al. 2004), with a total of 1047 observations worldwide, identified costs varying between €35 and €199/tC and, when opportunity costs (see Glossary) of land use were taken into account, between €89 and €1069/tC.

Tree planting is costly, and opportunity costs exist for converting existing non-forest land into new forests. Marginal cost estimates of carbon mitigation by forests can be compared with market prices of carbon, for example prices in the EU's Emissions Trading Scheme (ETS): these currently (August 2009) stand at around €15/tCO$_2$. However, it is important to remember that prices in carbon markets do not necessarily reflect the true social value of carbon reductions, but rather current demand and supply within carbon markets, and the institutional aspects of such markets (Defra, 2008). EU ETS prices are expected to rise over time, implying an improving competitive position for forests as a mitigation option, although carbon price volatility will also be important to forest managers (Turner et al., 2008).

It is argued by Van Kooten and Sohngen (2007) that if carbon sequestered through tree planting was to be traded in markets alongside credits through emissions reductions, it would be relatively economically attractive if traded at US$50/tCO$_2$. Assuming a threshold of about US$30/tCO$_2$, tree planting activities are generally competitive with emissions reductions, particularly in tropical and boreal regions (Table 12.1). The costs of carbon sequestration in forestry also compare well with those of emerging technologies for carbon capture and storage. However, if the opportunity cost of land is fully taken into account (and if emissions reduction credits can be purchased for US$50/tCO$_2$ or less), tree planting appears less attractive. Hanley (2007) shows that forests are cost-effective sources of mitigation for Scotland, relative to wind energy investments, but agricultural land-use changes are also relatively cost-effective.

Preliminary work has been commissioned by the Department for Environment, Food and Rural Affairs (Defra) to examine marginal carbon abatement costs for a range of land-use activities in the UK (Moran et al., 2008). Carbon sequestration costs through forestry were estimated to range from £8 per tCO$_2$ (afforestation of sheep grazing areas) to £48 per tCO$_2$ (for afforestration of agricultural land), using a discount rate of 3.5%. The implications of such results are that there is evidence to support woodland creation on some marginal land rather than for afforestation on a larger scale, although much depends on whether agricultural subsidies continue to hold up land prices. Ongoing reforms (de-coupling) of the Common Agricultural Policy may have significant effects on agricultural land prices, which will change the net cost of woodland creation as a mitigation option. World food price changes will also have major effects. More complete costings have recently been estimated by ADAS (2009) and are described in Chapter 8.

Tree species and management regimes are also important factors in minimising economic costs. Prioritisation of areas

Table 12.1
Sustainable development implications of forestry actions for mitigation and adaptation ($/tCO$_2$).

Activity	Global	Europe	Boreal	Tropics
Planting	22–33	158–185	5–128	0–7
Planting and fuel substitution	0–49	115–187	1–90	0–23
Forest management	60–118	198–274	46–210	34–63
Forest management and fuel substitution	48–77	203–219	44–108	0–50
Forest conservation	47–195	N/A	N/A	26–136

Source: Adapted from van Kooten and Sohngen (2007).

that offer the most potential for sequestration through forestry would be greatly assisted by the development of maps providing indicative figures of how such costs vary spatially across the UK. Such an approach would provide the basis for a spatial cost-benefit analysis of forestry-based policy options on climate change. The analysis could identify:

- which options are economically sound for implementation, and where and how; and,
- which regions are likely to benefit most from forestry development, as well as those regions that may be adversely affected by forestry projects.

Key scenarios that merit attention are:

- carbon sequestration and storage in forests;
- production of wood for energy (when trees are cut and wood is used to substitute for fossil fuels – see Chapter 7);
- the use of wood products as substitutes for more carbon-intensive materials; e.g. in construction and furniture (see Chapter 7);
- tree planting for adaptive purposes such as on floodplains; and,
- tree planting/growing for the provision of multiple ecosystem services, including carbon.

The analysis of these scenarios should take account of relevant price signals including those in the agricultural and emissions trading sectors because these prices will affect the relative returns to forestry projects. The economically optimal level of mitigation through forestry, and the actual uptake of mitigation by private forest owners, will depend also on what we assume about the global price of carbon (Van't Veld and Plantinga, 2005).

Longer rotations can delay opportunities for using wood for energy generation or/and substituting wood for materials whose production is more intensive in GHG terms (Pajot and Malfait, 2008; Nijnik et al., 2009). Studies in Canada (Van Kooten and Bulte, 2000; Van Kooten, 2004) have suggested that a continual forest cycle in which trees are harvested and re-planted or regenerated, and in which substitution benefits are provided through the use of wood fuel and wood products, provides a sustainable means of sequestering carbon, storing it and avoiding emissions from other more damaging activities. The benefits of wood products and wood energy scenarios in the long-run are higher than under a strategy of carbon sequestration alone.

Analysis by Van Kooten (2009) has found that cost-effective emissions reductions might be created when short-rotation plantations are established for bioenergy. Evidence for this was earlier provided in Canada where hybrid poplar planted on marginal land appeared to be cost-effective (Van Kooten et al., 1993). An economic assessment of willow production in the UK (Boyle, 2004) also demonstrated that this can be economic if planting takes place on set-aside land with grants of £1600 per hectare and annual yields above 10 tonnes (oven-dried) per hectare ae obtained. Work by Dawson et al. (2005) and Galbraith et al. (2006) provides similar findings.

There is also evidence that forestry projects combined with use of wood products and renewable energy strategies offer economic opportunities in rural (and urban) areas through innovation, employment and the development of markets (EC, 1997; Van Kooten, 2004; Freer-Smith et al., 2007; Brainard et al., 2009). It is imperative, therefore, that measures for carbon sequestration in forests are considered within the context of policies for spatial planning, and of forestry, agricultural and rural policies and sustainable energy systems (Nijnik and Bizikova, 2006). This may save costs and assist in dealing with environmental problems associated with the changing climate.

12.2 Institutional aspects of forest carbon markets

The institutional framework relating to forest carbon markets is complex. The flexible mechanisms under the Kyoto Protocol (UNFCCC, 1998) – the Clean Development Mechanism and Joint Implementation – provide opportunities for countries to tackle climate change while making judgements on the economic feasibility of different courses of action. However, evidence suggests that the CDM and JI mechanisms are unlikely to create credit and permit (allowance) trading on a large scale, despite the growth of trading in CDM and JI credits globally (IPCC, 2007). Some studies suggest that such regulatory trading schemes fail, not because of a lack of interest, but primarily because of high transaction costs (Chomitz, 2000; Van Kooten, 2004). It appears that the complexity of the institutional arrangements for the flexible mechanisms have been a disincentive for action. To date, there have been only eight forestry projects approved under the CDM (see: http://cdm.unfccc.int).

A critical element of the institutional arrangements concerning carbon trading is their capacity to ensure that carbon benefits are delivered as stated. This applies to both regulatory and voluntary markets. In the UK, the Government has established a Quality Assurance Scheme for Carbon Offsetting that allows consumers to identify good quality offsets in voluntary markets (Defra, 2009). However, apart from the CDM, forestry is currently excluded from international regulatory markets in carbon. Therefore, at this time, voluntary markets are the principal means of generating carbon benefits from forestry in the UK. As described in Chapter 1, the Forestry Commission is establishing a Code of Good Practice for Forest Carbon Projects to ensure appropriate standards of delivery. This and the Government's Quality Assurance Scheme are intended to provide a framework to support the development of robust, transparent, reliable and timely carbon benefits that offer consumers genuine value for money, as well as achieving carbon savings.

The inclusion of forestry in regulatory emissions trading schemes has been impeded by a number of factors. These have been widely examined (see Chomitz, 2000; Marland et al., 2001; Subak, 2003; Van Kooten, 2004 and Nijnik et al., 2009) and include:

* establishing baseline emissions data;
* coping with 'leakages' (these may arise where the CO_2 emissions that a project is meant to sequester are displaced beyond its boundaries[2]);
* providing assurance of 'additionality' and of permanence of projects;
* establishing reliable measurement and monitoring of carbon sequestration and of costs;
* verifying that carbon sequestration has taken place;
* avoiding double counting;
* devising a process for certifying carbon credits and 'converting' them into emission permits;
* establishing property rights and institutions for exchanging carbon credits;
* putting in place appropriate legal arrangements and data requirements to allow schemes to operate.

Many of these challenges are also pertinent to voluntary carbon projects. However, the voluntary carbon market is less regulated and thus tends to have lower transaction costs. Some have argued that voluntary carbon trading is relatively successful (Taiyab, 2006). For example, it can comprise 37% of total voluntary transactions by the forestry sector (Hamilton et al., 2007). Across the world, schemes have been founded by governments, NGOs,

businesses and individuals. Types of projects include tree planting and conservation of forests, and in the majority of cases they offer 'cheap' carbon savings (House of Commons, 2007).

Uncertainty also plays a key role in the development of carbon markets for forestry. This is underlined in work by Turner et al. (2008) on forests in New Zealand. Estimating future benefits of carbon sequestration and storage is complicated by uncertainties in forest carbon dynamics. Estimates must determine how much carbon is sequestered and stored (and for how long) and assess how much carbon will be sequestered in the future under a changed climate. These uncertainties affect how many carbon credits a forest investment will earn. Uncertainties also exist in relation to carbon prices, the permanence of forest carbon stocks (their susceptibility to wind or fire risk for example), and concerns over double-counting and additionality of carbon credits. Assuring market confidence in the capacity of forestry investments to deliver mitigation benefits is essential if the future potential of forestry in this area is to be fulfilled. The Code of Good Practice for Forest Carbon Projects (see 1.5.4, Chapter 1) is intended to provide this assurance. For example, one way of dealing with uncertainties of fire or wind damage is to establish 'buffers' whereby a proportion of the anticipated carbon is set aside as an insurance. The Code also sets out proposals for proper procedures for registration, monitoring and verification of forest carbon projects. Private sector instruments for offsetting risks in carbon markets can also be expected to develop without government intervention.

12.3 Rural policy signals

Our review of the evidence indicates that the potential for the UK's forests and woodlands to contribute to climate change mitigation and adaptation is shaped by wide-ranging factors. These include important influences beyond the control of forestry policy. Some of these have an international dimension, such as the EU's Common Agricultural Policy (CAP), EU directives on the natural environment and multilateral climate change agreements (see 1.5.2, Chapter 1). Others have a domestic focus, such as UK designations on the natural environment. The breadth of factors that influence forestry's role in helping to tackle climate change show that a more integrated approach to planning involving forestry, agriculture and other land uses would bring benefits.

[2] It is possible to cope with 'leakages', for example by expanding the scope of the system to 'internalise' the 'leakages' or to design the project so as to be 'leakages' neutralising (Chomitz, 2000).

In the UK, forestry and farming have become competing land uses. According to Taylor *et al*. (1999), agricultural subsidies have been a significant deterrent to new forest planting, due to their effects on relative returns and thus on land rents. Recent reforms of the CAP have de-coupled support from production, leading to a large change in returns from certain farming activities, particularly in the uplands (Acs *et al*., 2008). These changes – such as falling returns from livestock grazing – can be expected to increase incentives at the margin to convert land to forestry, especially if Single Farm Payment is retained on planted land.

Policy support for renewable energy that increases the demand for wood energy and taxes on non-renewable forms of energy may also influence planting and management practices in forestry. Indeed, there is policy support for the development of woody biomass, with grant aid (Defra Energy Crop Scheme) currently available in all parts of the UK for household and community schemes (LUPG, 2004; see also the Bioenergy Infrastructure Scheme, Defra, 2007). As indicated in 1.5.3, Chapter 1, the UK Renewables Obligation (RO) also provides an incentive for the development and use of wood energy.

Forestry delivers a greater range of ecosystem services than carbon sequestration alone, and many of these benefits are highly spatially variant. This creates not only a need for accurate assessment of non-market benefits, but also for the design of policy instruments that take the full array of forest services into account. Again, it is important to examine the extent to which maximising the carbon sequestration benefits of forestry is consistent with delivering other ecosystem services. Government incentives for providing multiple ecosystem services including sequestration will also impact in complex ways on decision-making by private forest owners (see Caparros *et al*., 2009, for an example relating to new forest planting in Spain).

Apart from maximising monetary returns from land, land-use change decisions involve long-term investments that bring uncertainty (Schatzki, 2003) and are affected by other unquantified benefits and costs of alternative land uses (i.e. aesthetic values and recreation) (Ovando and Caparros, 2009). They may also be affected by liquidity constraints and decision-making inertia (Stavins, 1999). These considerations merit attention in the UK, as they could constrain the amount of new land that can be devoted to forestry based climate mitigation, and consequently the carbon sequestration benefits obtained.

12.4 Stakeholder attitudes

Economic analysis of forestry and climate change has been complemented by analysis of stakeholder attitudes (including land managers and the general public). For example, various cultural values affect the propensity of land managers to plant trees and to develop forest-based activities to tackle climate change.

Public attitudes to forestry in the UK are assessed in biennial surveys commissioned by the Forestry Commission. The 2007 and 2009 surveys include a section on climate change and forests. They tend to show growing awareness of and support for the role of forests in tackling climate change. General support for afforestation has been shown. A significant proportion of respondents wishing to see twice as many forests in their part of the country, primarily in the form of broadleaved and mixed forests. The proportion of people emphasising the role of forests as a source of renewable energy rose from 20% of the sample in 1999 to 50% in 2009. In the 2009 survey, 68% of respondents thought that using public money to manage existing woodland 'to help tackle climate change' was a good reason for such spending. The surveys also suggest that other forest amenities and benefits are important and, therefore, should be considered when forest strategies are to be implemented.

Studies of farmers' attitudes to trees and land conversion to forestry show reluctance to plant trees (Tiffin, 1993; Williams *et al*., 1994). Some of the work in this field is dated, although the findings are consistent with more recent work (see Burton, 2004 and Towers *et al*., 2006). Other authors have argued that there are important psychological, cultural and institutional barriers to afforestation in the UK. In particular, it is argued that the UK has a weakly developed forest culture (Mather *et al*., 2006; Nijnik and Mather, 2008). Land tenure has also been cited as a barrier to afforestation (Warren, 2002).

12.5 Conclusions

The acceptance of sustainable development as an over-arching objective requires forestry measures to be cost-effective, ecologically sustainable and socially desirable. Appropriate economic incentives and institutional and governance capacities are required to achieve this.

Forests can play a cost-effective role in a country's overall mitigation strategy, although the costs of CO_2

sequestration vary over a considerable range, according to land quality, alternative land uses, forest management option, and costs of alternatives.

Overall, tree planting for carbon mitigation is economically desirable either when combined with bio-energy production, or when afforestation provides environmental and/or social co-benefits.

Adaptation and mitigation activities are linked together, and the knowledge built up in the UK and beyond should be used to facilitate more successful interaction between mitigation and adaptation in the forestry and land-use sectors, and in the wider context of sustainable development and rural livelihoods.

Major problems arise concerning the inclusion of carbon credits from forestry into regulatory emission trading schemes because of the temporary nature of terrestrial carbon sinks, and issues such as 'leakages', double-counting and high transaction costs associated with measuring, assessing and monitoring of carbon. Opportunities to increase the cost-efficiency of climate change mitigation via the private sector rest in finding solutions to these problems.

The extent to which additional private sector forests are planted as part of a UK mitigation strategy will depend on the evolution of agricultural and renewable energy policy, as well as on the extent to which landowners can be rewarded for carbon sequestration.

12.6 Research priorities

- More research is needed to develop our understanding of the circumstances under which the forestry sector can offer sustainable, socially acceptable and low-cost opportunities for carbon sequestration. This includes analysis of the cost-effectiveness of forest-based carbon sequestration and storage and comparison of the marginal costs of carbon mitigated through different forestry management options (in different localities) compared to other possible alternatives for reducing net emissions (e.g. in agriculture, in housing, transport or industry).
- Further analysis is needed of the trade-offs and synergies between managing forests for carbon, compared with other public goals such as managing for biodiversity and recreation. This research should aim to quantify these trade-offs and synergies, and to design

mechanisms to maximise net benefits. This will help to provide a more thorough assessment of the effects of managing forests for carbon on indicators of sustainable development.
- Further work is needed to investigate the barriers (economic, institutional and cultural) to large-scale afforestation projects in the UK.
- The economics of forests for bio-energy needs further work. Moreover, it is important to improve understanding of the behavioural, social and economic barriers to the development of wood energy supply chains and the relative advantages of different wood energy supply systems (chip, pellet, CHP).
- Spatially explicit modelling of carbon and other non-market benefits of forests and woodlands including, *inter alia*, habitat networks and flood alleviation remains a research priority. This should include spatial modelling of both cost-effective mitigation, and the cost-benefit analysis of management alternatives.
- Further investigation is required into the nature of risk and uncertainty in developing forest carbon credit markets, and how this risk can best be managed.

References

ACS, S., HANLEY, N., DALLIMER, M., GASTON, K.J., ROBERTSON, P., WILSON, P. and ARMSWORTH, P.R. (2008). *The effect of decoupling on marginal agricultural systems: Implications for farm incomes, land use and upland.* Stirling Economics Discussion Paper. Online at: www.economics.stir.ac.uk/DPs/SEDP-2008-18-Acs-Hanley-et-al.pdf

ADAS. (2009). *Analysis of policy instruments for reducing greenhouse gas emissions from agriculture, forestry and land management – forestry options*. Report to Defra.

BOYLE, S. (2004). Royal Commission on Environmental Pollution. Biomass report: second consultant's report. Online at: www.rcep.org.uk/papers/general/119.pdf

BRAINARD, J., BATEMAN, I.J. and LOVETT, A.A. (2009). The social value of carbon sequestered in Great Britain's woodlands. *Ecological Economics* **68**, 1257–1267.

BURTON, R. (2004). Establishing "Community Forests" in England: can public forests be provided through private interests? In: Fitzharris, B and Kearsley, J. (eds) *Glimpses of a Gaian World: Essays on geography and senses of place*. University of Otago Press, Dunedin.

CAPARROS, A., CERDA, E., OVADO, P. and CAMPOS, P. (2009). Carbon sequestration with reforestation and biodiversity-scenic values. *Environmental and Resource Economics*, (in press).

CHOMITZ, K.M. (2000). *Evaluating carbon offsets from forestry and energy projects: how do they compare?* Development research group. The World Bank, 2357. Online at: www.worldbank.org/research

CRABTREE, R. (1997). Carbon retention in farm woodlands, In: Adger, W.N., Pettenella, D. and Whitby, M. (eds) *Climate change mitigation and European land-use policies*. CABI, Wallingford. pp. 187–197.

DAWSON, M., HUNTER-BLAIR, P., MULLAN, O. and CARSON, A. (2005). *Comparative costs and returns from short rotation coppice willow, other forestry plantations and forestry residues and sawmill co-products in small-scale heat and power and hear-only system*. The DARD Renewable Energy Study. Online at: http://dardni.gov.uk/file/con05026h.pdf

DEFRA (2007). *Bioenergy Infrastructure Schemes*. Online at www.defra.gov.uk/foodfarm/growing/crops/industrial/energy/infrastructure.htm

DEFRA (2008). *The social cost of carbon and the shadow price of carbon: what they are, and how to use them in economic appraisal in the UK*. DEFRA, London. Online at: www.defra.gov.uk/environment/climatechange/research/carboncost/index.htm

DEFRA (2009). *Carbon offsetting: Government Quality Assurance Scheme*. Online at: www.defra.gov.uk/environment/climatechange/uk/carbonoffset/assurance.htm

ENKVIST, P-A., NAUCLER, T. and ROSANDER, J. (2007). A cost curve for greenhouse gas reduction. *The McKinsey Quarterly 2007 (1)*, 35–45.

EUROPEAN COMMISSION (1997). Energy for the future: renewable sources of energy. White Paper for Community Strategy and Action Plan, COM(97)599 FINAL. EC, Brussels.

FORESTRY COMMISSION (2007). UK Public Opinion of Forestry 2007. Forestry Commission, Edinburgh.

FORESTRY COMMISSION (2009). UK Public Opinion of Forestry 2009. Forestry Commission, Edinburgh.

FREER-SMITH, P.H., BROADMEADOW, M.S.J. and LYNCH, J.M. (eds) (2007). *Forestry and climate change*. CABI, Wallingford.

GALBRAITH, D., SMITH, P., MORTIMER, N., STEWART, R., HOBSON, M., MCPHERSON, G., MATTHEWS, R., MITCHELL, P., NIJNIK, M., NORRIS, J., SKIBA, U., SMITH, J. and TOWERS, W. (2006). *Review of greenhouse gas life cycle emissions, air pollution impacts and economics of biomass production and consumption in Scotland*. SEERAD Project FF/05/08.

HAMILTON, K., BAYON, R., TURNER, G. and HIGGINS, D. (2007). *State of the voluntary carbon market. Picking up steam*. Washington D.C. and London. Online at: http://ecosystemmarketplace.com/documents/acrobat

HANLEY, N. (2007). *What should Scotland do about climate change?* Online at www.davidhumeinstitute.com/DHI%20Website/Events,%20transcripts%20&%20presentations/Events%202007/Hanley%20presentation.pdf

HOUSE OF COMMONS, ENVIRONMENTAL AUDIT COMMITTEE (2007). *The voluntary carbon offset market*. HC 331. Sixth report of session 2006–2007. House of Commons, London.

INTERGOVERMENTAL PANEL ON CLIMATE CHANGE. (2007). *Climate Change 2007: Synthesis Report*. Online at www.ipcc.ch/pdf/assessment-report/ar4/syr/ar4_syr.pdf

LAND USE POLICY GROUP (LUPG). (2004). *Cap reform (June 2003)/ implications for woodlands*. Issues paper from the Woodland Policy Group with input from the Forestry Commission. Online at: www.lupg.org.uk/pdf/pubs_Woodland_and_CAP_reform[1].pdf

MARLAND, G., FRUIT, K. and SEDJO, R. (2001). Accounting for sequestered carbon: the question of permanence. *Environmental Science and Policy* 4, 259–268.

MATHER, A., HILL, G. and NIJNIK, M. (2006). Post-productivism and rural land use: cul de sac or challenge for theorisation?. *Journal of Rural Studies* 22, 441–455.

MORAN, D., MACLEOD, M., WALL, E., EORY, V., PAJOT, G., MATTHEWS, R., MCVITTIE, A., BARNES, A., REES, B., MOXEY, A. and WILLIAMS, A. (2008). *UK marginal abatement cost curves for the agriculture and land use, land use change and forestry sectors out to 2022, with qualitative analysis of options to 2050*. Final Report to the Committee on Climate Change, RMP4950.

NEWELL, R.G. and STAVINS, R.N. (2000). Climate change and forest sinks: factors affecting the costs of carbon sequestration. *Journal of Environmental Economics and Management* 40, 211–235.

NIJNIK, M. (2005). Economics of climate change mitigation forest policy scenarios for Ukraine. *Climate Policy* 4, 319–336.

NIJNIK, M. and BIZIKOVA, L. (2006). The EU sustainable forest management and climate change mitigation policies from a transition countries perspective. In: Reynolds, K (ed.) *Sustainable forestry: from monitoring and modelling to knowledge management and policy science*. CAB International, Wallingford. pp. 56–66.

NIJNIK, M. and BIZIKOVA, L. (2008). Responding to the Kyoto Protocol through forestry: a comparison of opportunities for several countries in Europe. *Forest Policy and Economics* 10, 257–269.

NIJNIK, M. and MATHER, A. (2008). Analysing public preferences for woodland development in rural landscapes in Scotland. *Landscape and Urban Planning* 86, 267–275.

NIJNIK, M., PAJOT, G., MOFFAT, A. and SLEE, B. (2009).

Analysing socio-economic opportunities of British forests to mitigate climate change. European Association of Environmental and Natural Resource Economics, Amsterdam.

OVANDO, P. and CAPARROS, A. (2009). Land use and carbon mitigation in Europe: A survey of the potentials of different alternatives. *Energy Policy* **37**, 992–1003.

PAJOT, G. and MALFAIT, J.J. (2008). Carbon sequestration in wood products: Implementing an additional carbon storage project in the construction sector. *The European forest based sector: bio-responses to address new climate and energy challenges,* 6–8 November, Nancy, France.

STAVINS, R. and RICHARD, K. (2005). *The cost of US forest based carbon sequestration.* Arlington, VA, Pew Center on Global Changes.

SCHATZKI, T. (2003). Options, uncertainty, and sunk costs: an empirical analysis of land use change. *Journal of Environmental Economics and Management* **46**, 86–105.

STERN, N. (2006). *Stern review: the economics of climate change.* Online at: www.hm-treasury.gov.uk/stern_review_report.htm

STAVINS, R.N. (1999). The costs of carbon sequestration: a revealed-preference approach. *American Journal of Agricultural Economics* **89**, 994–1009.

SUBAK, S. (2003). Replacing carbon lost from forests: an assessment of insurance, reserves, and expiring credits. *Climate Policy* **3**, 107–122.

TAIYAB, N. (2006). *Exploring the market and voluntary carbon offsets.* International Institute for Environment and Development, London.

TAYLOR, J.E., YÚNEZ-NAUDE, A. and HAMPTON, S. (1999). *Agricultural policy reforms and village economies: a computable general-equilibrium analysis from Mexico.* Journal of Policy Modelling 21(4), 453-480.

TIFFIN, R. (1993). *Community forests: conflicting aims or common purpose?* Unpublished M.Phil. thesis, University College, London.

TOWERS, W., SCHARWZ, G., BURTON, R., RAY, D., SING, L. and BIRNIE, R. (2006). *Possible opportunities for future forest development in Scotland.* A scoping study report to the Forestry Commission.

TURNER, J.A., WEST, G., DUNGEY, H., WAKELIN, S., MACLAREN, P., ADAMS, T. and SILCOCK, P. (2008). *Managing New Zealand planted forests for carbon – a review of selected management scenarios and identification of knowledge gaps.* Report to the Ministry of Agriculture, New Zealand.

UNITED NATIONS FRAMEWORK CONVENTION FOR CLIMATE CHANGE (1998). *The Kyoto Protocol to the Convention on Climate Change.* UNEP/IUC, Bonn.

VAN'T VELD, K. and PLANTINGA, A. (2005). Carbon sequestration or abatement: the effect of rising carbon prices on the optimal portfolio of greenhouse-gas mitigation strategies. *Journal of Environmental Economics and Management* **50**, 59–81.

VAN KOOTEN, G.C., THOMPSON, W. and VERTINSKY, I. (1993). Economics of reforestation in British Columbia when benefits of CO_2 reduction are taken into account. In: Adamowicz, W., White, W. and Phillips, W. (eds) *Forestry and the environment: economic perspectives.* CABI, Wallingford. pp. 227–247.

VAN KOOTEN, G.C. and BULTE, E. (2000). *The economics of nature: managing biological assets.* Oxford, Blackwell.

VAN KOOTEN, G.C., EAGLE, A.J., MANLEY, J. and SMOLAK, T. (2004). How costly are carbon offsets? A meta-analysis of carbon forest sinks. *Environmental Science and Policy* **7**, 239–251.

VAN KOOTEN, G.C. (2004). *Climate change economics.* Edward Elgar, Cheltenham.

VAN KOOTEN, G.C. and SOHNGEN, B. (2007) Economics of forest carbon sinks: a review. *International Review of Environmental and Resource Economics* **1**, 237–269.

VAN KOOTEN, G.C. (2009). Biological carbon sink: transaction costs and governance. *Forestry Chronicle* **85**, 372–376.

WARREN, C. (2002). *Managing Scotland's environment.* Edinburgh University Press, Edinburgh.

WILLIAMS, D., LLOYD, T. and WATKINS, C. (1994). *Farmers not foresters: constraints on the planting of new farm woodland.* Department of Geography Working Paper 27. University of Nottingham, Nottingham.

HUMAN BEHAVIOURAL AND INSTITUTIONAL CHANGE	Chapter 13

A. Lawrence and C. Carter

Key Findings

Human behaviour needs to change, to both mitigate and adapt to climate change. There is a scarcity of social science research into climate change and much of the section draws on findings from relevant research linking social values, beliefs and knowledge with attitudes and behaviour towards the environment or sustainable consumption.

Many people find it hard to make sense of information about climate change, with its complexity and uncertainty. The ways in which people understand the role of trees and forests in this varies within society. Information and knowledge are not in themselves sufficient to change attitudes and behaviour, as personal and cultural values, experiences and beliefs also have a strong influence. If intervention is desired to bring about behavioural change, this will need to be tailored to the knowledge, values and experiences of specifically defined groups; one single approach will not suit all.

Change at the individual level is not adequate without institutional change. Institutions need to be adaptive. Characteristics of adaptive organisations are that they incorporate organisational learning, enhance social capital through internal and external linkages, partnerships, and networks, and make room for innovation and multi-directional information flow.

It appears that trees and forests can have a strong role in the way that people make sense of their environment and how it is changing. This suggests a particularly significant role for woodland management and the engagement of forestry with the public, in contributing to societal understandings and responses to climate change.

The success of a climate change policy in which forests play an important part will depend, to a degree, not just on economic issues – as outlined previously – but also on public attitudes and behaviour, and on institutions.

Institutions can be laws, conventions, cultural practices and/or organisations. Institutions affect how we think about, frame and regulate problems and how society and lifestyles develop. In order to facilitate an individual's ability to adapt, institutions also need to adapt.

Specific research on the social and institutional aspects of climate change in the UK is only slowly emerging, so this review of the evidence also draws on studies conducted in Europe and further afield. Even less research has been specifically conducted on the perceived role and significance of trees, woodlands and forests in climate change from a socio-cultural perspective. The need for

social research has been highlighted by experts (see Box 13.1).

This chapter first reviews evidence on people's attitudes and beliefs and highlights the fact that changing behaviour needs more than improved information and knowledge. This leads us to look at the wider context for adapting societal structures and behavioural patterns by looking at the role of institutions in directing, facilitating or constraining change. The concepts of adaptiveness and resilience are as important in the social and policy arena as they are in forest management. Finally, we review evidence that trees, woodlands and forests have a symbolic role which makes them potentially a powerful means of helping people understand and adapt to climate change.

13.1 Attitudes and beliefs

Several, largely quantitative, studies exist on the public's knowledge and perceptions of, and attitudes to, climate change (Defra, 2001–2008; Downing, 2008; Downing and Ballantyne, 2007; Maibach *et al.*, 2008). The findings generally indicate a high level of awareness of terms such as 'climate change' but not a clear understanding of the processes and causes of climate change. In particular, recent studies show a lack of understanding of how trees may contribute to climate change mitigation (Forestry Commission, 2007a,b,c) although a recent study with young people and children showed that they had a good understanding of the potential of trees to reduce levels of carbon dioxide in the atmosphere (Lovell, 2009).

Data from the Public Opinion of Forestry survey data for 2007 (Forestry Commission, 2008) show a modest level of awareness of forestry's value in the context of climate change (51%). Awareness is higher among rural and ethnically white British people than others. The data also show that there is variation among economic classes in terms of their knowledge about causes of climate change and mitigation, and beliefs about the future impact of climate change. This variation also correlates with frequency of woodland visit – those who visit more often tend to be more knowledgeable and less pessimistic.

This variation reflects a more general disparity between social classes and ethnic groups in terms of access to and appreciation of woods and forests (Forestry Commission, 2008).

13.2 Information, experience and behavioural change

More information does not necessarily change people's behaviour, especially where complex issues are concerned such as climate change (Sturgis and Allum, 2004; Ockwell *et al.*, 2009). Instead we need to better understand and consider the role of different influences affecting choices and behaviour. Without the appropriate emotional, cultural or psychological disposition, information will make no difference.

For example, research in Australia found that public understanding of global environmental issues drew not only on scientific information, but also on local knowledge, values, and moral responsibilities (Bulkeley, 2000). People who lack immediate, sensual engagement with the environmental consequences of their actions display greater destructive tendencies; again, awareness is not enough to curb destructiveness (Worthy, 2008). Emotional connection to the environment tends to be greater amongst those who have grown up in rural areas than in urban (Hinds and Sparks, 2008; Teisl and O'Brien, 2003). Engendering greater empathy towards nature tends to increase the level of connectedness people feel towards it (Schultz, 2000). Emotional affinity with nature is able to predict nature protective behaviour, such as public commitments to environmental organisations and the use of public transport (Kals *et al.*, 1999). Some of these links are stronger than others and all bear further research.

More specifically in relation to trees and forests, Nord *et al.* (1998) found strong correlations between frequency of visits to forest areas and self-reported pro-environmental behaviours. Emotional connection has been rated as more important than knowledge, in forming attitudes to environmental issues such as logging native forests (Pooley and O'Conner, 2000). A survey of nearly 2000 Swedish private individual forest owners showed that strength of belief in climate change and adaptive capacities were found to be crucial factors for explaining observed differences in adaptation among Swedish forest owners (Blennow and Persson, 2008).

13.3 Adaptive capacity

Adaptive capacity can be defined as the characteristics of organisations, communities, or societies which enhance their ability to adapt to environmental change. Adaptive forest management is an approach which recognises that complexity and uncertainty require us to treat forest management as experimental, requiring enhanced monitoring and feedback to decision makers. Given the range of forest and woodland ecosystems, and uncertainty about how climate change will affect them, no single approach to mitigation and adaptation will suit all situations. Forest managers, therefore, need to have sufficient flexibility to choose locally appropriate management practices, and to work with other stakeholders, especially local people, to systematically improve these practices by means of observation, analysis, planning, action, monitoring, reflection and new action (Seppälä et al., 2009).

Studies of the social and institutional requirements for adaptive forest management are scarce compared with more technical studies. One such in Ontario, Canada (MacDonald and Rice, 2004) showed that:

- institutional barriers are more limiting than technical barriers;
- most conflict is in the assessment and design steps of the adaptive management cycle;
- the process needs flexibility, trust, and consensus-building;
- wider application of active adaptive management requires staff retraining, cooperation among management agencies, encouragement of innovation and regular adjustment of policies and practices.

This last point links to the need for adaptive capacity in the wider context in which forest management takes place. Work on adaptation published by IUFRO notes that:

'The predominant hierarchical, top-down style of policy formulation and implementation by the nation state and the use of regulatory policy instruments, such as forest laws, are likely to be insufficiently flexible and may stifle innovative approaches in the face of climate change... Given the uncertainties surrounding the impacts of climate change, a more flexible and collaborative approach to forest governance is needed that can respond more quickly to policy learning. Policies will need to place greater emphasis on financial incentives for individual and cooperative/partnership to

forest management, supported, where necessary, by appropriate regulations.' (Seppälä et al., 2009)

An adaptive policy context will not focus on forestry alone but recognise that many drivers of change originate in other sectors (agriculture, energy, transportation and land use). The IUFRO report argues that market-based instruments such as forest certification, and approaches such as criteria and indicators for the monitoring and reporting of sustainable forest management, are more likely than regulatory approaches to serve this purpose (Seppälä et al., 2009). However, economic incentives and regulation are not mutually exclusive and in the UK context with a legacy of regulatory approaches (Kitchen et al., 2002), it may be necessary to build in an adaptive approach to regulation. Others also argue that the separation of mitigation and adaptation in policy processes may be counterproductive (Swart and Raes, 2007).

Finally, if more adaptive communities also have more in-built ecological resilience, there will be a need for higher levels of tree planting in urban, peri-urban and targeted privately owned rural areas. This means that the literature on motivations for tree planting, and engaging with spatial planning systems, is relevant (Götmark, et al., 2009; Hauer and Johnson, 2008; Pauleit et al., 2002; Ross-Davis et al., 2005; Saavedra and Budd, 2009; Siry et al., 2004; Van Herzele and Van Gossum, 2008).

13.4 Need for new approaches to knowledge generation and use

The high profile of climate change knowledge means that knowledge claims become politically contested (Bäckstrand and Lövbrand, 2006). The result is a lack of consensus about knowledge, methods and ethics around climate science (particularly in the context of forestry) (Lövbrand, 2009), and a perceived split between technocratic knowledge ('science can fix the problem') and locally relevant knowledge (Adger et al., 2001). Such situations can provide governments with the scope for a more participatory interpretation and assessment of knowledge, credibility and authority, and some authors argue that this makes climate change knowledge potentially more inclusive and open-ended (Demeritt, 2006; Lövbrand, 2009). This is supported by citizen science networks recording changes in seasonal behaviour of species in connection with climate change, which suggest that reflexivity (awareness of own basis for knowledge) and credibility (awareness of others' basis for knowledge) contribute to personal and societal meaning-making around climate change. Public knowledge about climate change can include training in environmental monitoring (Lawrence, 2009a,b) as well as involving interest groups and others in consultation and decision-making on land use planning and management.

Forest management has for centuries relied on a linear model of knowledge generation and communication (i.e. research and extension). Adaptiveness requires a different approach which responds to complexity by drawing on a range of knowledge types, and which builds in monitoring for learning in the face of uncertainty. North American literature in particular indicates that this requires a radical adjustment of knowledge and strategies to adequately plan and enhance the uses of trees and forests in line with ongoing and future climatic changes (McKinnon and Webber, 2005; Ohlson *et al.*, 2005; Spittlehouse, 2005). This requires forest managers to embrace a more process-based management approach that balances careful long-term 'design' with maintaining the capacity to 'adapt' (Fürst *et al.*, 2007).

13.5 The particular role of trees and forests in social change

Trees and woods have a significant role in many people's life and could thus potentially help in people's understanding of climate change. For example Henwood and Pidgeon (2001) found that woods are key features in defining place, and people see them as symbolic of nature itself. Many people value the contribution of woods to human well-being (through knowledge, experience and sense of relationship with woodlands) more highly than other forest ecosystem functions (Agbenyega *et al.* 2009). This suggests a particularly significant role for woodland management and the engagement of forestry with the public, in contributing to societal understandings and responses to climate change.

13.6 Conclusions

The evidence given above suggests the following:

- Many people find it hard to make sense of scientific information about climate change, and the associated complexity and uncertainty.
- Factors other than knowledge or the simple provision

of information are important in achieving behavioural change in relation to global warming.
- Social behaviour is influenced by institutions. These institutions need to be able to change and to continue to be able to do so; in other words, to be adaptive.

These findings come largely from areas other than forestry or woodland management. We can hypothesise that they are relevant to forestry, but these hypotheses need to be tested. The following conclusions can be drawn specifically in relation to trees woods and forests:

- Knowledge about the role of trees in climate change mitigation and adaptation varies within society, and is often confused.
- Trees, woods and forests have a special symbolic value in many people's sense of place and understanding of the environment.

These findings emphasise both the potential and the need for further research.

13.7 Research priorities

- We can hypothesise that trees have a significant role in influencing people's understanding of and responses to climate change. This needs to be further tested across wider geographical areas and among different social groups, and through action research, which explores the effects of developing people's experience of and emotional connection with trees and forests, for example through art or education activities.
- The climate change debate presents a particular opportunity for forestry experts to engage people in environmental analysis. Research highlights the gulf between expert and local knowledge. Given the special symbolic value of trees and forests in the climate change debate, forestry knowledge could be received and trusted differently from some other forms of technical expertise. This needs to be tested.
- Research in other countries suggests that adaptive forestry organisations need to develop new approaches to using knowledge (research and innovation). This needs to be tested in the UK forestry context, for example by understanding how forest managers make sense of new information about climate change/species suitability/silvicultural practices, and how this affects decision-making. Such adaptiveness also includes working co-operatively with other organisations, and the processes and outcome of such partnerships needs to be researched in the UK context.

References

ADGER, W.N., BENJAMINSEN, T.A., BROWN, K. and SVARSTAD, H. (2001). Advancing a political ecology of global environmental discourses. *Development and Change* 32, 681–715.

AGBENYEGA, O., BURGESS, P.J., COOK, M. and MORRIS, J. (2009). Application of an ecosystem function framework to perceptions of community woodlands. *Land Use Policy* 26, 551–557.

BÄCKSTRAND, K. and LÖVBRAND, E. (2006). Planting trees to mitigate climate change: Contested discourses of ecological modernization, green governmentality and civic environmentalism. *Global Environmental Politics* 6, 50–75.

BARNETT, A. (2009). *"IHDP: should 90% of climate change research be social science?,"* Climate feedback. Online at: http://blogs.nature.com/climatefeedback/2009/04/ihdp_should_90_of_climate_chan.html

BLENNOW, K. and PERSSON, J. (2009). Climate change: motivation for taking measure to adapt. *Global Environmental Change* 19, 100–104.

BULKELEY, H. (2000). Common knowledge? Public understanding of climate change in Newcastle, Australia. *Public Understanding of Science* 9, 313–333.

DEFRA (2001–2008). *Attitudes to climate change*. Various reports. DEFRA, London.

DOWNING, P. (2008). *Public attitudes to climate change, 2008: concerned but still unconvinced*. Ipsos MORI. Online at: www.ipsos-mori.com/researchpublications/researcharchive/poll.aspx?oItemId=2305

DOWNING, P. and BALLANTYNE, J. (2007). *Tipping point or turning point?* Ipsos MORI. Online at: www.ipsos-mori.com/Search.aspx?usterms=tipping%20point

FORESTRY COMMISSION. (2007a). *UK public opinion of forestry 2007: England*. Forestry Commission, Edinburgh.

FORESTRY COMMISSION. (2007b). *UK public opinion of forestry 2007: Scotland*. Forestry Commission, Edinburgh.

FORESTRY COMMISSION. (2007c). *UK public opinion of forestry 2007: Wales*. Forestry Commission, Edinburgh.

FORESTRY COMMISSION (2008). *Public opinion of forestry*. Forestry Commission, Edinburgh. Online at: www.forestry.gov.uk/forestry/infd-5zyl9w

FÜRST, C., VACIK, H., LORZ, C., MAKESCHIN, F., PODRAZKY, V. and JANECEK, V. (2007). Meeting the challenges of process-oriented forest management. *Forest Ecology and Management* 248, 1–5.

GÖTMARK, F., FRIDMAN, J. and KEMPE, G. (2009). Education and advice contribute to increased density of broadleaved conservation trees, but not saplings, in young forest in Sweden. *Journal of Environmental Management* 90, 1081–1088.

HAUER, R.J. and JOHNSON, G.R. (2008). State urban and community forestry program funding, technical assistance, and financial assistance within the 50 United States. *Arboriculture and Urban Forestry* 34, 280–289.

HENWOOD, K. and PIDGEON, N. (2001). Talk about woods and trees: threat of urbanization, stability, and biodiversity. *Journal of Environmental Psychology* 21, 125–147.

HINDS, J. and SPARKS, P. (2008). Engaging with the natural environment: the role of affective connection and identity. *Journal of Environmental Psychology* 28, 109–120.

KALS, E., SCHUMAKER, D. and MONTADA, L. (1999). Emotional affinity toward nature as a motivational basis to protect nature. *Environment and Behavior* 31, 178–202.

KITCHEN, L., MILBOURNE, P., MARSDEN, T. and BISHOP, K. (2002). Forestry and environmental democracy: the problematic case of the South Wales Valleys. *Journal of Environmental Policy and Planning* 4, 139–155.

LAWRENCE, A. (2009a). The first cuckoo in winter: phenology, recording, credibility and meaning in Britain. *Global Environmental Change* 19, 173–179.

LAWRENCE, A. (2009b). Taking stock: learning from experiences of participatory biodiversity assessment. In: Lawrence, A. (ed.) *Taking stock of nature: participatory biodiversity assessment for policy and planning*. Cambridge University Press, Cambridge. (in press)

LÖVBRAND, E. (2009). Revisiting the politics of expertise in light of the Kyoto negotiations on land use change and forestry. *Forest Policy and Economics*, (in press.)

LOVELL, R. (2009). *Wood you believe it? Children and young people's perceptions of climate change*. Forest Research, Farnham. Online at: www.forestresearch.gov.uk

MACDONALD, G.B. and RICE, J.A. (2004). An active adaptive management case study in Ontario boreal mixedwood stands. *Forestry Chronicle* 80, 391–400.

MAIBACH, E., ROSER-RENOUF C., WEBER, D. and TAYLOR, M. (2008). *What are Americans thinking and doing about global warming?* George Mason University/Porter Novelli. Online at: www.climatechangecommunication.org/images/files/PN_GMU_Climate_Change_Report.pdf

MCKINNON, G.A. and WEBBER, S.L. (2005). Climate change impacts and adaptation in Canada: Is the forest sector prepared? *Forestry Chronicle* 81, 653–654.

NORD, M., LULOFF, A.E. and BRIDGER, J.C. (1998). The association of forest recreation with environmentalism. *Environment and Behavior* 30, 235–246.

OCKWELL, D., WHITMARSH, L. and O'NEILL, S. (2009). Reorienting climate change communication for effective mitigation. *Science Communication* 30, 305–323.

OHLSON, D.W., McKINNON, G.A. and HIRSCH, K.G. (2005). A structured decision-making approach to climate change adaptation in the forest sector. *Forestry Chronicle* 81, 97–103.

PAULEIT, S., JONES, N., GARCIA-MARTIN, G., GARCIA-

VALDECANTOS, J.L., RIVIARE, L.M., VIDAL-BEAUDET, L., BODSON, M. and RANDRUP, T.B. (2002). Tree establishment practice in towns and cities - Results from a European survey. *Urban Forestry and Urban Greening* 1, 83–96.

POOLEY, J.A. and O'CONNER, M. (2000). Environmental education and attitudes: emotions and beliefs are what is needed. *Environment and Behavior* 32, 711–723.

ROSS-DAVIS, A.L., BROUSSARD, S.R., JACOBS, D.F. and DAVIS, A.S. (2005). Afforestation motivations of private landowners: An examination of hardwood tree plantings in Indiana. *Northern Journal of Applied Forestry* 22, 149–153.

SAAVEDRA, C. and BUDD, W.W. (2009). Climate change and environmental planning: Working to build community resilience and adaptive capacity in Washington State, USA. *Habitat International* 33, 246–252.

SCHULTZ, P.W. (2000). Empathizing with nature: the effects of perspective taking on concern for environmental issues. *Journal of Social Issues* 56, 391–406.

SEPPÄLÄ, R., BUCK, A. and KATILA P. (2009). *Making forests fit for climate change: a global view of climate-change impacts on forests and people and options for adaptation*. IUFRO. Online at: www.iufro.org/download/file/3581/3985/Policy_Brief_ENG_final.pdf

SIRY, J., ROBISON, D. and CUBBAGE, F. (2004). Economics of hardwood management: Does it pay to plant? *Forest Landowner* 63, 32–34.

SPITTLEHOUSE, D.L. (2005). Integrating climate change adaptation into forest management. *Forestry Chronicle* 81, 691–695.

STURGIS, P. and ALLUM, N. (2004). Science in society: re-evaluating the deficit model of public attitudes. *Public Understanding of Science* 13, 55–74.

SWART, R. and RAES, F. (2007). Making integration of adaptation and mitigation work: mainstreaming into sustainable development policies? *Climate Policy* 7, 288–303.

TEISL, M.F. and O'BRIEN, K. (2003). Who cares and who acts? Outdoor recreationists exhibit different levels of environmental concern and behavior. *Environment and Behavior* 35, 506–522.

VAN HERZELE, A. and VAN GOSSUM, P. (2008). Typology building for owner-specific policies and communications to advance forest conversion in small pine plantations. *Landscape and Urban Planning* 87, 201–209.

WORTHY, K. (2008). Modern institutions, phenomenal dissociations, and destructiveness toward humans and the environment. *Organisation and Environment* 21, 148–170.

SECTION 6
CONCLUSIONS
Chapter 14 - Overview and research priorities

OVERVIEW AND RESEARCH PRIORITIES

Chapter

14

Sir D. J. Read FRS and P. H. Freer-Smith

This Assessment presents a science-based analysis of the current and potential capabilities of the UK's forests and forest products to contribute to the mitigation of climate change.

It shows unequivocally that: (1) a significant contribution to mitigation could be made by maintaining and increasing the rates at which CO_2 is removed from the atmosphere by the UK's forests – the abatement of emissions; (2) there are other major contributions that forests and trees can make, for example in urban environments, and (3) that there are areas where substantial research is required to assess and reduce impacts and to develop the contribution of UK forestry to the mitigation of, and adaptation to, climate change.

The report follows the IPCC convention of identifying key findings at the start of each Chapter. Authors have also identified research priorities at the end of their respective chapters and these proposals require serious and urgent consideration. Gaps in our understanding are identified and, in some instances, there is a clear requirement for more scientific evidence in order that uncertainties in the projected impacts of climate change can be reduced. The driver in the new research programmes will be the need to enhance the contribution that UK trees and woodlands can make to a low carbon economy. It is the expressed intention of the report to provide the detailed evidence required to achieve sustainable forest management under what will be complex and changing climatic circumstances.

14.1 Overview

The UK faces wetter winters and hotter, drier summers with more drought, heat-waves and flooding. As the future emissions' trajectories cannot be forecast precisely, climate projections are based upon three possible scenarios: high, medium and low emissions. They show considerable regional variation in climate sensitivity. For example, the 2009 UKCP09 projections for a medium-emissions scenario suggest that by the 2050s, average summer temperatures in southern England will rise by 2°C above

the 1961–90 average; summer rainfall in the south west will fall by 20%, and winter rainfall in the north west will increase by 15%. Much of our existing woodland, together with the stands being established now, will experience these changes. If global emission control measures do not result in significant decreases in GHG emissions, the UK climate trends will become even more pronounced by the time that new woodlands planted over the next 10–15 years are felled. As a result of the inevitable uncertainties involved in climate predictions, recommendations for improving the contribution of UK forests and urban trees to mitigation of climate change will themselves carry some uncertainty. This in turn necessitates that planting scenarios are subject to risk assessment. As well as research directed towards identification of the major areas of uncertainty, it will be essential to pursue programmes that continuously monitor the effectiveness of actions taken. Sufficient flexibility must be retained in all programmes to enable changes of direction to be achievable where and when they are deemed necessary.

On-going climate change is impacting our woodlands now and will influence their ability to provide environmental, economic and social benefits in the future. Furthermore, pests and diseases of forest trees, both those that are already present in the UK and those that may be introduced, currently represent a major threat to woodlands by themselves. When combined with the direct effects of climate change, these threats are likely to become even more serious. Identification of the nature of current and predicted climate change impacts will provide those responsible for the management of our forest resource with the opportunity to adapt their practices in such a way as to limit the severity of impacts. Action is required to change the extent, composition and structure of our woodlands in order to avoid future serious limitations of the goods and services they provide and also to prevent wildlife losses. Indeed, the contribution which the UK's forests and woodlands can make to abatement of GHG emissions cannot be achieved unless effective measures are taken to ensure their adaptation. A number of actions are set out which are designed to achieve such adaptation. These range from the creation of appropriate new woodland to management of some existing woodlands in such a way as to optimise diversity and encourage natural regeneration. In addition it will be necessary to engage in selective restocking of some current stands using species and provenances better suited to the changing climate.

In the urban environment, trees will play an increasingly important role in helping society adapt to the changing climate. It is recognised that all measures taken to provide mitigation and adaptation should be both socially and environmentally acceptable, as well as being cost-effective.

A number of key features of UK forestry are a product of past forest policy. On the positive side, the UK has increased its forest area from a low of c. 5% to the current level of nearer 12%. This includes the establishment of about 1 million ha of new fast-growing conifers that currently represent a significant carbon sink. Provided that this important resource is managed with emissions abatement as one of the primary management objectives, it can continue to provide an effective carbon sink. On the negative side the strength of the carbon sink provided by UK forests could decline from a value of about 16 MtCO$_2$ year^{-1} in 2004 to as little as 4.6 MtCO$_2$ year^{-1} by 2020. This is because of ongoing harvesting and the age distribution of the whole UK forest resource, which is a consequence of the decline in afforestation rates since 1989. This decline has serious implications for future UK GHG inventories and in the timescale of the UK's first three carbon budgets (to 2022), can only be modified to a limited extent by increased woodland planting. This is a major concern at a time when the UK Government is seeking to establish a low carbon economy and to take a lead on climate change internationally. However, what emerges from this assessment is that woodland creation over the next 40 years could significantly benefit the UK GHG inventory by 2050. For example, by planting c. 23000 ha per year over the next 40 years the UK could, by the 2050s, be locking up on an annual basis, an amount of carbon equivalent to 10% of our total GHG emissions. Such a programme would restore the UK's annual woodland creation rates to values similar to those seen in the 1970s, 80s and 90s. It would increase the proportion of woodland cover in the UK landscape to 16%. Trees and woodlands across the UK contribute to numerous policy objectives, including the provision of recreation and amenity, the conservation

of biodiversity and water management. Clearly, the establishment and management of the new woodlands must be planned in such a way that the required emissions abatement can be achieved without compromising multifunctionality.

Our analysis shows that UK forests and forest soils contain significant amounts of carbon and that the strength of the carbon sink they provide will decline over the next few years unless practice and policy change. Economic analysis suggests that woodland creation is a cost-effective approach for abatement of GHG emissions. This approach is most cost-effective where land opportunity costs are lowest. Currently, UK conifer plantations are providing cost-effective abatement. If the social, economic and environmental co-benefits of woodland creation are also considered, then GHG abatement through woodland creation appears even more cost-effective. The maintenance of a useful carbon sink in existing UK forests and any significant woodland creation programme would require the existing regulatory framework and sustainability standards to be maintained and developed (e.g. the UK Forestry Standard, FC Guidelines on Climate Change, the new Code of Practice for Forest Carbon Projects). Effective standards, guidance and management plans will be essential to ensure that climate change objectives are achieved. These need to be underpinned by strong evidence appropriate to UK conditions. Maximising the capacity of forests to mitigate and adapt to climate change requires actions that are tailored to local and regional conditions. Woodland creation will also require careful spatial planning and targeting in order that it contributes effectively to the full range of forestry objectives.

Success in achieving delivery of the full abatement potential of the UK forestry sector requires an integrated approach involving consideration of not only the carbon stocks in forests but also the roles of wood and timber products directly in substituting for fossil fuels and indirectly as components of the built environment. Forest management has to be effectively co-ordinated with the full timber supply and utilisation chain to ensure that the flow of wood from the forest continues.

Woodlands need to be managed as part of the landscape. Inevitably there are demands on land for other purposes – notably food and energy production and urban and infrastructure development – which affect the economic potential for land to be allocated to forestry. Policies and practices in agriculture, planning development and other urban and rural activities will affect the capacity of woodlands to deliver climate change mitigation and adaptation objectives. Policy incentives need to be re-designed so that adequate reward is given to the provision of non-market benefits, including those relating to the climate change mitigation and adaptation functions of forests. The knowledge built up in the UK and beyond should be used to facilitate more successful mitigation–adaptation interactions in the forestry/land-use sectors in the wider context of sustainable development and the promotion of rural livelihoods.

14.2 Identification of research requirements

The major research requirements which are identified at the end of each chapter are drawn together and summarised below.

The UK has only two forest sites at which CO_2 and energy balances are being recorded continuously; one upland conifer site and one lowland broadleaved woodland. It is important that these sites are maintained and that the measurements should be extended to allow continuous recording of other GHG, particularly CH_4 and N_2O fluxes. Our current understanding of the effects of geographical location, natural events, site disturbance, species, stand age, Yield Class and management operations on GHG exchange is very limited and more research is required. Collaborative experimental programmes, for example, between Defra, CEH, SEPA, University teams and the FC would allow tall towers and aircraft to be used to evaluate emissions and removals of trace gases across landscapes to define major sources and sinks of GHG in relation to forestry, agriculture and other land uses at regional and district scales. This would enable the daily and seasonal sources and sinks of forested and agricultural landscapes to be evaluated at regional scales.

There is evidence, particularly from continental Europe, of increasing forest productivity, perhaps resulting from changes in forest management and increased nitrogen availability. Rising CO_2 concentrations, warmer temperatures, longer growing seasons and in some places increased rainfall could also contribute to the enhancement of forest growth. The absence of a framework for collecting and evaluating detailed information on forest growth and productivity has contributed to the lack of clear evidence of growth trends of British forests. This could delay the implementation of adaptation measures and is a serious disadvantage for those confronted with the need to design new woodland creation programmes. The development of

a framework for monitoring and evaluating growth changes on an annual basis is therefore required. Such a framework should be developed as a UK Climate Change Indicator for the forestry sector, and reported on annually. We need to determine whether or not increased forest productivity is occurring in the UK. If increases are detected, the extent to which these will be sustained under projected future climate conditions should be estimated. Conversely, we must recognise that in those areas subject to the deleterious impacts of climate change, the decline and dieback of some of our major species could be envisaged. This would have serious consequences, among which the net release of carbon from forests and forest soils would be the most threatening for the UK GHG balance. The impacts of extreme events on forest ecosystems need to be better understood, particularly the effects of prolonged drought on tree physiology and mortality. Climate projections show that the risks of drought are likely to increase in the UK. Since soil moisture deficits are already serious limiting factors for some tree species in some UK sites, the threats posed by even greater rainfall deficits are significant indeed. A number of other important issues of this type are identified in the chapters of this assessment, many of which require well focused research programmes.

It is important to ensure that the monitoring systems in place are adequate for the new policy and management challenges posed by climate change. The UK's National Forest Inventory is currently under development and will produce a digital map of British woodlands in 2010. In broad terms, the procedures for monitoring and predicting timber production from the UK's forests place us in a good position to plan and review. The National Forest Inventory provides important information on forest cover and woodland type which should help with action on climate change. However, most UK local authorities lack the basic inventories describing the nature and extent of urban trees and woodlands in their districts. This information gap needs to be addressed urgently, with the resulting information on the trees and woods of urban and peri-urban areas being added to the national forest inventory.

We need to continue to improve the forestry-related information that underpins the GHG inventory, and it is essential that there is confidence in the assessment of the potential for forestry to contribute to the abatement of net UK GHG emissions. The conclusions of this review rely heavily on the GHG accounting models used in Chapter 8. While there is sufficient evidence and agreement to justify the recommendations which are made, it is also important to improve these models further, and to underpin them with measurement programmes. The analyses presented in Chapters 7 and 8 show the importance of CO_2 emission abatement achieved through the use of wood in place of fossil fuels and of wood products in substitution for building materials. A comprehensive evaluation of life cycle analyses from a wide range of wood products compared with alternative materials is required to better demonstrate the role of forest management and product displacement in GHG abatement. An operational decision support system needs to be developed to downscale national level assessments of abatement potential through changes in forest management, and to aid the implementation of appropriate abatement measures at regional and local scales.

In the face of the uncertainties inherent in all climate change impact assessments, our modelling capabilities have enabled valuable indications of the likely roles and responses of forests under a range of scenarios. However, we require the development of a hierarchical modelling system combining the practical applicability of knowledge-based decisions support approaches with more theoretical process-based models. Such a system should be designed to represent the effects of changing atmospheric composition and be extended outside the evidence-base of empirical models. Decision support systems are required to integrate understanding and to characterise the structure, function and economic benefits of urban and peri-urban trees and woodlands.

'Climate matching' analysis can identify broad regions that currently experience a climate similar to that projected for the UK in the future. This provides an opportunity to explore the impacts of likely climate change predictions based on model simulations. The approach should be used to explore likely changes in woodland ecosystems, the suitability of tree species for commercial forestry and to inform alterations in forest management that might be required in response to the changing climate. It must, however, be understood that such an approach can only provide broad guidance as complete analogues of future conditions do not exist. We need to improve our understanding of which factors will become limiting for which species at a regional level. Forest trials of potential species that may be suitable for the current and projected British climate are required. This will allow the scope for species translocation, genetic improvement and the use of new provenances and species to be examined. These approaches are important for developing our understanding of how different species will respond to climate change.

Forest planning faces difficult decisions on how to address the many objectives of forestry, in a changing climate.

Managers will require more advice and information from the research community in order to make rational decisions when faced with unknown or unfamiliar conditions, and with multiple demands and objectives. We must develop and maintain databases describing how different species (both trees and other woodland species) are predicted to respond to climate change (e.g. using knowledge-based systems, such as Ecological Site Classification, and empirical climate space/envelope modelling). These tools need to be informed by strengthened forest monitoring for early detection of change. This will be essential to allow damage-limiting adaptation measures to be imposed in a timely manner. We need to improve our understanding of how climate change will influence disturbance regimes of wind, fire, pests and pathogens and to develop methodologies to help forest managers identify sites and stands most vulnerable to climate change.

There is evidence of an increased number of pest and pathogen outbreaks both in UK forests and globally over recent years. Whether or not these are related to climate change, such outbreaks pose a serious threat because they could compromise both the growth and resilience of forests. The extent to which increased world trade in plants and forest products and climate change contribute to current threats is uncertain, but there is a need to reduce the future risks and to manage the existing outbreaks. The temperature response of growth should also be determined for a range of tree pests and pathogens to provide the basis for epidemiological modelling of future outbreaks under a changing climate. It is essential that appropriate and effective interception and monitoring systems are in place to prevent the introduction of pests and pathogens. Early identification of impending threats and of new outbreaks can prevent their establishment. Scientific analysis and awareness are the keys to preventing new outbreaks from becoming established. The management of pests and diseases, if they do become established, must also be predicated upon scientific understanding of the outbreaks concerned.

We must improve our quantitative understanding of the impacts of forest management alternatives on the carbon and nutrient budgets and yield of plantations, particularly for new species and management methods. It is known that land preparation, thinning, harvesting and windthrow can have important effects on soil carbon stocks but a better understanding of rates of carbon sequestration and stocks in older forest stands is required because these may need to be retained for landscape and biodiversity reasons. Information of this kind would also allow an improved reporting of forest carbon stocks, including those present in soils, in the National Forest Inventory and would underpin accounting models for forest carbon.

In spite of the considerable potential for using trees as a sustainable source of energy and thereby to contribute to emissions abatement, there are a number of barriers to greater use of woody biomass particularly fast growing species, and these are considered in Chapter 7 of this Assessment. To date there has been a failure to achieve significant planting of woody energy crops in the UK. Estimates provided by the directorate of the Energy Crops Scheme indicate that so far, there has been only limited adoption of short rotation crops (SRC) and short rotation forestry (SRF) practices in the UK. To underpin the wider use of energy crops, more research is required on choice of species, future yields, and on potential diseases and pests. The values of the ecosystem services provided (including the likely carbon benefits) by these schemes also must be evaluated. More extensive implementation of farm-scale trials would fill some of these knowledge gaps. There is considerable potential for the future of new bioscience technologies to improve the photosynthetic gains of bioenergy systems. Other research needs identified include an evaluation of minimal input systems, analysis of energy crops with different carbon qualities (e.g. increased lignin for calorific combustion) and of those with improved resistance to biotic and abiotic stresses. Poplar is currently the model bioenergy tree, and its various genotypes are an important resource from which enhanced traits including improved carbon sequestration and energy production should be obtainable in second generation crops. In addition, new technologies for conversion of biomass to fuel are likely to be developed and, by 2020, gasification and other technologies may be deployed to improve the efficiency with which wood-based energy supplies are processed and delivered.

Initiatives are urgently required to stimulate a step change in the extent to which UK-grown forest products are used in our buildings. Here it must be acknowledged however that the case for increased use of wood is hampered by incomplete and fragmented evidence of the qualities of wood products. In order to highlight the benefits of wood relative to those of other construction materials, GHG balances and energy efficiencies for different construction systems using consistent assessment methods are required. Life cycle analyses of wood products and of biofuel energy systems are essential. The turnover rates of carbon in different wood product pools must be better quantified. It is also important to understand better the behavioural, social and economic barriers to the development of wood energy supply chains and the relative advantages of different wood energy supply systems (chip, pellet, CHP). Furthermore, research on the optimal adaptation of our woodlands needs to take into account the increased requirement for sustainable wood products and woodfuel as the climate change mitigation role of forestry increases.

Further analyses of the cost-effectiveness of forest-based carbon sequestration and emissions abatement programmes are required. These should include comparisons of the marginal costs of abatement through different forestry management options relative to those of other possible alternatives for reducing net emissions (e.g. in agriculture, in housing, transport or industry). Mechanisms enabling the maximisation of the net benefits of managing forests for abatement need to be described. Spatially explicit modelling of carbon and non-market benefits of forests and woodlands remains a research priority. The economic value of ancillary benefits of woodland creation (biodiversity, water quality, recreation, soil protection) need to be incorporated both within cost-effectiveness assessments and in marginal abatement cost curve analyses carried out for the forestry sector. Further investigation is required into the nature of risk and uncertainty in developing forest carbon credit markets. It is not yet clear how this risk can best be managed. The extents of the trade-offs and synergies between managing forests for carbon, relative to other public goals such as recreation, biodiversity conservation, timber supply and water management must be quantified. The benefits of trees and woodlands in helping businesses and people to adapt to the impacts of climate change require to be evaluated.

There is an increasing awareness among the public of the threats to their economic security, lifestyles and wellbeing posed by climate change. It is our intention that this Assessment should highlight the potential for trees and forests to mitigate climate change, so reducing these impacts. It should stimulate greater engagement by individuals, businesses and government in consideration of the future role of trees and forests in the UK landscape. Undoubtedly, some of the measures shown in this study to have significant mitigation potential may not in the first instance receive universal approval. Progress towards broadly acceptable strategies for reducing the impacts of climate change will depend upon cooperative working between organisations, interest groups and individuals, and an understanding of the need to identify widely acceptable solutions. What is very clear is that inaction is no longer an option.

Glossary

Abatement: to decrease GHG emissions including the reduction of net GHG emissions by increasing GHG removal (or uptake) from the atmosphere.

Adaptation: a process by which organisms become better suited to their habitat.

Adaptation to global warming: initiatives and measures to decrease vulnerability of natural and human systems to climate change effects.

Annex B country: one of the 39 industrialised countries under the Kyoto Protocol which has accepted targets for limiting or reducing emissions. These targets are expressed as percentages of each country's GHG emissions as estimated for a base year of 1990 (also known as the assigned amount for the country).

Carbon dioxide equivalent (CO_2e): Over a 100-year time interval the Global Warming Potential (GWP) of methane (CH_4) is 23 times that of CO_2 and nitrous oxide (N_2O) is 296 times that of CO_2. Therefore total GHG amounts are sometimes expressed as CO_2e where more than one GHG is being considered.

Carbon leakages: losses of forest carbon (or increases in GHG emissions) that can occur as an inadvertent result of implementing an emissions reduction or sequestration project. For example, the effectiveness of a project aimed at preventing deforestation in a particular area of forest may be reduced if the causes of the deforestation are simply 'displaced' to a different area of forest. Similarly, the effectiveness of an afforestation project may be reduced if it leads to loss of interest and support for other afforestation projects already in existence.

Carbon offsetting: a way of compensating for GHG emissions by making an equivalent carbon dioxide saving elsewhere. Carbon offsetting involves calculating emissions and then purchasing 'credits' from emission reduction projects see: www.defra.gov.uk/environment/climatechange/uk/carbonoffset/index.htm

Carbon pool or reservoir: a component of the earth system in which a greenhouse gas is stored. Trees, deadwood, debris, litter and soils forming forests are all examples of carbon pools or reservoirs, as are any harvested wood products that retain carbon in wood during their use. Even wood disposed of in landfill could be regarded as a carbon pool.

Carbon sink/source: the carbon balance of a forest is often described as a sink if there is a net transfer of carbon from the atmosphere to one or more of the carbon pools in the forest (resulting in carbon sequestration). When a forest is described as a carbon source there is a net transfer of carbon to the atmosphere.

Clean Development Mechanism (CDM) - The CDM, as part of the Kyoto Protocol, allows industrialised countries with a greenhouse gas reduction commitment (called Annex B countries) to invest in projects that reduce emissions in developing countries as an alternative to more expensive emission reductions in their own countries.

Cost-effectiveness: cost-effectiveness analysis compares the relative expenditure (costs) and outcomes (effects) of two or more courses of action. Regarding carbon sequestration in forestry, cost-effectiveness is expressed in terms of a ratio where the denominator is the carbon stock changes and the numerator is the Net Present Value of the expenditure on the forest measure.

Discounting: is a process used to give different values to costs and benefits occurring at different times. Formally, the Net Present Value (NPV) of a project generating incomes at different times is:

$$NPV = -c + Y1/(1+r) + Y2/(1+r)^2 + ... + YT/(1+r)^T$$

Where c is the initial cost, $Y1,....,YT$ - the revenues in real terms at year-zero prices occurring at times $1,....,T$; and r is the discount rate.

The effect of discounting physical carbon is to increase the costs of creating carbon offset credits because discounting effectively results in 'less carbon' attributable to a project. Discounting financial outlays, on the other hand, reduces the cost of creating carbon offsets. Since most outlays occur early on in the life of a forest project, costs of creating carbon offsets are not as sensitive to the discount rate used for costs as to the discount rate used for carbon.

Emissions trading: a process to control pollution whereby a central authority (e.g. a regional or national body, or an international body) sets a limit or cap on the amount of a pollutant (e.g. carbon dioxide) that can be emitted. It is sometimes called cap and trade. Companies or other groups are issued emission permits and are required to hold an equivalent number of *allowances* (or credits) that represent the right to emit a specific amount. The total amount of allowances and credits cannot exceed the cap, limiting total emissions to that level. Companies that need to increase their emission allowance must buy credits from those who pollute less. The transfer of allowances is referred to as a trade. In effect, the buyer is paying a charge for polluting, while the seller is being rewarded for having reduced emissions by more than was needed. Thus, in theory, those who can easily reduce emissions most cheaply will do so, achieving the pollution reduction at the lowest possible cost to society.

Externalities: an effect of a purchase or use decision by one set of parties on others who did not have a choice and whose interests were not taken into account. In such a case, prices do not reflect the full costs or benefits in production or consumption of a product or service. A positive impact is called an *external benefit*, while a negative impact is called an *external cost.*

Flexibility mechanisms: market-based mechanisms in the Kyoto Protocol which are intended to support Annex B countries in meeting their emissions limitation/reduction targets at least costs. These mechanisms are Emissions Trading, the Clean Development Mechanism and Joint Implementation. Annex 1 countries can invest in J1 and CDM projects as well as host J1 projects. Non Annex 1 countries can host CDM projects. Note that Belarus and Turkey are listed in Annex 1 but not Annex B; and Croatia, Lichtenstein, Monaco and Slovenia are listed in Annex B but not Annex 1.

Forest-based sector Technology Platform (FTP): has defined and is currently implementing a research and development roadmap for the European forest-based sector.

Global Warming Potential (GWP): a measure of the contribution made by a combination or mix of greenhouse gases to global warming, which involves converting quantities of non-CO_2 greenhouse gases to an equivalent amount of CO_2. Over an assumed 100-year time frame, the global warming potential of 1 tonne of methane is equivalent to 23 tonnes of CO_2 (or 24.5 tCO_2e) while 1 tonne of nitrous oxide is equivalent to 296 tonnes of CO_2 (or 320 tCO_2e).

Greenhouse gases (GHG): gases that have significant infrared radiation absorption bands in the troposphere. In order of importance, the relevant *natural* GHG in the troposphere are: water vapour, carbon dioxide (CO_2), methane (CH_4), ozone (O_3) and nitrous oxide (N_2O). UK forests exchange all of these GHG with the troposphere to a larger or smaller extent.

Joint Implementation (JI): as part of the Kyoto Protocol, allows industrialised countries with a greenhouse gas reduction commitment (called Annex B countries) to earn emission reduction units (ERUs) from an emission-reduction or emission-removal project in another Annex B country.

Marginal abatement cost (MAC): in economics, marginal cost is the change in total cost that arises when the quantity produced changes by one unit. It is the cost of producing one more unit of a good. A marginal abatement cost curve represents the marginal economic cost of one extra tonne of GHG mitigation.

Market price of carbon: reflects the value of traded carbon emissions rights to those in the market given the constraints on supply of these rights to emit imposed by current policy. If the carbon market covers all emissions and is competitive, then the market price will be equal to the MAC for a given target.

Maximum mean annual increment (MMAI): the average rate of volume growth from planting/regeneration to any point of time is known as the 'mean annual increment'. This value will increase during the early phases of stand development (i.e 'stand initiation' and 'stem exclusion') before reaching a peak and then declining. This peak value is known as the 'maximum

mean annual increment' (MMAI) and will vary between species and sites. It provides an estimate of the maximum average rate of volume production (and hence carbon sequestration) that can be maintained on a site in perpetuity. In conifer species, the age of MMAI usually occurs towards the end of the 'stem exclusion' phase.

Mitigation: (of global warming) actions to decrease GHG emissions, to enhance sinks, or both, in order to reduce the extent of global warming.

Opportunity costs: income foregone – for example, the net income which would have been achieved by a crop now replaced by an alternative land use. For example, the costs of woodland creation may not only be the establishment costs. This value has to be considered in the economic calculations.

Sequestration: (of C or CO_2) the removal from the atmosphere of carbon or carbon dioxide through biological or physical processes and their retention in living biomass or wood products.

Shadow price of carbon (SPC): a shadow price in that it is the marginal abatement cost (MAC) to reach a target concentration, although that choice of target concentration depends on *social* valuation. Then the SPC might be measured by price that we *might* agree should be the carbon (dioxide) price that all (or most) countries would set internally (via taxes or cap and trade mechanisms) and which ought then to be equal to the MAC. The SPC is intended to guide decisions if incorporated into a market price, i.e. as a tax or allowance price, with rebates for other carbon pricing instruments (like the EU ETS). In practice, decisions as well as investment are guided by market prices. Whereas the social cost of carbon (SCC) is determined purely by our understanding of the damage caused and the way we value it, the SPC can adjust to reflect the policy and technological environment. This makes the SPC a more versatile concept in making sure that policy decisions across a range of government.

Social cost of carbon (SCC): measures the full global cost today of an incremental unit of carbon (or equivalent amount of other GHG) emitted now, summing the full global cost of the damage it imposes over the whole of its time in the atmosphere. It measures the scale of the externality, i.e. *social* cost of the damage caused by releasing a tonne of CO_2. Its magnitude depends on the choice of the social welfare function and its implied social weighting scheme (which is that adopted by the *Stern Review* and the Government *Green Book*) and on the rate of pure time preference. The SCC matters because it signals what society should, in theory, be willing to pay now to avoid the future damage caused by incremental carbon emissions. The SCC is conceptually different from the market price of carbon.

Transaction costs: (in economics) is a cost incurred in making an economic exchange.

Yield Class: in the British system Yield Classes are created by splitting the range of possible MMAI (see definition above) into steps of two cubic metres per hectare. Thus a stand of YC 14 has a maximum MMAI of about 14 m^3 ha^{-1}, i.e. greater than 13 but less than 15 m^3 ha^{-1}.